PLANT METABOLISM

PLANT METABOLISM

H. D. Kumar
Professor of Botany
Banaras Hindu University
Varanasi

and

H. N. Singh
Reader in Botany
Banaras Hindu University
Varanasi

M

First published in India 1976 by
Affiliated East-West Press Private Limited

First published in Great Britain 1979 by
THE MACMILLAN PRESS LTD
London and Basingstoke
Associated companies in Delhi Dublin
Hong Kong Johannesburg Lagos Melbourne
New York Singapore and Tokyo

Printed in India at Raj Bandhu Industrial Co., New Delhi–110064

British Library Cataloguing in Publication Data

Kumar, Har Darshan
Plant metabolism.
1. Plants – Metabolism
I. Title II. Singh, Hriday Narain
581.1'33 QK881

ISBN 0-333-25638-7

Contents

Preface

Research in plant metabolism during the last few decades has revealed diverse aspects of interlocking metabolic cycles. Living cells which, with their subcellular organelles, are the ultimate units of function, are now best understood as molecular machines dispensing energy in a highly ordered and controlled manner for various biochemical and biophysical reactions involved in anabolic and catabolic processes. It should be noted that inside the living cell these reactions do not occur in isolation, but rather are interconnected in complex pools and cycles. However, in a book, they have to be arbitrarily classified and described under discrete chapter headings, such as photosynthesis and respiration, erroneously making the processes appear independent of each other. Further, some overlap in the distribution of topics under different chapters being inherent in the very nature of the subject cannot be entirely avoided. Since most metabolic processes are highly complicated and interconnected, they may best be studied through the application of computer-based system analyses, but this has not so far been done to any appreciable extent. Our aim has been to provide a simple and straightforward account of otherwise complex metabolic events. We therefore suggest that, after having grasped the essence of a process, a beginning student consult some original papers and the more advanced recent publications listed at the end of this book.

Two of the most important new contributions to plant physiology in the last decade are: (1) the detection of a major new pathway of photosynthetic CO_2 assimilation, and (2) the discovery of the process of photorespiration in some green plants. We find that these metabolic activities, and their interrelations with the anatomical, morphological, economic, and ecological characteristics of the plants, have not received the coverage they deserve in the available books on plant physiology. These exciting discoveries have a strong bearing on plant productivity and photosynthetic efficiency.

Many succulent plants exhibit a pronounced variation in their cell sap pH which rises during the night and falls during the day. Recent findings have disclosed a remarkable similarity between the carbon fixation reactions of CAM plants and those of plants with the C_4 dicarboxylic acid pathway. However, a major difference between the two is that reactions that are spatially separated in C_4 plants (as between the chloroplasts of bundle-sheath cells and those of mesophyll cells) are temporally (night and day) isolated in CAM plants (see Chapter 7).

We have included some other topics of current interest which we fee' reflect the new emerging trends of research in this fascinating discipline biology. These are the dynamic nature of cell membranes; the isolation of plant cell protoplasts; the multiplicity of form of several biological macromolecules and compounds, such as chlorophylls and enzymes; the endosymbiotic theory of plastid origin and its economic potential; photosynthesis *in vitro*; and lipid transfers *in vitro*. Also we have briefly covered sulphur compounds and sulphur-selenium interrelations because of their importance in metabolism. And, instead of highlighting the effects of various factors on different metabolic phenomena, we have preferred to discuss the regulation and control of such processes as photosynthesis and respiration.

We wish to acknowledge the valuable help we received in the preparation of this text. We are grateful to Professor W. M. Laetsch (Berkeley, California) for his critical reviewing of the manuscript and his many useful suggestions for its improvement. We are greatly obliged to him as well as Professor C. C. Black and Dr W. T. Hall for supplying the electron micrographs and reprints of their latest papers and for their interest in this venture. We thank our friends and colleagues, notably Dr A. R. Gopal-Ayengar, Professors Mahatim Singh, B. M. Johri, and H. Y. Mohan Ram for encouraging, criticizing, and discussing our efforts from time to time. We are also indebted to Professor G. E. Fogg, Sc.D., F.R.S., and Professor W. D. P. Stewart for their constant help in various ways. Finally, we thank Dr A. K. Kashyap and Mr Subhash Bhardwaj for their pains in preparing some of the illustrations.

We request our readers to send us their criticisms of this volume and their suggestions for its improvement.

Although the book has resulted from the joint effort of the two authors, the first drafts of Chapters 1, 6, 7, 9, and 13, and "The Chloroplast" in Chapter 4 were prepared by H. D. Kumar and those of the remaining chapters by H. N. Singh. Chapters 6 and 9 and "The Chloroplast" in Chapter 4 are based on lectures delivered by H. D. Kumar at various Indian universities under the University Grants Commission (UGC) Scheme of National Lectureships (1972–73). Professor Kumar is indebted to the UGC for their permission to incorporate these lectures (now modified and updated) in this text.

H. D. Kumar
H. N. Singh

Varanasi
September 1975

1: Structure and Function of Plant Cells

INTRODUCTION

Current biological usage generally recognizes the division of living organisms into two main categories, viz., noncellular and cellular. The former includes viruses which do not have a cellular construction, i.e., the viruses contain either DNA or RNA (not both) and have only a very limited number of enzymes. In contrast, a cell constitutes the basic structural and functional unit of all organisms (protists, plants, and animals), has both DNA and RNA, and also a wide variety of enzymes.

Cellular organisms are further divided into procaryotes (bacteria and blue-green algae) and eucaryotes (higher plants, animals, fungi, algae, protozoa, and nonvascular as well as vascular cryptogams). Procaryotes lack discrete, well-organized nuclei, chloroplasts, and mitochondria, and have chemically distinctive cell walls, largely composed of mucopolysaccharides containing amino sugars and muramic acid whereas eucaryotes have several kinds of discrete subcellular organelles, including mitochondria, endoplasmic reticulum, microbodies, Golgi bodies (dictyosomes), lysosomes, ribosomes, and, in green tissues of plants, also chromatophores or plastids. Some organelles, e.g., nucleus, mitochondria, and chloroplasts, are bounded by double membranes, others, e.g., microbodies, Golgi bodies, and lysosomes, are bounded by single membranes, and yet others, e.g., ribosomes, are membrane-free structures. When present, the cell wall of eucaryotes does not contain amino sugars and muramic acid. Further, the procaryotic cell is monogenomic whereas the eucaryotic cell is polygenomic. This means that the procaryote has a single genome in its nucleoplasm. (It also has extragenomic elements called plasmids which enhance its survival value.) In the eucaryotic cell, at least two (three in plant cells) different genomes, essential for the primary life processes, occur, one in the nucleus and the other in the mitochondria (the third in the chloroplast), the biological expression of the cell being a function of coordinated interaction between the different genomes.

Plant cells differ from animal cells in many ways, the most important difference being that plant cells have a conspicuous cellulosic wall which is distinct from, and located external to, the cell membrane whereas animal cells have no cell walls but only cell membranes. Another noteworthy distinction is that most mature plant cells have large vacuoles whereas animal cells generally do not have any vacuoles. Most animal cells, and also some lower plants and algae, have small dark bodies, called centrioles, located just outside the nucleus, but no such centrioles are found in most higher plant cells. The photosynthetic cells of plants also have plastids which are lacking in animal cells. Despite these differences, plant and animal cells share certain common properties, such as a basic similarity in the mode of storage, transcription, and translation of genetic information, a universally applicable genetic code, and a basically uniform biochemical and metabolic machinery, especially that relating to the metabolism of foods and to the biosynthesis of metabolic intermediates and biologically important macromolecules. It is an established fact that a living cell is the smallest unit capable of self-assembly and self-replication; yet, in some laboratories, efforts (so far largely unsuccessful) are under way to culture isolated chloroplasts or mitochondria. It has been confirmed by the tissue culture technique that plant cells are totipotent, i.e., a single vegetative plant cell possesses the potentiality to divide and redivide and give rise, after appropriate differentiation, to an entire plant. In many animals (except insects), only the germ cells (fertilized egg) possess this capacity. But, though plant cells are totipotent, the actual functioning and expression of the individual cells, when these occur in a multicellular plant, is governed and regulated by control mechanisms to achieve the balanced and coordinated functioning of the entire plant. Further, although all the cells of a multicellular plant contain the same genetic information, all such information is not expressed in an identical manner. There are variations in form and function among the cells constituting the different tissues or organs of the plant; these variations are a consequence of the regulated, partial expression (or repression) of the potentially complete genome.

A general feature of most living cells is that they are not static, as may be erroneously inferred from textbook diagrams, but in a state of constant flux (see Haggis *et al.*, 1964). This has been proved by observations on living cells under the phase-contrast microscope. For example, fairly continuous undulating variations have been noticed along the edges of the *Paramecium*.

In this book, we shall mostly discuss the structural and functional aspects of the cells of higher plants (see Fig. 1-1). These cells exhibit a pronounced diversity in form and function, each constituting an extremely complex unit which in turn is made up of a number of different kinds of highly complex components or organelles. The whole cell is much more complex than the sum of complexities of its component parts and this is perhaps one of the most outstanding attributes of life and its self-perpetuation.

The vast majority of plant cells arise only from pre-existing cells, mostly through the process of cell division but sometimes, as in angiosperms, by fertilization and triple fusion. The various cells of multicellular plants are not equally active in cell division which is often largely restricted to certain growth zones or meristems. Such meristematic regions of actively dividing cells are commonly located apically, e.g., at root and shoot apices, but they may sometimes be located intercalary (i.e., between different tissues, such as the vascular cambium) and rarely basal.

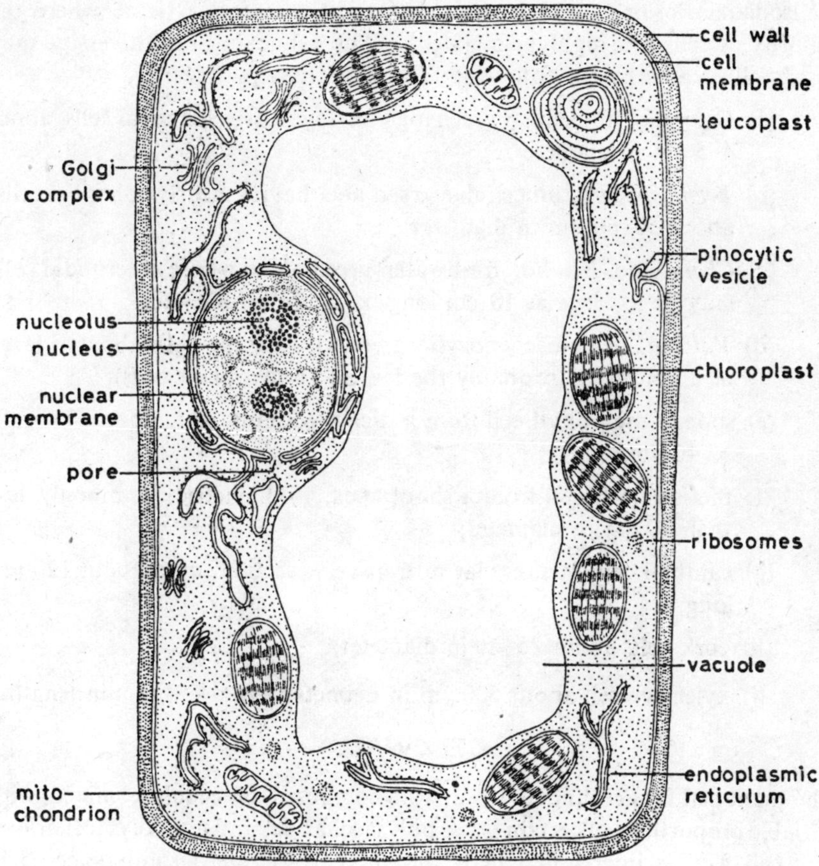

Fig. 1-1 Diagrammatic illustration of a mature, green plant cell. (Leucoplast and chloroplast do not occur in the same cell.)

GENERAL STRUCTURE

A green plant is commonly composed of a large number of different kinds of cells, such as those found in epidermis, periderm, parenchyma, collenchyma, sclerenchyma, endodermis, xylem, and phloem. All these cells are derived through enlargement, differentiation, and maturation of the products of cell division of certain meristematic cells which are much smaller than the mature cells of nonmeristematic tissues.

A young cell consists of a cell wall and a protoplast. The latter harbours the organelles and includes the ground cytoplasm in which the organelles occur. Young cells normally do not have any large central vacuole, but they may have a number of tiny vacuoles distributed throughout the cytoplasm which coalesce to form a large central vacuole in the mature cell. Each cell can be visualized as a compartment with many sides (usually 14, but this number varies from cell to cell), and may be in contact with several adjoining cells. The dimensions of the different faces of a cell may be either relatively similar (as in parenchyma where the cells are isodiametric) or extremely variable (as in the vascular tissue where the majority of cells are highly elongated). The range of such diversity may best be illustrated by the following representative examples:

(a) *Chlorella*, unicellular green alga having nearly spherical cells, about 4–5 μm in diameter;

(b) *Eremosphaera*, unicellular green alga having nearly spherical cells, about 50–60 μm in diameter;

(c) *Chara* and *Nitella*, freshwater green algae whose internodal cells may be as large as 10 cm long × 1 mm in diameter;

(d) *Valonia*, marine coenocytic green alga having cells about as large as a hen's egg (probably the largest known plant cells);

(e) spongy mesophyll cell from a dicotyledonous leaf, about 80 × 40 × 40 μm;

(f) meristematic cell from a shoot apex, isodiametric, commonly less than 10 μm in diameter;

(g) cambial cell, rectangular in transverse section, upto about 120 μm long;

(h) cork cell, about 25 μm in diameter;

(i) xylem vessel, about 500 μm in diameter and a few mm in length.

CELL WALL

The plant cell is surrounded by a wall composed of cellulose mixed with variable proportions of certain other polysaccharides and nonpolysaccharides. The cell wall is interrupted here and there by numerous interstices, i.e., small spaces between closely set cellulose microfibrils, which are finally packed by noncellulosic components. Walls of many cells have also minute pores through which fine protoplasmic connections, called plasmodesmata, are seen. The functional significance of plasmodesmata is not clear, but they interconnect adjoining cells and may participate in transporting materials. Chemically, the cell walls are made up largely of cellulose, but usually they have also pectin and hemicelluloses. The cellulose is in the form of long thread-like structures, called microfibrils; in very young (primary) walls, these fibrils are commonly spread out in the form of a loose network. Primary walls are fairly elastic and flexible. Adjacent cells are connected by a common abutting wall, called middle lamella, which is largely made up of pectin and serves to bind the cells together. In older cells, additional layers of wall material are laid down to form the secondary wall which is often much thicker than the primary wall, very rigid and firm, and composed of a succession of compact layers of cellulose microfibrils disposed in a parallel fashion. Secondary walls usually contain also certain stiffer materials, e.g., lignin, which increase the mechanical strength of the cells. Many plant cell walls have also one or the other (or both) of two water-impermeable chemical components, called suberin and cutin.

Research carried out by Setterfield and Bayley (1961) on the structure and physiology of plant cell walls has provided a rough indication of the chemical composition of primary cell walls, viz., 10–15% of cellulose, 5–15% of hemicelluloses, 2–8% of pectic substances, 1–2% of proteins, 0.5–3.0% of lipids, and about 60% of water, all values being expressed as average percentages of fresh weight.

PROTOPLAST

CELL MEMBRANE

Within the cell wall lies the living material or protoplast which is enveloped by a cell membrane or plasma membrane, called plasmalemma. Unlike the cell wall, the plasmalemma does not permit free entry of materials into or out of the cell cytoplasm; it is differentially or selectively permeable, thereby regulating the entry and exit of various materials. In contrast to the cell wall which is an extracellular secretion of the protoplast, lies external to the plasmalemma, and is nonliving, the cell membrane constitutes an integral part of the living cell and encloses the fluid substance, cytoplasm, along with its associated organelles. Other similar membranes found within the cell are the tonoplast which delimits the vacuoles from the cytoplasm and the membranes surrounding the intracellular organelles, such as the nucleus, mitochondria, and plastids. Plasmalemma and other membranes are also involved in maintaining the optimum environment for such processes as photosynthesis and respiration. In addition, they are the sites of many complex biochemical phenomena, e.g., energy conversion in photosynthesis, respiration, and protein synthesis.

Under an electron microscope the cell membrane commonly appears as a lighter line (electron transparent) sandwiched between two darker (electron opaque) lines. The entire membrane system is thought to consist of a central bimolecular layer of fatty material placed between two monomolecular layers of proteins of equal or unequal thickness. The composite protein-lipid-protein layer is known as a unit membrane and is comparable to a very fine sieve through which only small-sized molecules can pass.

Recent studies on the chemistry and functions of cell membranes have tended to attach greater biological significance to their protein, rather than lipid, components. Besides serving as permeability barriers delineating microenvironments within the cell, the membranes are involved in energy transduction in chloroplasts and mitochondria, in active transport by means of carrier proteins, in pinocytosis, and in cell fusion. In view of these multifarious functions, it is not proper to regard them as static structures based on the classical bilayer model. In fact, they are highly dynamic structures, and Cook's (1971) subunit model of membrane structure is based on this consideration. According to this model, membranes are considered to be composed of lipoprotein subunits in which two kinds of bonding occur, viz., the inter-subunit and the intra-subunit bonding. The former force is weaker than the latter, thus allowing for selective and preferential breakage of inter-subunit bonds. Although there are variants of the subunit model, the essence of all is that the changing pattern of proteins within the subunits constitutes the membrane specificity and that the weak nature of the inter-subunit bonds makes the membrane readily submit to pinocytosis and fusion.

Investigations on membrane organization constitute a fascinating and active area of research in modern cell biology, and a large number of electron micrographs of various cell membranes have been published during the past two decades. Based on such studies, conflicting views have been put forward to explain the general structure of membranes. These electron micrographs, though useful, can provide only a rather static view. In the living cell, the membranes, far from being static, are constantly undergoing alterations, breaking, and rejoining in various ways.

THE CYTOPLASM

Internal to the plasmalemma is the protoplasm which is made up of (1) the cytoplasm, and (2) the nucleus. The cytoplasm is the metabolic centre of the cell, consists of a granular matrix or molecular aggregates of various sizes, and is generally thought to be colloidal in nature. In mature cells, the cytoplasm is commonly present in the form of a thin layer just beneath the plasmalemma and rarely exceeds 7–8% of the total cell volume, but some cytoplasmic threads intersect the cell. The cytoplasmic matrix consists largely of water (about 85–95%), the balance (about 5%) being made up of globular and fibrous proteins.

The cytoplasms of different cells may vary greatly in viscosity. An elegant method for estimating this viscosity consists of introducing fine particles of iron or nickel into the cytoplasm of certain cells (e.g., those of Myxomycetes), and then placing the cells in a magnetic field; the amount of magnetic force required to produce a certain movement of the particle can then be determined and from this the viscosity estimated. Observing the streaming movement in many living cells and determining the rate of plasmolysis are two more useful methods of estimating relative viscosities.

Many researchers have attempted to analyze the protoplasm (or cytoplasm) chemically. This is, of course, very difficult in plants since the protoplasts cannot be separated easily from the cell walls. Animal cells which lack cell walls, and the plasmodia of certain slime moulds (Myxomycetes) which lack cell walls as well as vacuoles, are favourite experimental subjects for gross analyses of protoplasm (Table 1-1). However, such analyses should not create the impression that protoplasm may be synthesized by mixing substances in the proportions stated in Table 1-1. The living protoplasm is something more than these substances and therein lies the mystery of life.

Table 1-1 Analyses of plasmodia of *Fuligo* and *Reticularia* (Based on data of Lepeschkin, 1924 and Kiesel, 1925)

Substance	Percentage of dry weight
Fats	14–18
Proteins	8–29
Nucleoproteins	25–32
Free nucleic acids	2.5–3.7
Monosaccharides	8–14
Polysaccharides	1.8–3.0
Minerals	3–5
Unidentified	4–6

Cytoplasmic Organelles and Inclusions

Endoplasmic reticulum and ribosomes The cytoplasm of eucaryotic cells is made up of a complex system of submicroscopic, anastomosing membranes, known as endoplasmic reticulum (ER), which form a network and probably serve to provide structural or skeletal support to the cytoplasm. In some cells, especially those active in protein synthesis, the ER membranes are lined externally by small particles, called ribosomes, and this kind of ER is described as 'rough' ER; in other cases, the ribosomes are not disposed along the ER which is then qualified as 'smooth'. The ER is frequently in intimate contact with the outer membrane surrounding the nucleus and such ER-nuclear envelope connections probably serve as a channel, facilitating the exchange of materials or information between the nucleus and the cytoplasm. Likewise, the ER may also sometimes be connected with the plasmalemma (though this is rare in plant cells) and then may constitute similar channels for the transport of substances between the cytoplasm and the external environment. Strands of ER membranes may also occasionally interconnect adjoining cells through plasmodesmata. A good example of such intercellular connections is furnished by the sieve tube element and its associated companion cell in the phloem of angiosperms. Strands of ER have been seen to pass through the companion cell into the sieve tube (or vice versa) in some electron micrographs.

Since the ER membranes are composed of proteins (and fats), it is conceivable that some of these proteins may be endowed with enzymatic function, in which case the ER may well function as a biosynthetic site for certain chemical substances in the cytoplasm.

The ribosomes are submicroscopic granules composed of protein and RNA, either occurring freely in the cytoplasm or being attached to the ER. On the ribosomes, amino acids are assembled to form enzymes and other proteins. Further details about the ribosomes are given in Chapter 12.

Microtubules Electron microscope studies of many eucaryotic cells have revealed the widespread presence of microtubules, about 250 A thick. These constitute a sort of microskeleton for the cell and may play a significant role in the determination and specification of the cell shape, chromosome movement, and motility. They also impart rigidity, flexibility, and firmness to the protoplast.

Microtubules from many diverse kinds of cells seem to be identical as regards their chemistry, micromorphology, and drug sensitivity (see Margulis, 1973). A good example of microtubules is the mitotic spindle which is responsible for the distribution and migration of the chromosomes to the two poles during nuclear division. The rate of such migration is rather slow and ranges between 0.2 μm and 5.0 μm per minute. The driving force of such a slow movement is not known.

Chemically, microtubules are largely composed of certain proteins, called tubulins, a major component of the vertebrate brain tissues. These proteins are sensitive to low concentrations of colchicine which has proved to be very useful in analyzing the assembly of microtubular proteins into tubules. Several morphologically distinguishable microtubule proteins may be found in a cell.

A few mutant strains of certain flagellates (e.g., the paralyzed flagellar mutants) are thought to represent potential alterations in microtubules. Suspected microtubular mutants have been successfully isolated in *Chlamydomonas reinhardii*, *Paramecium aurelia*, *Saccharomyces pombe* (temperature-sensitive), *Schizosaccharomyces pombe* (colchicine-resistant), *Triticum aestivum* (somatic association suppressor or promotor mutants), and *Drosophila melanogaster* (sterile, Y mutants). Most of these mutants exhibit a chromosomal pattern of inheritance. They are now being used for genetical studies and analysis of microtubules.

Plastids These are discoidal, globular, or oval bodies found in the cytoplasm of most plant cells. They consist of nucleoproteins, lipids, carbohydrates, and various kinds of small molecules. Plastids may be classified according to their colour and function. The coloured plastids containing chlorophyll and other photosynthetic pigments are called chloroplasts and those containing certain orange, red, or yellowish pigments are known as chromoplasts. Amyloplasts (colourless plastids) probably serve as centres for the storage of starch.

The chloroplasts which may vary from about 5 μm to 20 μm in diameter are the site of photosynthesis. Each chloroplast is enveloped by two membranes, but mature chloroplasts are not connected to them. The inner membrane usually forms lamellae, called thylakoids, as shown in Fig. 1-2. Many thylakoids may get stacked together to form what is known as a granum, and a single chloroplast may contain as many as 60 grana. Under the electron microscope the membranes show a globular micromorphology and seem to be embedded with certain particles. Chemically, the membranes consist of roughly equal proportions of lipids and proteins to which the chlorophylls and other pigments are bound. They contain also various enzymes and other components of the photosynthetic electron transport chain.

The ground matrix of the chloroplast is called stroma. It contains the enzymes concerned with photosynthetic CO_2 fixation and also ribosomes and DNA fibrils (not associated with histones).

Further details about chloroplasts are given in Chapter 4.

Mitochondria These are filamentous or ovoid structures (0.2–2.0 μm in diameter) bounded by two unit membranes. The inner membrane is variously infolded into numerous parallel, flattened, tubular (mostly in plants), or interdigitating membranes, called cristae, which project as minute finger-like lobes from the periphery to the core of the organelle. These cristae are generally covered with stalked granules (6–9 nm in diameter), called oxysomes, which are thought to contain certain enzymes. However, many researchers consider the oxysomes as an artefact of fixation.

The general ground space (granular material) of the mitochondrion is known as the mitochondrial matrix.

Chemically, mitochondria contain roughly 35% proteins, 25% phospholipids, and 5–10% nucleic acids. They are the site of oxidative metabolism or aerobic respiration and contain the Krebs cycle enzymes. Since they also contain some ribosomes, they can synthesize some proteins. Like chloroplasts, new mitochondria arise through the division of pre-existing mitochondria and are endowed with a partial genetic autonomy.

Phylogenetically, the mitochondria of eucaryotic cells are often thought to have been derived from ancient bacterial cells (bacteria do not have mitochondria) which invaded some larger archaic cells and, during the course of evolution, took up a symbiotic mode of life as the mitochondria of the larger cells (see Raven, 1970). This is analogous to the independent, endosymbiotic origin of the chloroplasts from ancient blue-green algae (discussed in Chapter 4). However, many workers do not agree with such a view of mitochondrial origin.

Golgi bodies Also known as dictyosomes, these organelles (about 2–5 μm in diameter) may be located around the nucleus or scattered throughout the cytoplasm. They are specially abundant in meristematic cells. Each Golgi body (a collection of such bodies is often called the 'Golgi apparatus') consists of packed arrays, called cisternae, or stacks of flattened vesicles which are slightly dilated at the edges. The membranes composing the stacks seem to have a smooth surface.

The Golgi apparatus seems to have a well-defined polarity in flagellate algae, with its one end (called basal) lying deep in the cytoplasm near the nucleus, and the other end lying near the surface of the plasma membrane. Its vesicles continually change in shape and size as they move to the apical pole. Golgi bodies probably originate from the ER by budding and later merge with the plasmalemma. It seems that the Golgi apparatus functions like a vesicle for packaging secretory materials, characteristic of each eucaryotic cell type, and transporting them to the cell surface. Manton (1967) has shown that the Golgi apparatus in the flagellate algae *Prymnesium* and *Chrysochromulina* is involved in the synthesis, storage, and deposition of wall materials on the surface of the cell. Similarly, the coccoid chrysophyte, *Pleurochrysis*, seems to be dependent on its Golgi apparatus for the formation of the cell wall (Brown, 1969). Morr *et al.* (1965) have made a detailed study of Golgi bodies in higher plants. They have seen structural polarity in Golgi bodies in which an individual body is visualized to have a forming face and a maturing face. New cisternae are added to the forming face, probably from the ER, and on maturation they become part of the maturing face.

It has also been suggested that the Golgi apparatus functions in a way that can be characterized as 'reverse pinocytosis' or 'exocytosis' as opposed to pinocytosis or endocytosis. In endocytosis, large solid particles are transported from the outside to the interior of the cell. Procaryotic cell membranes do not show any such properties, i.e., they are incapable of endocytosis or exocytosis. Indeed this constitutes another major distinction between procaryotes and eucaryotes.

Peroxisomes These are a single-membrane type of microbody (0.2–2.0 μm in diameter), occurring in animal cells and in leaf mesophyll cells of plants. They function in close cooperation with chloroplasts in photorespiring plants in reactions involving the light-stimulated uptake of oxygen and glycollate oxidation (see Chapter 9). Another similar type, glyoxysomes, are known to occur only in plants where they are abundant in germinating fatty seeds. They contain enzyms for the transformation of fatty acids into sugars.

Vacuole Young meristematic plant cells usually contain a number of small vacuoles which later coalesce to form a single large vacuole. This large vacuole often occupies as much as 85–95% of the volume of the older, mature cell. Sometimes portions of plasma membrane become invaginated into the cell cytoplasm and form small vacuoles. Their formation constitutes one means by which water and mineral salts (or small food particles in animal cells) may enter the cell.

The vacuole plays a major role in cell physiology, especially in the uptake and regulation of water and salts. It is filled up with a dilute solution, vacuolar sap, containing dissolved gases, mineral salts, sugars, various kinds of acids, amides, glycosides, water-soluble proteins, alkaloids and pigments such as anthocyanins, and certain crystalloidal substances. Some of these materials may be absorbed passively from ambient solutions whereas others, e.g., mineral salts, are accumulated actively, i.e., at the expense of metabolic or respiratory energy. The sugars and pigments are synthesized in the cytoplasm from where they migrate through the vacuolar membrane (tonoplast) into the vacuole.

Healthy cells of procaryotes do not have any true vacuoles, but they may have other vacuole-like structures, called gas vacuoles or pseudovacuoles. These procaryotic vacuoles do not have unit membranes which are typical of eucaryotic vacuoles. The membrane of the gas vacuole is mainly proteinaceous and devoid of lipids and carbohydrates.

Spherosomes These are spherical organelles, less than 1.5 μm in diameter, and widespread in many plant cells. Each spherosome is enveloped in a single unit membrane and is thought to develop by the accumulation of oily material at the end of a membrane of the ER. The delimited part is then constricted to form a prospherosome which finally matures into a spherosome. Chemically, spherosomes are rich in proteins and lipids. Their function is not known, but there is some indication that they are involved in synthesis and storage of lipids.

Lomasomes Located at the cytoplasmic periphery and associated with cell walls, especially in fungi, algae, and higher plants, these are believed to function in cell wall synthesis. Their ontogeny and precise function are not clearly known.

Lysosomes These are very common in animal cells but rare in plants. They are found in algae and fungi and have also been reported to be present in aleurone grains of cotton seeds and pea seeds (Hughes *et al.*, 1970). Structurally similar to peroxisomes, the lysosomes contain certain hydrolytic enzymes, e.g., acid phosphatase, which, when released into the cytoplasm, can result in its lysis and destruction; in normal cells, these enzymes are trapped within the lysosomes by the barrier of lysosomal membrane. Because of their lytic action, the lysosomes are commonly regarded as 'suicide bags'.

Many leading cell biologists are of the opinion that the ER, Golgi bodies, lysosomes, and other such organelles, e.g., peroxisomes and spherosomes, are ontogenetically related. This view certainly assumes an amazing degree of dynamism generated by a basically unit-membrane organization of the subcellular organelles of eucaryotic cells. However, mitochondria and chloroplasts are double-membrane structures with their own genetic apparatus. They have not been found to bear any ontogenetic relation with single-membrane organelles.

THE NUCLEUS

The second major portion of the protoplast is the nucleus (the other being the cytoplasm) which is spherical or oval and usually 5–50 μm in diameter. A plant cell may be uninucleate or multinucleate whereas the mature sieve tube cells in the phloem tissue of higher vascular plants are enucleate, though they have a nucleus when young. The nucleus is surrounded by its own double membrane, called the nuclear membrane, whose continuity is broken here and there by minute pores.

The ground matrix of the nucleus is rather granular and semifluid and is generally termed the nucleoplasm. Embedded in the nucleoplasm are three prominent and relatively distinct types of structural entities: chromosomes, nucleoli, and ribosomes. Chromosomes are threadlike structures, visible clearly only during cell division. They are largely composed of protein plus DNA, and bear genes in a linear arrangement. Nucleoli are denser than the nucleoplasm and are more or less oval. One or more nucleoli may occur within a single nucleus. They vary greatly in their dimensions but are commonly 5–15 μm in diameter. Nucleoli are rich in RNA content and seem to function mainly in the synthesis of ribosomal RNA. The nuclear ribosomes appear to be formed within the nucleolus and, like cytoplasmic ribosomes, are concerned with protein synthesis.

The nucleus is undoubtedly the most important control centre of the cell, with less important control centres being located in chloroplasts and mitochondria (and perhaps also in other organelles). The nucleus plays a significant role in cellular reproduction and differentiation, and directs the various metabolic activities of the living cell. It issues appropriate signals or instructions (the genetic information) which guide the life processes of the cell. Nevertheless, if cells are deprived of their nucleus, they can sometimes continue their normal activities for a certain period of time. Very interesting experiments on nuclear-cytoplasmic interrelations have been performed on the marine green alga *Acetabularia* which is commonly two inches tall. When young, it has a single conspicuous nucleus lying in one of the several lobes of the attaching organ (holdfast). The nuclear lobe can be amputated and the remaining, enucleated *Acetabularia* can be kept alive for a few weeks. These experiments indicate that the presence of the nucleus may not be absolutely necessary in this case and also that, perhaps, the cytoplasm itself contains sufficient genetic information in one form or the other to sustain life for some time. Recent researches have confirmed that at least some of the subcellular organelles of plant cells do have their own genetic information and can synthesize certain enzymes.

Oscar Miller of Oak Ridge National Laboratory, USA, has succeeded in actually observing, photographing, and studying the action of single genes under the electron microscope. For these studies, Miller chose to work on nucleolar genes of amphibian oocytes which have unusually large nuclei, each containing perhaps hundreds of nucleoli. Clusters of genes, coding for ribosomal RNA, are contained in these nucleoli. It was further shown that the nucleolar core axis consisted of DNA coated with protein and that the matrix consisted of ribonucleoprotein microfibrils. Elegant micrographs of various nucleolar genes were successfully taken (Fig. 1-3). Some of these showed a cluster of neighbouring genes which are separated from each other by intervening 'spacers' of transcriptionally-inactive DNA; others showed a single nucleolar gene with RNA polymerase molecules on the gene at attachment points for the ribonucleoprotein microfibrils. It has been estimated that each such gene contains 80–100 polymerases simultaneously transcribing ribosomal RNA precursor molecules (Miller, 1972).

ISOLATION OF PROTOPLASTS

A very significant development in modern cell biology has been the successful isolation of plant cell protoplasts through enzymatic digestion of the cell wall. Under suitable experimental conditions, treatment of cells

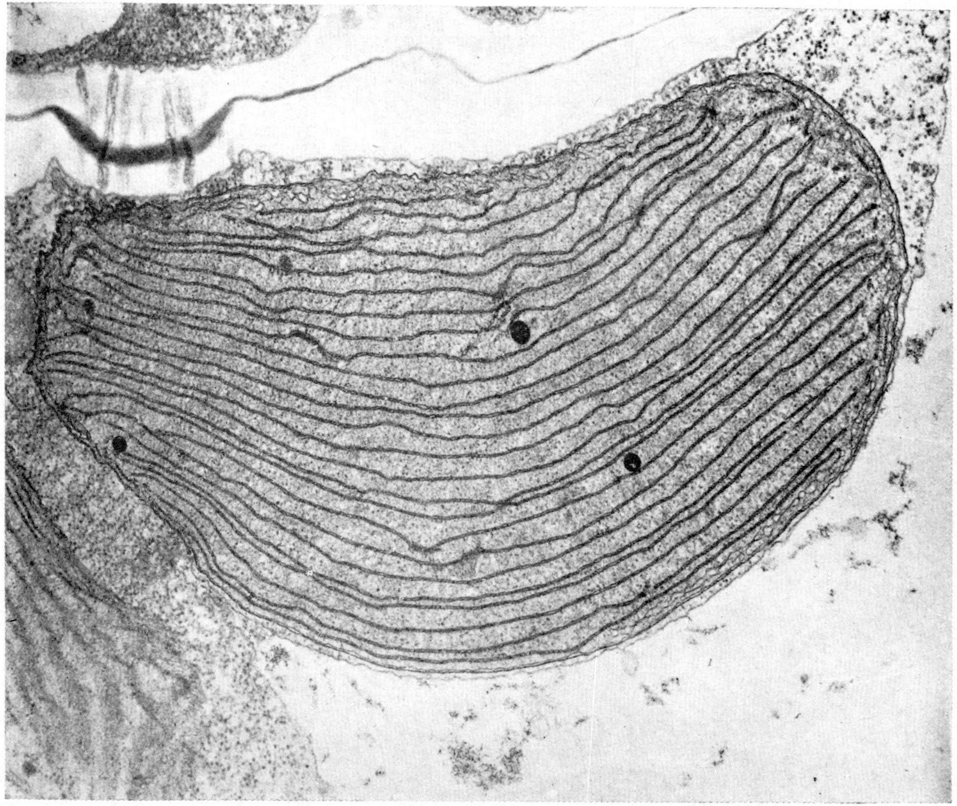

Fig. 1-2 Fine structure of a chloroplast from a mature bundle-sheath cell of sugarcane leaf. (Courtesy : Professor W. M. Laetsch, ×24,000.)

tandteile des Plasmodiums von *Reticularia lycoperdon*. *Zeit. f. physiol. Chemische.*, **150**, 149–75 (1925).

Lepeschkin, W. W., Über die chemische Zusammensetzung des Protoplasmas des Plasmodiums. *Ber. deut. bot. Ges.*, **41**, 179–87 (1924).

Manton, I., Further observations on the fine structure of *Chrysochromulina chiton*, with special reference to haptonema, peculiar golgi structure and scale formation. *J. Cell Sci.*, **2**, 265–75 (1967).

Margulis, L., Colchicine-sensitive microtubules. *Internat. Rev. Cytol.*, **34**, 333–61 (1973).

Miller, O., Genes in flagrante transcripto. *New Scientist*, 677–80 (1972).

Morr, D. J., Mollenhauer, H. H., Chambers, J. E., Glutaraldehyde as an aid to golgi apparatus isolation. *Exptl. Cell Res.*, **38**, 672–78 (1965).

Raven, P. H., A multiple origin for plastids and mitochondria. *Science*, **169**, 641–46 (1970).

Setterfield, G., Bayley, S. T., Structure and physiology of cell walls. *Ann. Rev. Plant Physiol.*, **12**, 35–62 (1961).

Villee, C. A., Biology, 6th edn. Saunders, Philadelphia (1972).

2: Bioenergetics

INTRODUCTION

Bioenergetics is the branch of biology concerned with the principle of energy transformation and migration in living cells (see Lehninger, 1971). Energy is defined as the capacity to do work and it may exist in various forms which may be potential, kinetic, thermal, chemical, electrical, radiant, etc. The various physical and chemical events occurring in living cells are always accompanied by energy changes and the science dealing with the laws of such energy changes is called thermodynamics or energetics in nonliving systems and bioenergetics in biological systems. Before we discuss the laws of bioenergetics we shall deal with some of the more common events concerning energy and matter migration in the biological world.

FLOW OF ENERGY AND MATTER IN THE BIOLOGICAL WORLD

Living organisms grow, differentiate, and reproduce by performing three main kinds of work, viz., biosynthetic, osmotic, and mechanical. All these works involve the expenditure of only one kind of energy, called chemical energy, stored in the form of cellular carbohydrates, fats, or proteins and utilized, during their oxidation in respiration (adenosine triphosphate (ATP) can also be formed during photosynthesis), only through the generation of a common energy-labile intermediate ATP, also appropriately termed the biological currency of the cell. The carbohydrates, fats, and proteins are synthesized from CO_2, H_2O, and N_2 in photosynthesizing cells at the cost of solar energy. In fact, only the (green) photosynthesizing cells of plants are capable of harvesting the sunlight energy and storing it in the form of fuel (food) substances. This food-making ability divides the whole biological world into two broad groups, the autotrophs which are capable of synthesizing their own food, and the heterotrophs which are incapable of doing so. Photoautotrophs include higher plants, algae, and some photosynthetic bacteria; these can meet their energy requirements from sunlight. In contrast, chemoautotrophs, which include mainly bacteria, obtain their energy from oxidation of inorganic substances. In a photosynthetic eucaryotic cell, the chloroplast can be regarded as an organelle concerned with synthesizing fuel substances from CO_2, and H_2O (in the presence of sunlight); and the mitochondrion can be regarded as an organelle of respiration concerned mainly with the oxidation of the fuels to CO_2, H_2O, and the chemical form of energy (ATP).

The energy-storing efficiency of the two kinds of organelles is shown in

cribing the basic aspects of thermodynamics and then discussing its application to biological processes in general.

PRINCIPLES OF THERMODYNAMICS

Thermodynamics, one of the most basic and exact branches of physics, provides an easy and precise method for analyzing physical and chemical phenomena. Though it has a complex mathematical component, its philosophy and logic are simple and straightforward. It deals with gross properties of matter, such as pressure, temperature, volume, and composition, and the thermodynamic feasibility of a chemical process which can be understood without any prior knowledge of the structure of the matter or the atomic theory and the mechanisms and pathways of chemical reactions.

A thermodynamic study of energy exchanges proceeds primarily with the specification of a collection of matter which is undergoing some physical or chemical change. Such a collection of matter is termed a 'system' and everything else in the universe is termed the 'surroundings'. Next comes the determination of the total energy of the system before it has suffered any physical or chemical change. This is called the initial state of the system. It is followed by the determination of the energy content of the final state which the system reaches after the physical or chemical change has actually taken place. During the transition from the initial to the final state, the system may absorb energy from, or deliver energy to, the surroundings. The final state is also sometimes called the equilibrium state. The amount of energy lost to, or gained from, the surroundings during the transition corresponds to the amount gained or lost by the surroundings; that is, the total amount of energy in the universe remains the same. Experience suggests that it is practically impossible to determine the total energy content of the initial and final states of the system. However, the next best alternative, which merely measures the energy difference between the initial and the final state, is possible to achieve. In fact, for a thermodynamic analysis it is sufficient if the amount and kind of energy exchanges between the system and the surroundings as the system passes from the initial to the final state is known. Since thermodynamic studies involve the consideration of such factors as volume, pressure, and temperature, the investigator always tries to reduce these variables to the minimum in order to facilitate the accuracy of the results. Happily, the biological reactions take place mostly in aqueous solutions which are in equilibrium with the atmosphere, and under such conditions the three most important thermodynamic variables (pressure, volume, and temperature) remain fairly constant.

Heat is the most common form of energy and various other forms of energy can be easily transferred to heat, thus making it possible to quantitate any kind of energy by transferring it to heat. The basic unit of energy in chemical thermodynamics is the gram calorie which is defined as the amount of energy required to raise the temperature of one gram of water at 15°C by 1°C (i.e., to 16°C). One kilocalorie (abbreviated kcal) is equivalent to 1000 calories.

FIRST LAW OF THERMODYNAMICS

According to this law, energy can be neither created nor destroyed. In other words, the total energy content of a system plus the surroundings remains constant in any physical or chemical process. That different forms of energy are readily interconvertible is an established fact. For example, a steam engine converts thermal energy into mechanical energy which is transformed into electrical energy by a generator. Similarly, electrical energy can be converted into chemical form by electrical charging of a storage battery.

Another implication of the first law is that there is quantitative correspondence between different forms of energy. This has led to the establishment of quantitative equivalence between a unit of heat energy (calorie), a unit of mechanical energy (erg), and a unit of electrical energy (watt); thus, one calorie is equal to 4.185×10^7 ergs or 4.185 watts. However, it should be clearly understood that despite such numerical equivalence it is not possible to completely convert one form of energy into another form.

Every reaction is generally accompanied by heat changes which means that the progress of the reaction may consume or liberate heat. Heat-consuming reactions are termed endothermic, and heat-liberating reactions, exothermic. A heat-measuring instrument, called the calorimeter, is normally used to measure the amount of heat change (called heat of reaction) suffered by a reaction, and from this the amount of work the given chemical reaction is capable of doing can be easily inferred. Heat and energy exchanges are normally expressed in terms of kcal of energy exchanged per gram molecular weight of the substance transformed under standard conditions. The standard state of a reactant or product is defined as a concentration of one gram molecular weight per litre of water at a temperature of 25°C and a pH of 7.0. Various organic substances when burnt in a bomb calorimeter in the presence of oxygen liberate heat, called the heat of combustion. Combustion here means the complete oxidation of the substance with oxygen, leading to the formation of CO_2 and water. Table 2-1 lists the heat-of-combustion values of some common organic substances.

Table 2-1 Heat of combustion of some common foodstuffs

Substance	Molecular weight	Heat of combustion (kcal/mole)
Glucose (carbohydrate)	180	673
Tripalmitin (fat)	809	7510
Glycine (amino acid)	75	239

The general rule is that the more complex a molecule is structurally, the more potential energy it possesses. The determination of the heat of combustion of any compound indicates also that its production will require the same amount of energy as was liberated during its combustion. Heat energy can be converted into work in a system only if the system is maintained at two different temperatures (just as in a steam engine). But biological systems are isothermal and as such they are unable to convert heat energy into any kind of meaningful work.

SECOND LAW OF THERMODYNAMICS

Whereas the first law deals with the conservation of energy during any physical or chemical process, the second law provides a measure of predicting the direction of energy change in any given system. This law introduces in the realm of energy transformations another component, called *entropy*, which represents the state of matter and energy. Entropy is very difficult to define precisely, but some of its attributes will help in understanding its real significance. For instance, it may be regarded as a measure of disorder or randomness. Thus, greater entropy means greater disorder, and vice versa. All spontaneous processes, whether chemical or physical, progress in a way that leads to maximum increase in the randomness or entropy of the universe. Entropy represents that energy part of the total which is never available for doing work. This brings us to the second implication of the second law, viz., all real processes are associated with increase in entropy and thus constitute a class of irreversible reactions. For example, let us consider a spontaneous reaction in which *a* is converted into *b* with the liberation of energy which, if made available completely, should be sufficient to drive the reverse reaction with an efficiency equal to the forward reaction. In such a situation, the transformation of *a* into *b* and *b* into *a* should continue endlessly. But this does not actually happen because each time a part of *a* is changed into *b*, a part of the released energy becomes unavailable or dissipated, thus making the reverse reaction less efficient than the forward. The repeated continuation of this process will lead to more and more loss of energy, with the result that the whole reaction will finally come to a stop. Since the unavailable form of energy for performing work is termed entropy, it is easy to regard entropy as a reason for the irreversible nature of a real process. This also explains why entropy or disorderliness in the universe should increase with the passage of time. The entropy unit is the measure that has the dimension of calories per mole.

All events in nature tend to seek a state which is accompanied by maximization of entropy of the universe (see Morowitz, 1968, 1970). We can directly measure the total energy change a given system suffers during its change from the initial to the final state. The total energy change at constant pressure is termed *enthalpy* change and is symbolized by ΔH. Though there is no direct way to measure entropy (S) or entropy changes (ΔS), these changes are found to be related to the total energy change, i.e., enthalpy change, by a third term or function, called *free energy*.

Enthalpy change, free energy, and entropy change are related by an equation which combines both the first and second laws of thermodynamics, and this relation holds good under conditions of constant temperature, pressure, and volume:

$$\Delta G = \Delta H - T\Delta S \tag{1}$$

where G represents free energy, ΔG is the change in free energy, ΔH is the change in enthalpy (total energy), T the absolute temperature, and ΔS the change in entropy. Equation (1) can be arranged as

$$\Delta H = \Delta G + T\Delta S \tag{2}$$

As already stated, entropy represents a nonavailable form of energy for work whereas free energy represents the available form of energy for work. In such a relation, if entropy increases, free energy decreases, and vice versa. If entropy is a measure of disorder, its opposite is equivalent to order.

THIRD LAW OF THERMODYNAMICS

The third law of thermodynamics states that the entropy of a perfect crystal at absolute zero temperature will be zero which is quite apparent from equation (1). Any system at a temperature above absolute zero will always have a certain amount of entropy. If the fine structure of a cell is examined, a most orderly arrangement of the various macromolecules is found within it. Such a highly structured complex, possible only at the cost of a large amount of energy, represents a minimum amount of entropy. However, when cells die and their orderly structures become disorganized, there occurs a great loss of energy in the form of entropy. Thus, growing cells consistently create order by increasing the entropy of the surroundings. Even nongrowing cells need a constant supply of energy to maintain their orderly state because their day-to-day normal activity of survival involves also many reaction steps, each causing an irreversible loss of a part of energy to the environment. This means that order is being constantly siphoned off as cellular events take place, and the maintenance of order therefore requires continuing assimilation of energy throughout cell life. In any spontaneous process that is exothermic, the value of free energy is negative whereas in an endothermic process it is positive. Exothermic reactions proceed with decrease in their free energy value which, when the reaction is at equilibrium, reaches the lowest value at which the temperature, pressure, and all other measurable quantities remain uniform throughout the system and the surroundings. In such a state, the entropy of the surroundings plus the system reaches the maximum value.

We shall now examine the relation between free energy and chemical equilibria, to see why such phenomena as free energy and entropy are of interest to biologists.

FREE ENERGY AND EQUILIBRIA

Let us consider a hypothetical exothermic reaction in which A is transformed into B. It is important to note that any spontaneous reaction never reaches completion point, i.e., about 80% of A may be changed into B with the passage of time, but at a certain point no further transformation of A into B occurs. The energetically unfavourable reaction B → A is also possible, provided that energy is available to force it. Since, by starting with A, a transformation of 80% of A into B has occurred, at equilibrium no more than 20% of B can be transformed into A. In fact, the position of the equilibrium is a characteristic feature of each reaction and is independent of the rate at which it is attained, i.e., the equilibrium position is unalterable and is expressed by the equilibrium constant (K_{eq})

$$K = \frac{b}{a}$$

where a and b represent molar concentrations of A and B, respectively. The present hypothetical reaction will have $K_{eq} = 80/20 = 4$. This means that the reaction favours a transformation of more A into B rather than more B into A. If there is more than one reactant and product, the K_{eq} can be obtained in the same manner:

$$A + B \rightleftharpoons C + D$$

$$K_{eq} = \frac{cd}{ab}$$

For spontaneous reactions, the K_{eq} lies in the direction of the products and will therefore have a numerical value greater than one. The higher the numerical value, the more favourable is the progress of the reaction towards the side of the products. The reciprocal of the K_{eq} value gives the strength of the reverse reaction. In such situations, if the concentrations of the reactants are greater than their equilibrium values, the reaction will proceed with the liberation, rather than consumption, of energy.

The simple mathematical relation between ΔG and K_{eq} is given by

$$\Delta G = -RT \ln K_{eq}$$

where G is the free energy change, R the gas constant (1.987 cal/mole/degree), T the absolute temperature, and $\ln K_{eq}$ is the natural logarithm of the equilibrium constant. In this relation, if ΔG is positive, then K_{eq} must be negative, and vice versa. $\Delta G°$ is termed the standard free energy change and is defined as the gain or loss of free energy in calories as one mole of reactant is converted into one mole of product at 25°C and pH 7.0. Table 2-2 shows the relation between K_{eq} and standard free energy, and in so doing explains that the reactions having K_{eq} values between 0 and 1 are exothermic.

Table 2-2 Numerical relation between K_{eq} and $G°$ at 25°C and pH 7.0

K_{eq}	$G°$ (kcal/mole)
0.001	+4.09
0.01	+2.73
0.1	+1.36
1.0	0.0
10	−1.36
100	−2.73
1000	−4.09

Many life reactions are unfavourably situated with regard to their equilibria. In such situations, the best alternative is to link the life reactions with spontaneous reactions in a way that makes the product of the first reaction the reactant of a subsequent spontaneous reaction. This kind of coupling will pull the entire process in the forward direction. Such a device is very much at work in most biological reactions.

An important point to note is that the thermodynamic principles discussed so far are based on closed systems which do not exchange materials with the universe. On the other hand, biological systems are open and readily exchange materials with the universe. Biological systems never attain true and complete thermodynamic equilibria. In fact, they represent a state of dynamism, called the dynamic steady state, in which the rate of formation of a given component is exactly counterbalanced by an equal rate of its removal or breakdown. Another property of living systems is that they increase the entropy of the universe and thus represent a kind of irreversible process. Such considerations have led to the development of a new kind of thermodynamics, called nonequilibrium thermodynamics, applicable to open systems such as biological activities. However, for all practical purposes the laws of classical thermodynamics are still very useful from the bioenergetics viewpoint.

Living cells store energy from oxidation of substances or by making electrons move along their downhill energy gradients through a process called phosphorylation. In this process, ATP is generated from ADP (adenosine diphosphate) and inorganic orthophosphate. We shall now describe this process and its mechanism.

PHOSPHORYLATION

Adenosine triphosphate is the energy currency and the driving force for all biosynthetic reactions. It is formed by the esterification of inorganic phosphate with the terminal phosphate of ADP at the expense of exothermic energy-liberating reactions. Cells have evolved three main routes to ATP synthesis, including substrate-linked phosphorylation (as in glycolysis), oxidative phosphorylation (as in mitochondria), and photosynthetic phosphorylation. The most common way by which cells capture or store energy in the form of ATP is by dropping electrons from a high energy state to a lower one, i.e., by moving electrons along the thermochemical gradient of electron carriers which undergo alternate redox changes. The difference between the energy of the electron donor and that of the electron acceptor represents the energy available to make ATP. The coupling of electron transport to phosphorylation means that ATP will not be formed unless electron transport occurs, and vice versa.

The resolution of the coupling of phosphorylation to electron migration has employed two main techniques, viz., (1) examination of the flow of compounds through intact pathways *in vivo*, and (2) dissection of the pathways into individual enzymatic reactions by fractionation studies, or by using specific metabolic inhibitors, or by producing mutant strains defective in one or the other enzymic or electron transport chain component.

The widely used techniques in the demonstration of phosphorylation *in vivo* have been the estimation of the level of ATP increase, or labelled pyrophosphate uptake, or labelled ADP, and phosphate incorporation into ATP under conditions where the only source of ATP is glycolysis, or oxidative phosphorylation, or photophosphorylation. These studies have revealed that: (a) two phosphate ions disappear and give rise to two molecules of ATP during the conversion of a molecule of glucose into two molecules of pyruvate in glycolysis; (b) 15 phosphate ions are used up and 15 molecules of ATP formed during oxidation of one molecule of pyruvate by oxygen; (c) two phosphate ions are taken up and converted into two molecules of ATP during the transfer of four electrons from two molecules of water to two molecules of $NADP^+$; and (d) conditions that promote phosphorylation promote also electron transfer, and vice versa.

The genes controlling the formation of cytochrome-c of mitochondria are located in the nucleus. It has also been demonstrated that synthesis of cytochrome-c occurs on 80S ribosomes in the cytoplasm and that mitochondria acquire cytochrome-c in readymade form from the cytoplasm. Mutants showing loss of this enzyme also lack aerobic respiration and the associated phosphorylation. Another class of extranuclear mutants termed *petite*, and widely investigated in yeast, lacks aerobic respiration and oxidative phosphorylation, and can grow at the cost of energy generated during fermentation or supplied in the form of ATP from outside. Such petite mutants contain rudimentary mitochondria, and it is believed that they have led to the removal of some of the components essentially required to assemble the structural units of mitochondria in proper functional order.

Mutant strains of *Chlamydomonas* and *Scenedesmus* in which the photosynthetic electron transport chain or one of the energy-trapping centres is blocked provide useful information regarding the components of the chain or the reaction centre that appeared to be required for phosphorylation, and possible sites for its occurrence (Levine, 1969). For example, noncyclic photophosphorylation which occurs in the presence of the added electron acceptor or Hill oxidant (ferricyanide) proceeds in a normal way in a mutant strain of *Chlamydomonas* that lacks P700 or PSI. Similarly, PSII defective or PSI defective mutant strains of *Scenedesmus* carry out either cyclic or noncyclic photophosphorylation, but never both. Clearly, cyclic photophosphorylation seems to be mediated by PSI, and noncyclic photophosphorylation by PSII. Cytochrome-b or cytochrome-f blocked mutants of *Chlamydomonas* do not perform noncyclic phosphorylation but instead perform cyclic phosphorylation, thus suggesting that a site for noncyclic phosphorylation probably lies between cytochrome-b and cytochrome-f and that for cyclic phosphorylation lies beyond cytochrome-f of the electron transport chain. Mutants of *Chlamydomonas* lacking plastocyanin do not show cyclic phosphorylation. This means that plastocyanin is intimately involved in cyclic phosphorylation.

The herbicide dichlorophenyl dimethyl urea (DCMU) which inhibits

PSII or Hill reaction or oxygen evolution also inhibits noncyclic photophosphorylation but not cyclic photophosphorylation. Similarly, CO and cyanides which poison cytochromes also inhibit phosphorylation. Iodoacetate which inactivates dehydrogenases, thus preventing the removal of electrons from the respiratory substrates, stops the formation of ATP. All these studies suggest that ATP formation is heavily dependent on both the structural and functional integrity of the electron transport chain and the downhill migration of electrons along this chain.

SUBSTRATE-LINKED PHOSPHORYLATION

In glycolysis, fructose-1,6-diphosphate yields two molecules of phosphoglyceraldehyde which is oxidized to the corresponding acid with simultaneous phosphorylation of its carboxylic group by inorganic phosphate. A phosphate group attached to an atom (carbon) which also bears a double bond $\left(\begin{matrix}C\\ \|\\ C\end{matrix}\!-\!P\right)$ is converted into a high energy state. This high energy state is related to the electronic structure of the molecule, since the excess of electrons on the carbon atom due to the double bond imparts it a negative character. As phosphate is also negatively charged, the two negative charges tend to repel each other. The net result is that a high energy phosphate group is donated to ADP and thus a molecule of ATP is formed. Such a formation of ATP is distinctive of glycolysis and is not coupled to the migration of electrons along the downhill gradient.

In glycolysis, there are two molecules, 1,3-diphosphoglyceric acid and phosphoenolpyruvate, in which the carbon atom holding the phosphate group contains double bonds and therefore excess of negative charges. Each of these molecules gives rise to one ATP from ADP and inorganic phosphate. There are two molecules of diphosphoglycerate and two of phosphoenolpyruvate and, hence, a total of four ATP molecules is generated per molecule of fructose diphosphate used. Since two molecules of ATP are utilized for the conversion of glucose into fructose diphosphate, there is a net gain of only two molecules of ATP in glycolysis.

The fate of the two pairs of electrons or hydrogens removed during oxidation of two molecules of phosphoglyceraldehyde depends upon whether conditions are favourable for aerobic respiration or fermentation. In aerobic respiration, the electrons are passed on from $NADH + H^+$ to oxygen through the respiratory electron transport chain, and nearly three molecules of ATP are generated per pair of electrons consumed by oxygen. In fermentation, the electrons are used to reduce either pyruvate to lactate or acetaldehyde (formed by decarboxylation of pyruvate) to ethyl alcohol.

Since the respiratory electron transport chain is located in the mitochondria, the electrons from the glycolytic $NADH + H^+$ which are formed in the cytoplasm have to travel to the mitochondria. Virtually nothing is known about the mechanism of such migration from cytoplasm to mitochondria; this aspect therefore remains a problem for future research.

OXIDATIVE PHOSPHORYLATION

Oxidative phosphorylation occurs in mitochondria and is coupled to the oxidation of the intermediates of the tricarboxylic acid (TCA) cycle (which also include the precursors of fatty acids and amino acids, and therefore of fats and proteins, respectively). Consequently, this cycle provides a common link between carbohydrates, fats, and proteins, oxidizing them completely to H_2O and CO_2. Three molecules of ATP arise per pair of electrons removed from NADH or NADPH ($+H^+$) to oxygen. The first group arises between NADH and FP, the second between cytochrome-b and cytochrome-c, and the third between cytochrome-a and oxygen. This has been confirmed by the results of inhibitor studies as well as by the observed difference in redox potentials of the electron carriers. Since five pairs of hydrogen or electrons are removed per molecule of pyruvate oxidized or per complete turn of the TCA cycle, the total number of ATP molecules generated during the conversion of pyruvate into CO_2 and H_2O is 15. Since two molecules of pyruvate are produced per molecule of glucose during glycolysis, the total ATP yield in the TCA cycle is (15×2) 30 molecules. Under aerobic conditions, the two pairs of electrons produced during the oxidation of two molecules of glyceraldehyde in glycolysis also yield six molecules of ATP in addition to two molecules of ATP; thus, the final ATP yield during complete oxidation of glucose is ($30 + 6 + 2$) 38.

PHOTOPHOSPHORYLATION

One site of ATP formation is between cytochrome-b and cytochrome-f of the electron transport chain, and another between the hypothetic reductant X of PSI and plastocyanin. The results indicate that one molecule of ATP is produced per pair of light-induced electron transfer.

Photophosphorylation has recently been shown to consist of two stages, viz., light and dark, and this is collectively termed the two-stage phosphorylation. Such a temporal division has been achieved by demonstrating that chloroplasts, when illuminated in the absence of phosphorylating agents (e.g., manganese and phosphate), generate high energy intermediates which can initiate the formation of ATP, even in the dark, if provided with the required phosphorylating reagents. This indicates that the function of light is to produce some high energy intermediates which can then be utilized for ATP formation through dark enzymatic processes. Despite extensive and intensive research, no such high energy intermediates have actually been identified so far.

MECHANISM OF PHOSPHORYLATION

There are two main theories for coupling electron transport to phosphorylation, viz., the chemical and chemiosmotic theories.

The chemical theory postulates the formation of some high energy chemical intermediate at the cost of the energy of the moving electron along the electron transport chain. This intermediate is not a free entity; it remains complexed with one of the electron carriers. For example, if the sequence of electron carriers is X to Y to Z (Fig. 2-3), then according to this theory, a coupling protein intervenes during the transfer of electrons from Y to Z, resulting in the formation of a reversible complex compound between the reduced Y and the coupling factor. This constitutes the postulated high energy chemical intermediate visualized under this theory. The consequence of this kind of reaction is that electron transport will not occur unless Y is freed, and Y will not be freed unless the complex is broken down. It is presumed that during phosphorylation inorganic phosphate replaces Y, and ADP replaces the coupling protein in the complex. The final outcome would be the formation of ATP and the normal resumption of electron transport. The intimate coupling between electron transport and phosphorylation is thus explained. Uncoupling here will mean the hydrolysis of the high energy intermediates or the inhibition of the formation of ATP. Consequently, electron transport will go on in the absence of phosphorylation. That this is indeed the case has been shown by the use of dinitrophenol, which causes hydrolysis of ATP, or oligomycin which causes hydrolysis of the high energy intermediates.

The chemiosmotic theory, a radical theory proposed by Mitchell (1967), postulates (see Figs. 2-4 and 2-5) that: (a) there is an ATP hydrolyzing enzyme, ATPase, so situated in the organelle membrane that it is accessible to OH^- ions from within and to ATP, ADP, PO_4, and H^+ ions from outside; (b) the ATP hydrolyzing activity of the enzyme obviously requires water; since the membrane is impermeable to water molecules, the hydrolysis requires the entry of OH^- from within and H^+ from outside; (c) any process that lowers the concentrations of H^+ and OH^- at the enzyme will favour the reaction, i.e., the formation of ATP, whereas any process that lowers the OH^- concentration inside the membrane will favour the synthesis of ATP; both photosynthesis and aerobic respiration cause accumulation of H^+ or protons inside the membrane, thus reducing the effective level of OH^- and favouring ATP synthesis; and (d) the electron carriers and other components are so arranged in the organelle membrane that only the proposed kinds of reactions will occur; however, the precise manner of disposition remains conjectural.

The accumulation of H^+ or protons inside the membrane will cause the pH to fall inside and rise outside, i.e., a proton gradient or pH-gradient will be created across the membrane. It has been observed that illuminated chloroplasts or mitochondria show a significant rise in pH outside the membrane and corresponding decreases towards the inside of the membrane.

A third theory for energy coupling mechanism, called the conformational theory, is increasingly influencing researches in this fascinating field. The focussing point of this theory is the mitochondrial membrane. It is believed that energy coupling is somehow related to minute conformational or structural transitions or fluxes in membrane structure. The fine structure of the mitochondrial membrane has been studied by many workers but, unfortunately, the electron microscope can provide only a rather static view of the membrane. In the living cell, the membrane may actually exist in a state of dynamic flux, with different parts undergoing minute changes, such as

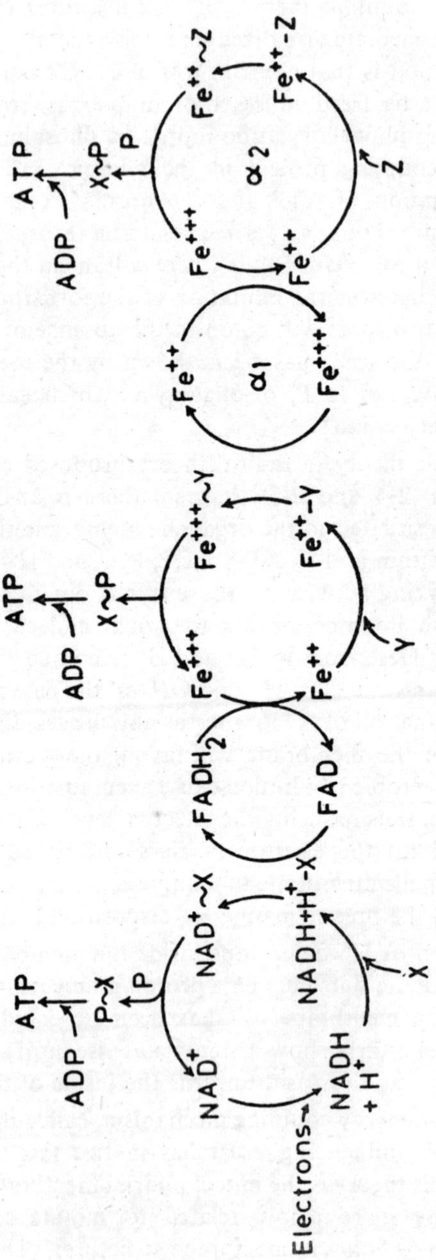

Fig. 2-3 Coupling of phosphorylation with electron transport in oxidative metabolism.

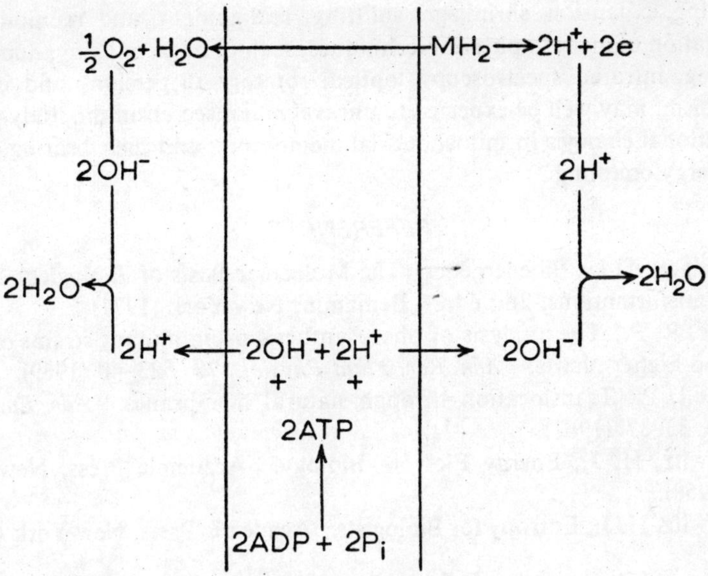

Fig. 2-4 General chemiosmotic model of Mitchell (1967).

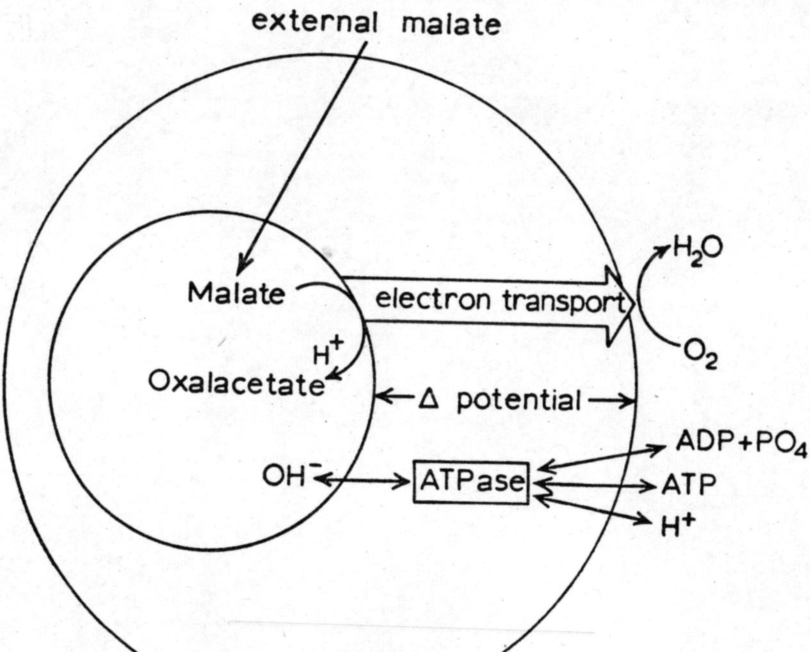

Fig. 2-5 Chemiosmotic model to explain mitochondrial phosphorylation. (After Mitchell, 1967.)

breaking, expansion, shrinkage, splitting, realignment, and reunion. The application of modern physical techniques, such as freeze etching, fluorescent probing, infrared spectroscopy, optical rotatory dispersion, and circular dichroism, may well be expected to unravel major secrets in the study of conformational changes in mitochondrial membranes, and their bearing, if any, on energy coupling.

REFERENCES

Lehninger, A. L., Bioenergetics: The Molecular Basis of Biological Energy Transformations, 2nd edn. Benjamin, New York (1971).

Levine, R. P., The analysis of photosynthesis using mutant strains of algae and higher plants. *Ann. Rev. Plant Physiol.*, **20**, 523–40 (1969).

Mitchell, P., Translocation through natural membranes. *Adv. Enzymol.*, **29**, 33–87 (1967).

Morowitz, H. J., Energy Flow in Biology. Academic Press, New York (1968).

Morowitz, H. J., Entropy for Biologists. Academic Press, New York (1970).

3: Enzymes

INTRODUCTION

The temperature at which most organisms carry on their life processes is too low to drive chemical reactions at any appreciable rate. The various activities of a plant cell, e.g., photosynthesis, respiration, growth, and reproduction, are manifestations of a complex series of biochemical reactions. Since ordinarily a chemical reaction that is thermodynamically possible cannot occur measurably at the temperature at which cells live, the question is: how do these cells manage to speed up their reactions at such physiological temperatures? The cells possess catalysts which make their biochemical reactions overcome the requirement for high temperatures. Whereas a requirement for high temperature (above 60°C) is the rule for all purely chemical reactions (though only for a rapid rate), the enzyme-mediated reactions are generally inhibited at temperatures exceeding about 45°C. This is because all enzymes are proteinaceous in nature and the proteins are denatured at high temperature. Thus, one of the unique characteristics of living cells is their ability to make complex reactions proceed rapidly at the temperature of the surrounding environment.

All enzymes are proteins, but all proteins are not necessarily enzymes. Some proteins are involved in the structural organization of various components of the cell. Such nonenzyme proteins are termed structural proteins. An enzyme, like structural proteins, is synthesized in a cell and speeds up a reaction that is thermodynamically possible and essential for the maintenance or the growth, or both, of the cell. The enzyme neither initiates the reaction nor alters its equilibrium or free energy status. It simply accelerates the reaction rate. Nonprotein or inorganic catalysts also accelerate a variety of chemical reactions. In contrast, a given enzyme catalyzes only a very few reactions (frequently, only one reaction). The ability of an enzyme to catalyze one specific reaction (and essentially no others) is perhaps its most significant property. This specificity is a function of the amino acid sequence which varies from enzyme to enzyme. Being proteins, enzymes are easily denatured by a variety of agents, such as heavy metal, strong salt solutions, heat, strong acids, and some organic solvents.

APOENZYMES AND COENZYMES

Many enzymes are composed of two parts, protein and nonprotein, both of which are essential for enzymatic activity. The nonprotein part may include flavin adenine dinucleotide (FAD), or nicotinamide adenine dinucleotide

(NAD), or metals such as Mo, Mn, and Zn. The nonprotein part is called a prosthetic group when it is *firmly* bound to the protein part, called apoenzyme. In this respect, FAD is the prosthetic group of succinic dehydrogenase. Another example is the porphyrin moiety of the haemoprotein peroxidases. The nonprotein part is termed a coenzyme when it is *loosely* bound to the apoenzyme. Two examples of this are coenzyme I (NAD) and coenzyme II, nicotinamide adenine dinucleotide phosphate (NADP). The metallic component of the enzyme may be firmly bound as in the case of iron in cytochromes, or loosely bound as in the case of nitrate reductase containing molybdenum. In general, the coenzyme or prosthetic group is heat-stable whereas the apoenzyme part is heat-labile. Vitamins constitute another group of coenzymes. Holoenzymes, which consist of both apoenzyme and prosthetic group or coenzyme, can be active only if both the components are present:

$$\text{Apoenzyme} + \text{Coenzyme or Prosthetic group} \rightarrow \underset{\text{Active form}}{\text{Holoenzyme}}$$

Although we have distinguished between the coenzyme and the prosthetic group on the basis of their individual strength of attachment with the apoenzyme, this distinction can sometimes be ambiguous or misleading, since strong or loose binding may vary with the conditions employed to separate the two components. A more unerring distinction is that the coenzyme acts as a cosubstrate for two reactions mediated by different enzymes whereas the prosthetic group reacts with two different substrates mediated by the same holoenzyme, of which the prosthetic group forms the integral part. For instance, the coenzyme NAD which acts as a hydrogen acceptor in most of the degradative oxidation reactions accepts the hydrogen removed by phosphoglyceraldehyde dehydrogenase from its substrate in a stoichiometric way, i.e., one mole of NAD is reduced for every mole of phosphoglyceraldehyde oxidized. The NAD is not then acting as a catalyst which in a very low concentration can transform a large amount of one substrate into another without suffering any chemical change or alteration. In this reaction, NAD violates both the conditions, as firstly, its reduction shows stoichiometric relation with the oxidation of glyceraldehyde and, secondly, it also undergoes chemical alteration by being converted into $NADH_2$. Its bonafide catalytic characters become apparent only when its reduction is coupled to the oxidation of its reduced form by another enzyme, i.e., the $NADH_2$ thus produced in glycolysis can act as a substrate for alcohol dehydrogenase or lactic dehydrogenase to produce alcohol or lactic acid. In this reaction, $NADH_2$ is produced in one kind of enzymatic reaction and is oxidized by oxygen with the help of respiratory cytochromes which constitute a series of entirely different enzymatic reactions (Fig. 3-1).

The prosthetic group does not act as a cosubstrate in the way NAD or NADP does. For example, alanine oxidase dehydrogenates alanine and passes the removed hydrogen to its prosthetic group FAD, and the latter transfers it to oxygen, thus forming H_2O_2. The resulting amino acid is then hydrolyzed to ammonia and the corresponding alpha-keto acid (Fig. 3-2). It can be seen that even the prosthetic group reacts stoichiometrically here, but it is returned to its original state by reaction with a second substrate (here, oxygen).

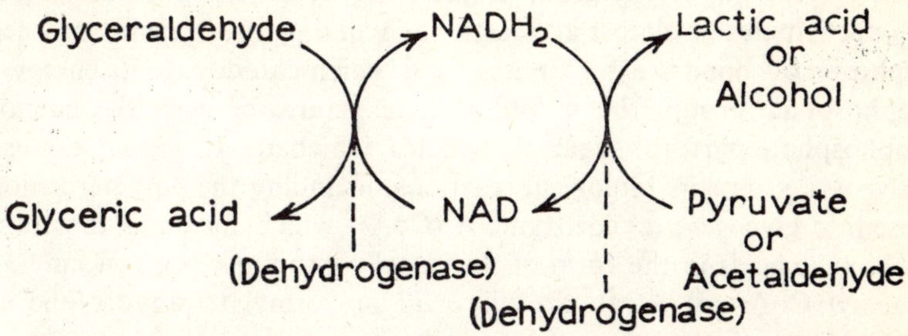

Fig. 3-1 Two different dehydrogenase-catalyzed reactions with participation of coenzyme I as cosubstrate.

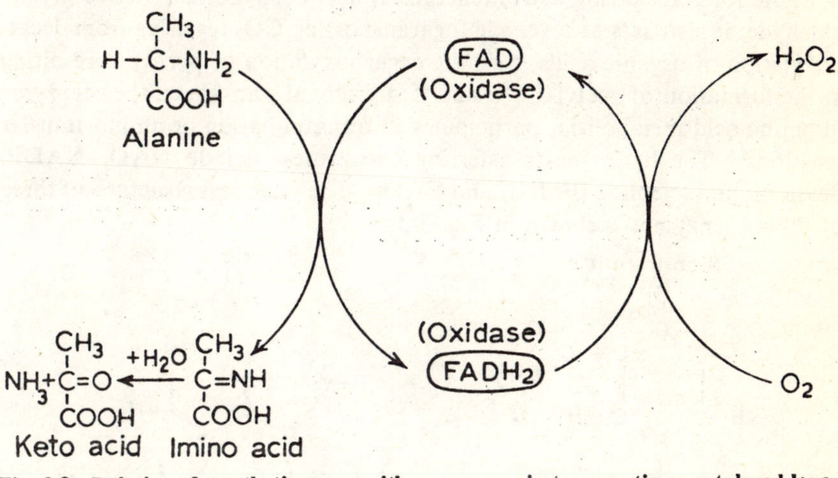

Fig. 3-2 Relation of prosthetic group with apoenzyme in two reactions catalyzed by a single enzyme.

The foregoing example illustrates the essential difference between the coenzyme and the prosthetic group and it would be quite appropriate to regard coenzymes as second substrates or cosubstrates between two coupled reactions, each catalyzed by a separate enzyme. The substrate specificity resides in the apoenzyme part of the holoenzyme, and denaturation of the apoenzyme part leads to its complete inactivation. Further, a single coenzyme can participate in many different reactions, but the holoenzyme, or, more appropriately, the apoenzyme, is normally specific for only one reaction, i.e., the choice of reaction is determined by the apoenzyme. Both substrate specificity and reaction specificity are functions of the apoenzyme part of the holoenzyme.

Coenzymes are extremely important because most of them are involved in group-transfer reactions without which no synthetic or physiological activity would be possible. They are the coordinators in various cellular functions. For instance, ATP is the coenzyme of one of the group-transferring holoenzymes known as kinases and, as stated in Chapter 2, the coenzyme ATP and the apoenzyme (kinase) are the core of the active absorption of sugars and other nutrients in glycolysis and aerobic respiration.

Some more important group-transferring coenzymes include: (a) ATP, which participates in almost all reactions involving the transfer of energy-rich phosphate bonds, e.g., kinases, and which can donate its energy-rich orthophosphate group, the pyrophosphate group, or even the adenosine monophosphate part; (b) sugar phosphates, which are the usual donors of the glycosyl group in biological reactions, including the polymerization of glucose into glycogen; (c) coenzyme A (CoA), which mediates the transfer of carboxylic acids in the form of their acyl compounds, e.g., succinyl CoA and acetyl CoA; (d) tetrahydrofolic acid or formyl tetrahydrofolic acid, which mediates the transfer of one-carbon compounds or fragments (C_1) from one reaction site to another; (e) thiamine pyrophosphate, which is responsible for transferring active (energized) acetaldehyde and active glycolaldehyde; it also acts as a vehicle for transporting CO_2 resulting from decarboxylation of organic acids, e.g., the decarboxylation of pyruvate resulting in the formation of acetyl CoA; and (f) pyridoxal phosphate, the coenzyme of amino acid metabolism, participates in transamination or amino-transfer reactions. The hydrogen-transferring coenzymes include NAD, NADP, flavin mononucleotide (FMN), and FAD. The chemical structure of three of these coenzymes is shown in Fig. 3-3.

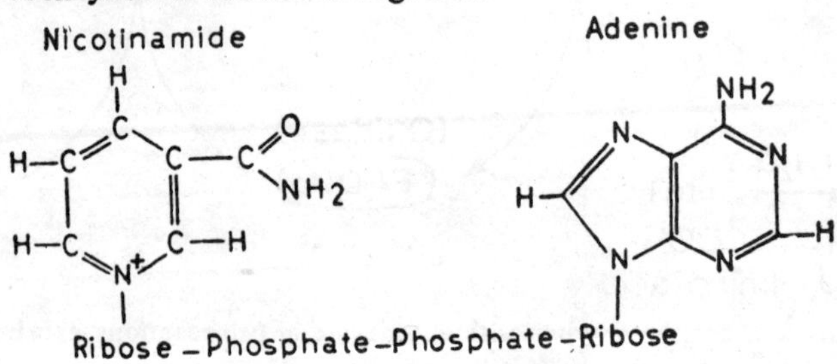

(NAD^+-oxidized form)

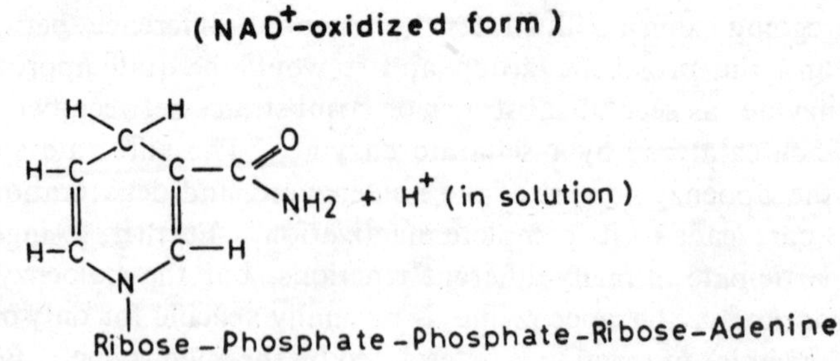

($NADH + H^+$–reduced form of NAD^+)

(a)

Fig. 3-3 Chemical structures of oxidized and reduced forms of coenzymes I and II, and of prosthetic group FAD (cont.).

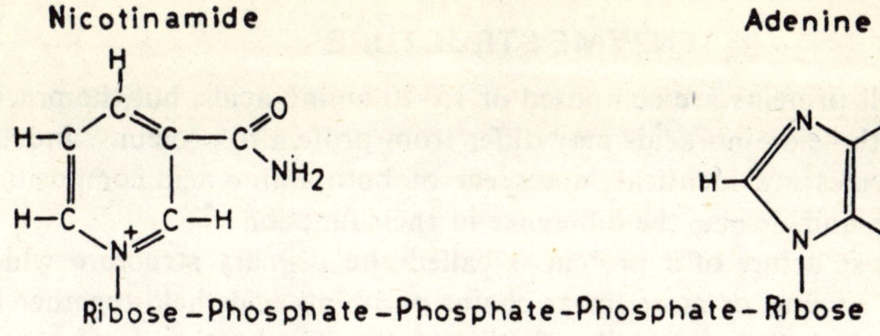

(NADP$^+$ – oxidized form)

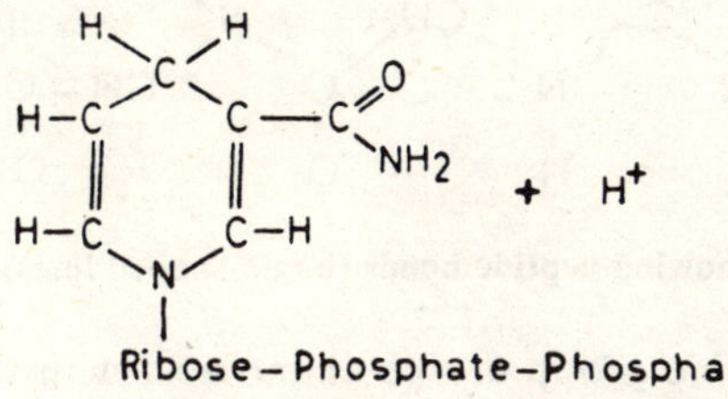

(NADPH + H$^+$ – reduced form of NADP$^+$)

(b)

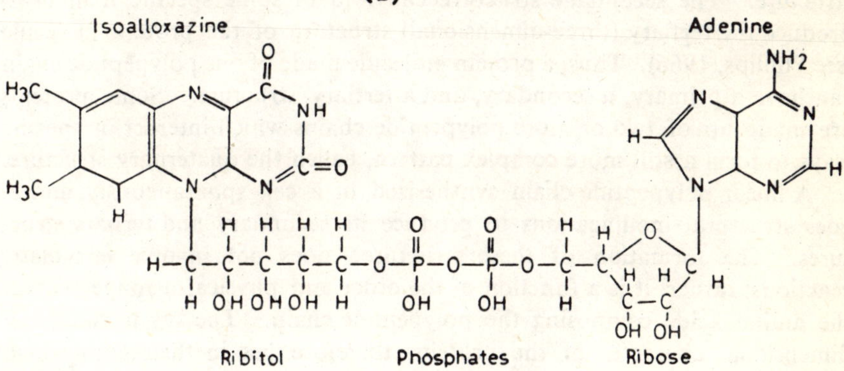

(FAD – oxidized form)

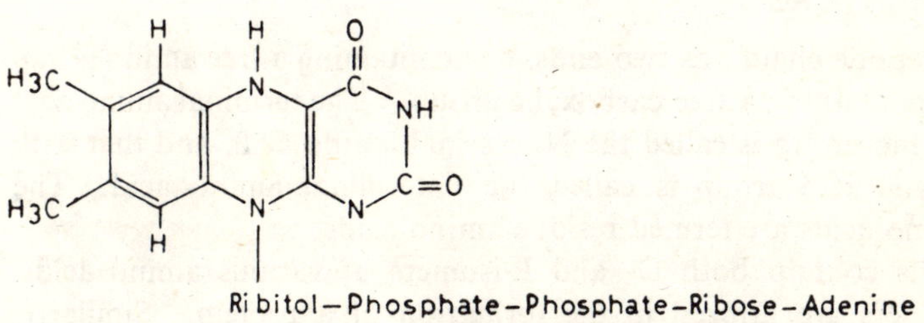

(FADH$_2$ – reduced form of FAD)

(c)

Fig. 3-3 Chemical structures of oxidized and reduced forms of coenzymes I and II, and of prosthetic group FAD.

ENZYME STRUCTURE

In general, all proteins are composed of 15–20 amino acids, but the precise sequence of these amino acids may differ from protein to protein. Indeed, no two enzymes are identical in respect of both amino acid composition and sequence and, hence, the difference in their functions.

The basic structure of a protein is called the primary structure which may consist of one or more linear chains of amino acids held together by covalent bonds, called peptide bonds (Fig. 3-4). The peptide bond results

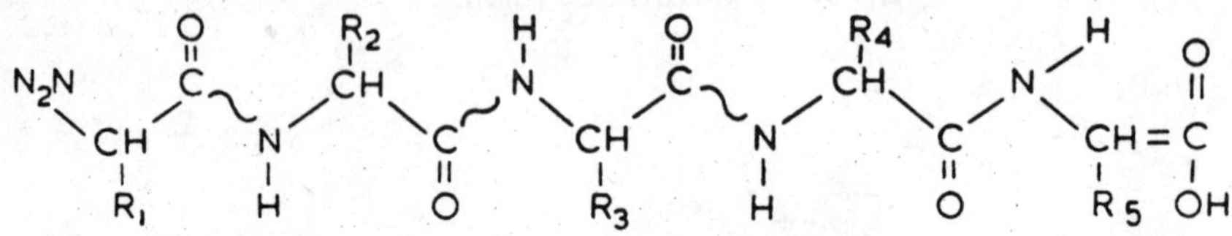

Fig. 3-4 Primary structure of a protein showing peptide bonds (bugle-shaped lines).

from the interaction between the amino group of one amino acid with the carboxyl group of another. Since a protein has many such peptide bonds, it is also called a polypeptide. When the linear (primary) structure assumes a corkscrew-like configuration or helical pattern, it is called a secondary structure. The secondary structure can fold in some specific manner to produce the tertiary (three-dimensional) structure of the protein molecule (see Phillips, 1966). Thus, a protein molecule made of one polypeptide chain can have a primary, a secondary, and a tertiary structure. Some proteins are made up of two or more polypeptide chains which interact in specific ways to form a still more complex pattern, called the quaternary structure.

A linear polypeptide chain synthesized in a cell spontaneously undergoes structural modifications to produce its secondary and tertiary structures. The formation of these structures does not involve enzymatic reactions; rather, it is a function of the order and physical arrangement of the amino acids composing the polypeptide chain. The key to the three-dimensional character of the proteins therefore lies in their amino acid sequence or primary structure.

PRIMARY STRUCTURE

A linear polypeptide chain has two ends, one containing a free amino group and the other containing a free carboxylic group. The terminal amino acid with a free amino group is called the N-terminal amino acid, and that with a free carboxylic acid group is called the C-terminal amino acid. The remaining amino acids are termed residue amino acids.

Though cells contain both D- and L-isomers of various amino acids, only the L-isomers are utilized in the formation of a protein. Similarly, though cells have more than 20 kinds of amino acids, the same 15–20 amino acids are used in the organization of widely different protein molecules. Why many different proteins prefer the same type of amino acids is still not understood. One reason may be that during the course of evolution cells may have tried and tested several kinds of amino acids and then discarded those found unsuitable.

Proteins are generally more than 20 amino acids long; this means that some amino acids must occur repeatedly for the enzyme to achieve its requisite length. Determination of the chemical structure of an enzyme must therefore proceed with the quantitative analysis of the amino acid composition of the polypeptide chain. This can be achieved by subjecting the pure crystalline enzyme to acid hydrolysis which fragments the polypeptide into its constituent amino acids. The relative amounts of the 20 amino acids thus liberated are assayed chromatographically. The two enzymes β-galactosidase and polypeptide chain A of tryptophan synthetase have been studied by this method and shown to differ considerably in respect of their relative amino acid content: the former contains 1173 amino acid residues and the latter has 267 amino acids. The sequence has also been studied and involves the following steps:

(a) fragmentation of protein molecule into a number of smaller chains of variable length and composition;
(b) separation of these fragments by the ion-exchange technique, countercurrent distribution method, or fingerprint technique; and
(c) sequencing the separated fragments by exopeptidases which split off amino acids one at a time from one end of the separated peptide or fragment.

Fragmentation of protein molecule can be achieved by the combination of a chemical and an enzymatic method. The chemical method involves the use of cyanogen bromide which, in slightly acidic solution, cleaves the peptide bond between methionine and the adjacent amino acid. Treatment of tryptophan synthetase A protein with cyanogen bromide yields four fragments, as the molecule contains four methionine residues. The fragments so obtained can be further broken down into smaller pieces by two successive enzymatic treatments; the proteolytic enzyme trypsin specifically splits polypeptide chains at the carboxylic side of arginine and lysine, and another proteolytic enzyme, chymotrypsin, hydrolyzes peptide bonds adjacent to the aromatic amino acids tryptophan, phenylalanine, and tyrosine. However, both these proteolytic enzymes fail to act if proline occurs next to the amino acids acted upon by the enzyme concerned.

The smaller fragments obtained from chemical and enzymatic treatments can be separated by using methods based on differences in solubility or electrical charge of the various fragments.

SECONDARY STRUCTURE

The primary structure of a protein is essentially one-dimensional. It has long been known that enzyme activity is associated with the three-dimensional character rather than the primary structure of the protein. Denaturing agents which destroy the three-dimensional configuration abolish also the enzymatic property of the protein. It has already been stated that the specificity of the protein resides in its primary structure. We shall now examine how the primary structure of any given protein determines its secondary, tertiary, or quaternary structure.

A secondary structure, also called α-helix, has greater structural complexity than the primary structure. Since proteins are very long polypeptide chains, they must have some means (in addition to the peptide bonds) to make them stable; α-helix is a step in this direction. It derives its stability from the formation of hydrogen bonds between the hydrogen atom of the α-amino group of each amino acid residue and the oxygen atom of the α-carboxylic group of every fourth amino acid residue along the chain. The amino acid proline does not favour the formation of α-helix because of the nonavailability of the hydrogen of its α-amino group for bond formation. Thus, a polypeptide will lack α-helical structures in those regions containing proline residues. Another reason for the polypeptides to have localized α-helix is the presence of cysteine residue whose —SH groups interact to form intramolecular disulphide bonds, thus preventing the intramolecular formation of hydrogen bonds between $—NH_2$ and C=O groups. However, the formation of the secondary structure and other intramolecular interactions depends upon the location of certain amino acids at strategic positions along the polypeptide chain. In other words, the α-helix formation in a protein is governed by its primary structure.

TERTIARY STRUCTURE

The tertiary structure depends largely upon the nature of the amino acid side chains and, hence, upon the primary structure. A polypeptide chain assumes the three-dimensional configuration by coiling, twisting, and bending in such a way as to bring towards its interior (centre) those amino acids (isoleucine, valine, proline, and phenylalanine) whose side chains are hydrophobic, and towards its exterior (periphery) those amino acids (glutamate, lysine, and threonine) whose side chains are hydrophilic. The orientation of hydrophilic groups at the periphery of the tertiary structure is important since most of the enzymes carry out their reactions in an aqueous environment. In addition, the disposition of hydrophobic groups to the interior provides a fairly high degree of stability to this structure.

QUATERNARY STRUCTURE

Some proteins, e.g., myoglobin, are composed of single polypeptide chains whereas others, e.g., β-galactosidase, are made up of more than one polypeptide chain. β-galactosidase is an aggregate of four identical chains, each being 1173 amino acids long. The quaternary structure results from the manner the chains interact to produce the complete molecule. It is generally believed that the kinds of interaction involved in the determination of secondary and tertiary structures are the same as those involved in creating the quaternary structure. Likewise, agents that break the secondary or tertiary structure are equally effective in damaging the quaternary structure. Denaturants do not destroy the primary structure since after removal of the agents from the treatment medium, the denatured proteins can regain their secondary, tertiary, and even quaternary configuration. This indicates that the primary structure is solely responsible for determining the higher-order structures; thus, the conclusion that the catalytic specificity resides in its primary structure is substantiated.

FACTORS AFFECTING ENZYME ACTIVITY

SUBSTRATE-ENZYME INTERRELATION

The substance acted upon by an enzyme is called substrate, and the enzyme activity is measured in terms of the molecules of substrate transformed per unit time. This can be illustrated with reference to the enzyme β-galactosidase which attacks the disaccharide substrate lactose and hydrolyzes it to glucose and galactose (Fig. 3-5). In fact, this enzyme attacks the β-galactoside linkage, and other substances having this linkage can also be used as substrates by this enzyme. One such substance is the colourless compound

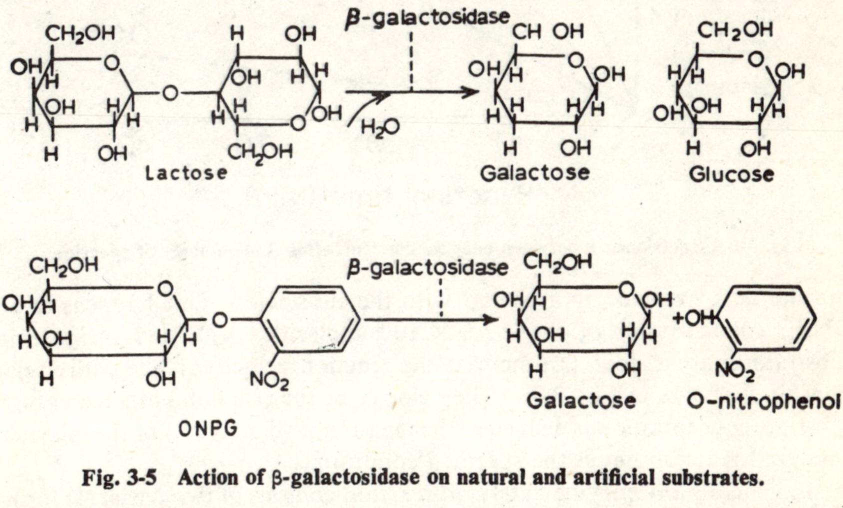

Fig. 3-5 Action of β-galactosidase on natural and artificial substrates.

orthonitrophenylgalactoside (ONPG) which when hydrolyzed yields galactose and an intensely yellow orthonitrophenol (Fig. 3-5). Thus, the intensity of yellow colour can be used as a measure of enzyme activity. When substrate concentration is fairly high and not limiting, the rate of an enzyme reaction depends mainly upon the enzyme concentration. Similarly, when the concentration of enzyme is not limiting, the reaction rate increases with increase in substrate concentration.

The effect of increasing the concentration of β-galactosidase enzyme on its substrate ONPG is shown in Fig. 3-6. It can be seen that the intensity of the yellow colour increases with increase in enzyme content to a certain extent. The ONPG is unable to undergo significant hydrolysis without the enzyme. This figure also explains the fast nature of enzyme-mediated reactions in relation to nonenzyme reactions.

Another aspect of the substrate-enzyme relationship becomes evident when the substrate concentration (ONPG) is raised from 0.00025 M to 0.001 M against a fixed concentration of the enzyme (10^{10} molecules of β-galactosidase). The rate of reaction now increases till every molecule of the

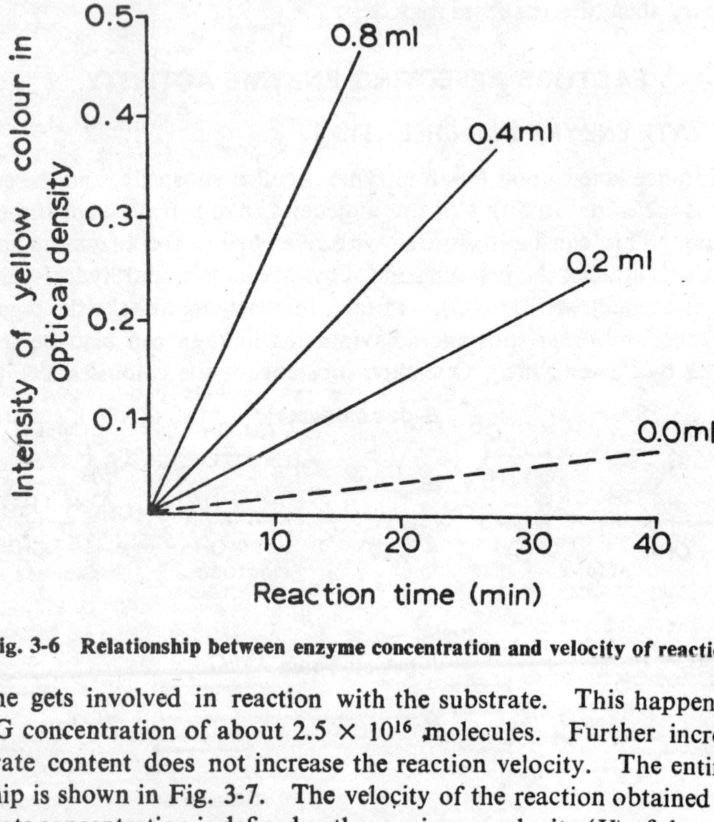

Fig. 3-6 Relationship between enzyme concentration and velocity of reaction.

enzyme gets involved in reaction with the substrate. This happens at an ONPG concentration of about 2.5×10^{16} molecules. Further increase in substrate content does not increase the reaction velocity. The entire relationship is shown in Fig. 3-7. The velocity of the reaction obtained at high substrate concentration is defined as the maximum velocity (V) of the enzyme-catalyzed reaction under the specified conditions.

In general, any enzyme-catalyzed reaction consists of two parts: (1) formation of an enzyme-substrate complex, and (2) dissociation of the complex into its products and the enzyme.

$$\text{ONPG} + \beta\text{-galactosidase} \rightleftarrows \beta\text{-galactosidase-ONPG} \rightarrow \text{ONP} + \beta\text{-galactosidase}$$

At maximum velocity, every enzyme molecule is saturated with the substrate. At half of the maximum velocity, one half of the enzyme molecules remains free and the other half is in the form of enzyme-substrate complexes. The substrate concentration required to yield half the maximum velocity is defined as the Michaelis constant (K_m), named after the originator of this theory, L. Michaelis. K_m is equal to the dissociation constant of the enzyme-substrate complex; it is also a measure of the affinity of the enzyme for the substrate. A high K_m value means that the enzyme has low affinity for the substrate, and vice versa. The K_m of β-galactosidase for ONPG is nearly

3×10^{-4} M, i.e., this is the concentration that saturates 50% of the enzyme molecules. The slope of the curve (Fig. 3-7) measures the rate of velocity of the reaction which reaches the maximum value with an ONPG concentra-

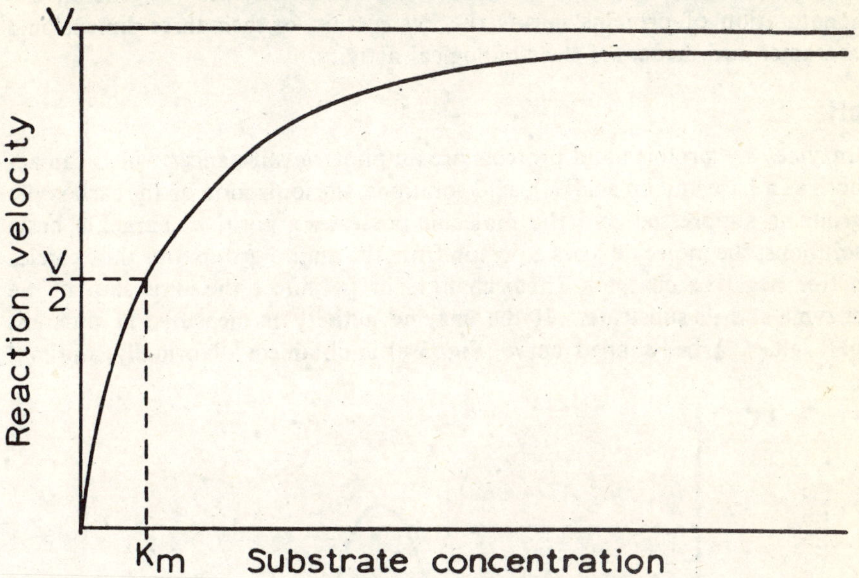

Fig. 3-7 Relationship between substrate concentration and velocity of reaction.

tion of nearly 2.5×10^{16} molecules. It should be noted that, in all enzyme-catalyzed reactions, the dissociation of the substrate-enzyme complex to generate the products and the free enzyme is always a rate-limiting step which is practically impossible to measure by any means. Since K_m is equal to the dissociation constant and since it can be conveniently determined experimentally, the normal practice is to use K_m as a measure of the rate of enzyme-substrate complex giving rise to the products.

TEMPERATURE

Within certain limits, the velocity of an enzyme-catalyzed reaction increases as the temperature increases. Suppose the velocity of the enzymatic reaction is V_1 at 10°C and V_2 at 20°C. The exact ratio (V_2/V_1) by which the velocity of the reaction changes for a 10°C rise in temperature is termed Q_{10} or the temperature coefficient. It is known that the velocity of many biological reactions roughly doubles with a 10°C rise in temperature (i.e., $Q_{10} = 2$) and is halved if the temperature is decreased by 10°C. This relation normally holds good upto 40°C, but not beyond it. Between 37°C and 40°C, the velocity of the reaction is maximum and this temperature is called the optimum temperature. The temperature required to start the reaction is called the minimum temperature. Beyond the optimal temperature, the reaction rate always decreases sharply, but below the optimal the reaction rate always increases with rise in temperature. This increase in rate results from the increased kinetic energy of the reacting molecules. Increase in temperature above the optimal level destroys the secondary, tertiary, and quaternary structures of enzymes with the consequence that enzyme activity is lost. This is called the temperature denaturation of the enzyme. In general, the denaturation of proteins means the destruction of their three-dimensional character and, hence, of their biological activity.

pH

Enzymes are proteins and proteins are amphoteric substances which can act both as a base and an acid. In acid solutions, the ionization of the carboxylic group is suppressed and the molecule possesses a positive charge; in basic solutions, the molecule loses a proton from the amino group and thus carries a free negative charge. Thus, changes in pH affect the ionic state of the enzyme and its substrate. If the enzyme activity is measured at different pH values, a bell-shaped curve (Fig. 3-8) is obtained. Normally, optimal

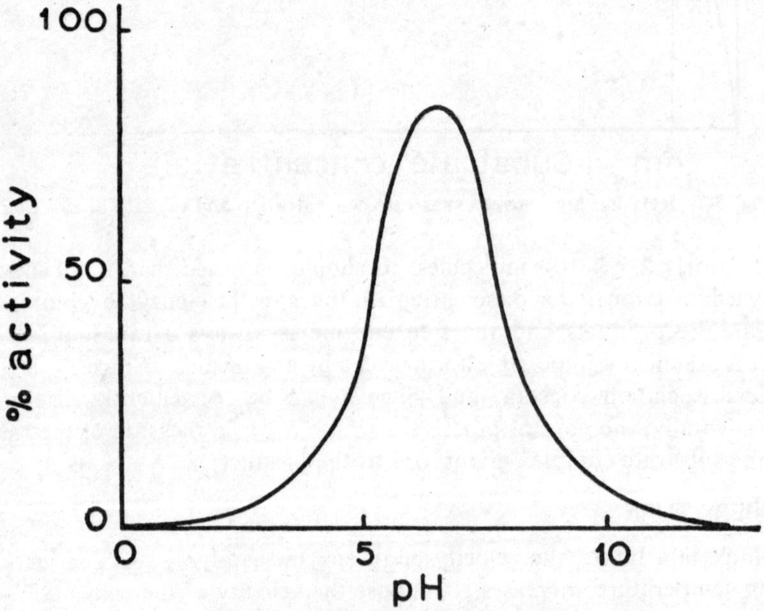

Fig. 3-8 General relation between pH and enzyme activity.

enzyme activity is observed between pH 5.0 and pH 9.0. At pH values outside this range, most enzymes become denatured. Also, the pH values affect the charged state of the enzyme and its substrate. Let us assume that a negatively charged enzyme E^- reacts with a positively charged substrate S^+ ($E^- + S^+ \rightarrow E{-}S$) to form the complex ES. Low pH will protonate the enzyme, thereby preventing it from reacting with the substrate:

$$E^- + H^+ \rightarrow E{-}H$$

High pH will ionize S^+, thus forcing it to lose its positive charge and, consequently, its reaction affinity for the enzyme.

OXIDIZING AGENTS

Most oxidoreductases (dehydrogenases) contain —SH groups which determine their catalytic activity. If —SH groups are oxidized, activity is lost. Oxidizing agents, such as O_2, inactivate cells by inactivating their dehydrogenases.

HEAVY METALS

Heavy metals, such as silver, inactivate enzymes by causing denaturation. They have positive charges which interact with the negative charges of the enzyme and result in the loss of negative charge of the enzyme molecule. The final effect is that all such enzymes are precipitated after treatment with heavy metals.

MECHANISM OF CATALYSIS

The ability of the enzyme to recognize its substrate and to form an enzyme-substrate complex, resulting in product formation, has led to the concept that enzymes have specific surface sites for binding their substrates. These sites are described as active, catalytic, or substrate. All enzyme molecules are enormous in comparison with their substrate molecules. Clearly then, only a part of the enzyme area would be engaged in the formation of the active site. In fact, researchers have noted the formation of an enzyme-substrate complex by the enzyme lysozyme which catalyzes the hydrolytic dissolution of bacterial cell walls. Molecular details about the mechanism of binding the substrate by this enzyme are now known. Approximately 20 amino acids are known to encircle a hollow area which constitutes the active site of the enzyme. The substrate is held on the active surface by forming weak bonds with the side chains of the 20 amino acids. The same mechanism is proposed for the hydrolysis of the lactose molecule by β-galactosidase. The most significant point is that it is the tertiary or quaternary structure that brings the widely scattered 20 amino acids together to constitute the active site. In other words, active site formation is a function of the three-dimensional configuration of proteins and this is why enzymes lose their activity after destruction of their three-dimensional structure.

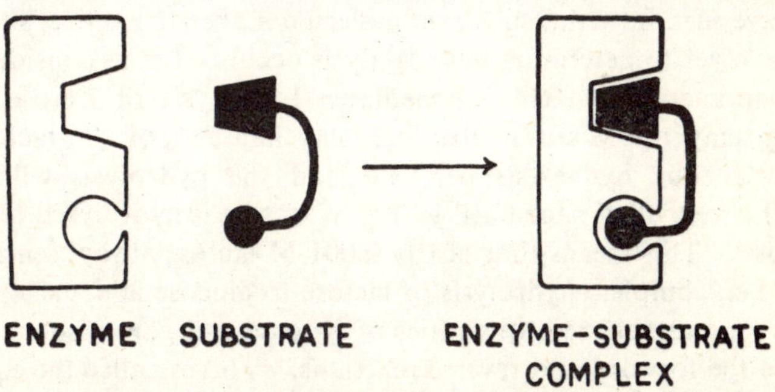

Fig. 3-9 Lock-and-key model illustrating enzyme-substrate interaction.

Until recently, the interaction of enzyme and substrate was visualized in terms of the 'template' or lock-and-key model (Fig. 3-9) of Fisher (see Moss,

1968; Bernhard, 1968). Another model, called the 'induced-fit' model, has been proposed by Koshland (see Moss, 1968) to account for the same enzyme-substrate interaction. Fisher's theory implies rigidity or inflexibility of the active site whereas Koshland's view permits a substantial degree of flexibility in the region of the active site. Since we know that conformation of the active site to the three-dimensional space greatly changes depending upon the surrounding conditions, Koshland's theory seems more logical than Fisher's. In the induced-fit model, the enzyme must suffer a conformational change to bring widely separated groups close enough for substrate binding, or for catalysis, or for both to take place. The substrate is the inducer of correct conformational change in the enzyme molecule, as explained in the hypothetical diagram (Fig. 3-10). The nonpolar zone

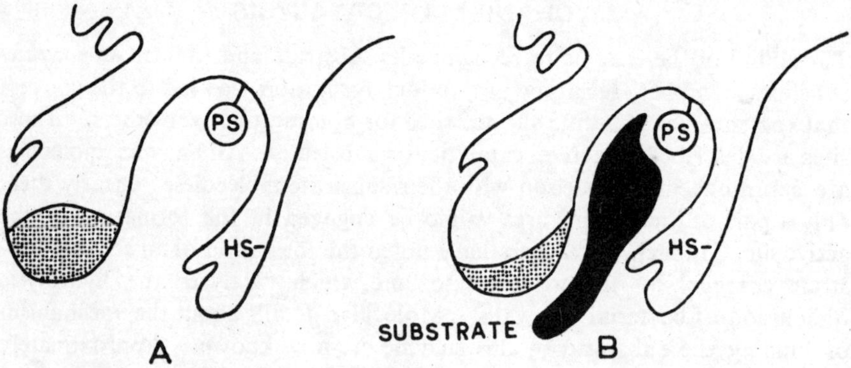

Fig. 3-10 'Induced-fit' theory explaining enzyme-substrate interaction.

(hydrophobic, shown as clear space), and polar region (charged groups, shown dotted) are both involved in substrate binding. A phosphoserine (PS) and the —SH group of a cysteine participate in the catalysis. In the absence of the substrate, the catalytic and substrate binding groups remain several bond distances away from each other. Approach of the substrate induces the conformational change in the enzyme molecule in such a way as to promote substrate binding and catalysis.

We have seen how the substrate molecule makes the active site hold it, but we have yet to determine how catalysis occurs. Let us consider the case of the spontaneous and enzyme-mediated hydrolysis of lactose. At the prevailing temperature and neutral pH, less than 50% of 1 M lactose undergoes spontaneous hydrolysis per year and the hydrolysis will continue steadily at a very slow pace until 99.9% of lactose is hydrolyzed to galactose and glucose. This means that nearly 0.001 M lactose always remains unhydrolyzed, i.e., complete hydrolysis of lactose to glucose and galactose never occurs as long as sufficient quantities of glucose and galactose are available to balance the forward and reverse reactions. This is called the equilibrium stage. If the same reaction is catalyzed by β-galactosidase, the equilibrium will be reached in just a few minutes though, in this enzyme-catalyzed reaction too, the 0.001 M lactose will remain unhydrolyzed because of a balance between the forward and reverse reactions. Thus, enzymes do nothing unusual other than speeding up the reactions without affecting the chemical equilibrium. A theory, called the transition state theory, has been postulated to explain why the spontaneous reaction proceeds much slower than the enzyme-catalyzed reaction. According to this theory, lactose (or any other substrate) in solution can have many conformational or transitional states, one of which represents the actual substrate for hydrolytic attack. This reactive state is very rare, but its frequency can be greatly increased by the presence of its enzyme. This explanation may be true for both forward and reverse reactions. One important point is that the proportion of the reactive state among different conformational states of the substrate depends upon substrate concentration. Accordingly, at equilibrium, both forward and backward reactions occur with the same speed; this is because the frequency of the reactive state attains the same numerical value.

ENZYMIC CONTROL OF METABOLISM

The most crucial factor for making the cells active in growth or metabolism is the maintenance of their enzyme machinery intact and functional. Cells synthesize or degrade those substances that are essential for their survival and growth under a defined condition. For example, if cells use glucose instead of lactose for energy and carbon source, they do not produce the enzyme β-galactosidase; instead, they form the glycolytic enzymes for glucose utilization. Similarly, if cells are provided with tryptophan exogenously, the endogenous production of tryptophan is stopped. In order to maintain a continuous balanced flow of energy and metabolites along various metabolic pathways, cells have evolved some regulatory processes, such as (a) induction and repression of enzyme synthesis in response to the low molecular weight metabolites; (b) activation of preformed enzymes by hydrogen ions or by their own active form; and (c) allosteric control. These are now briefly described.

INDUCTION AND REPRESSION

A substance such as lactose must enter the cell before it can be metabolized. Due to the permeability barrier of the cell membrane, a protein molecule, called permease, is generally involved in transporting the exogenous metabolites to the interior of the cells. Permeases are as specific in their function as any other enzyme. Lactose also gets into the cell interior through its own permease. β-galactosidase then makes a hydrolytic attack on the lactose and cleaves it into galactose and glucose. If lactose is not supplied to the cells, the permease and β-galactosidase are not synthesized. It is the presence of lactose that signals the cells to synthesize the two enzymes needed. This phenomenon of enzyme synthesis in response to substrate is called enzyme induction and such enzymes are termed inducible. Other examples of inducible enzymes are nitrate reductases, sulphate reducing enzymes, and glycolytic enzymes. In fact, all those enzymes that catabolize the substrates are generally inducible in nature. In contrast, the group of enzymes involved in the synthesis of a metabolite, e.g., tryptophan, isoleucine, and histidine, are repressible in nature because their synthesis is stopped in the presence of the metabolite they produce. Tryptophan, when supplied to the cells, signals the cells not to synthesize the enzymes concerned with its own production. Thus, in both induction and repression, *de novo* synthesis of proteins occurs, the main difference being that in induction the synthesis occurs only in the presence of the substrate whereas in repression it takes place when the end product of a biosynthetic pathway, say tryptophan, is missing from the cells.

Repression and induction controls are two different biological mechanisms for achieving the single aim of stopping the production of enzymes not required by the cells at some particular time. The biological advantage of such controls lies in the fact that cells are prevented from unnecessarily wasting their valuable raw materials (enzyme substrates) and energy in the synthesis of enzymes that are not immediately required.

ACTIVATION OF PREFORMED ENZYMES

Certain proteolytic enzymes which hydrolyze proteins are produced as inactive enzyme precursors. These so-called zymogens are subsequently converted into active enzymes by hydrogen ions or by the already active forms of proteolytic enzymes:

$$\text{Pepsinogen} \xrightarrow{H^{+} \text{ or Pepsin}} \text{Pepsin}$$

$$\text{Trypsinogen} \xrightarrow{\text{Trypsin}} \text{Trypsin}$$

Pepsinogen and trypsinogen are inactive forms of pepsin and trypsin enzymes, respectively.

There are indications for the presence of zymogen-like proenzymes in plants. Pea seeds contain amylopectin glucosidase in an inactive form. By proteolysis, this inactive form is converted into an active form. In yeast, the active form of a protease enzyme results from hydrolysis of its inactive proprotease (see Huffaker and Peterson, 1974).

ALLOSTERIC CONTROL

Tryptophan is produced from glutamine and chorismate by three enzymic steps as follows:

$$\text{Glutamine} + \text{Chorismate} \xrightarrow{\text{Anthranilate synthetase}} \text{Anthranilate} \xrightarrow{\text{IGP synthetase}} \underset{\text{IGP}}{\text{Indole glycerol phosphate}}$$

$$\text{IGP} + \text{Serine} \xrightarrow{\text{Tryptophan synthetase}} \text{Tryptophan}$$

We have already seen that one way of controlling tryptophan synthesis is through enzyme repression by tryptophan itself. This prevents the production of fresh enzyme molecules, but does not block the activity of preformed enzymes. Consequently, cells will synthesize tryptophan, though at a reduced rate, through the activity of these enzymes even if they receive readymade tryptophan. Fortunately, cells have an additional control mechanism to check this unwanted synthesis of tryptophan. This mechanism operates through inactivation of the first enzyme of the pathway by its end product. Thus, an excess of tryptophan prevents its further production by inactivating the enzyme anthranilate synthetase. Note that the substrate of this enzyme is glutamine and chorismate and not tryptophan.

Like anthranilate synthetase, the first enzyme of the metabolic pathway also has a site for the binding of the substrate or its structural analogue and an additional site for binding of the end product which is never a structural analogue of the substrate. The substrate-binding site is called the isosteric site and the end-product (inhibitor)-binding site is called the allosteric site. All such enzymes are collectively referred to as allosteric enzymes. The problem is to determine how the binding of the end product inactivates the first enzyme. It is proposed that binding of the end product on the allosteric site changes the conformation of the isosteric site in a way that renders it unfit for substrate binding and catalysis. This is how tryptophan prevents anthranilate synthetase from further synthesizing anthranilate. The low availability or complete lack of anthranilate will finally limit the IGP and tryptophan synthesis.

All enzymes forming the centre of branched metabolic pathways are believed to be of allosteric nature. Some agents, such as adenosine monophosphate (AMP), stimulate the activity of phosphofructokinase by allosteric mechanism. Since allosteric enzymes are inhibited or stimulated in their activity by substances structurally quite unrelated to their substrate, the general convention is to regard all such substances as modifiers. Accordingly, a substance that inhibits the activity of any enzyme by allosteric means, e.g., tryptophan which inhibits the activity of anthranilate synthetase, is called a negative modifier. In contrast, the positive modifier includes substances that promote the enzyme activity of allosteric means, e.g., effect of AMP on phosphofructokinase.

CLASSIFICATION AND NOMENCLATURE

Reactions and the enzymes catalyzing them have been divided into six major groups (see Anonymous, 1965), listed here with examples. Popular names of enzymes are given within parentheses. Technically, the name of an enzyme consists of two parts: the first denotes the name of the substrate, and the second which ends in *-ase* indicates the type of reaction catalyzed.

Oxidoreductases (Dehydrogenases) catalyze oxidation-reduction between two substrates A and B

$$A^{-}_{\text{reduced}} + B^{+}_{\text{oxidized}} \rightarrow A_{\text{oxidized}}\ B_{\text{reduced}}$$

Some examples of oxidation-reduction enzymes are triosephosphate oxidoreductase and pyruvic oxidoreductase.

Transferases (Transaminases) catalyze the transfer of groups other than hydrogen. They include enzymes such as transketolase (transfers the 2-carbon fragment containing keto group to, for instance, phosphoglyceraldehyde, thus forming xylulose phosphate) and kinase (transfers the terminal phosphate of ATP to other substances, or the energy-rich phosphate group of a compound to ADP to produce ATP; pyruvate kinase is an example of the latter reaction in which phosphoenolpyruvate is converted into pyruvate and ATP is formed).

Hydrolases (Sucrases, Lipases) catalyze the splitting or breaking up of ester, ether, peptide, glycosyl, and acid anhydride bonds by the addition of water molecules. Lipases are an example of ester bond hydrolysis.

Lyases (Decarboxylases) reversibly catalyze the removal of groups from substrates by mechanisms other than hydrolysis. For example, carboxylases and dicarboxylases remove (or add) carboxyl groups or dicarboxyl groups.

Isomerases (Triosephosphate isomerases) bring about the interconversion of optical, geometric, and positional isomers. Examples include triosephosphate isomerase, xylulose epimerase, and ribose isomerase.

Ligases (Thiokinases) participate in linking together two compounds coupled with the breaking up of a pyrophosphate bond in ATP or guanosine triphosphate (GTP). For example, succinate CoA ligase guanosine diphosphate (GDP) yields succinyl CoA:

$$GTP + Succinate + CoA \rightarrow GDP + Succinyl\ CoA + P_i$$

ISOENZYMES

Isoenzymes (isozymes) are multiple molecular forms of a protein. They catalyze a single reaction, but differ from one another in electrical charges. This property makes it easy to separate and identify them by electrophoresis. A good example of isozymes is aspartokinase, the enzyme catalyzing the first step of a branched biosynthetic pathway leading to the synthesis of lysine, methionine, and threonine, as follows:

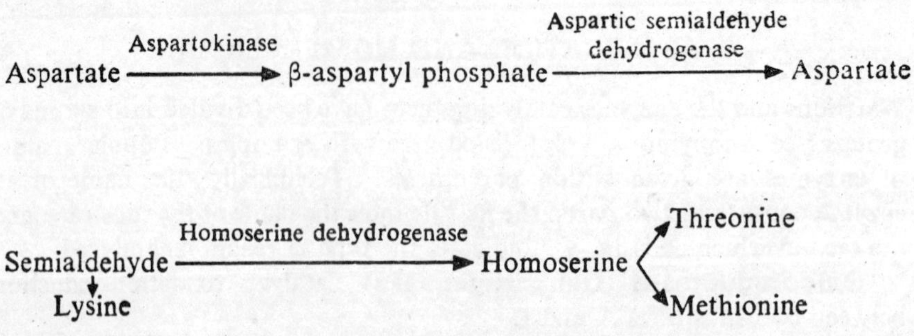

β-aspartyl phosphate is produced by an ATP-requiring reaction which is catalyzed by aspartokinase, and the aspartyl phosphate is subsequently reduced to homoserine in two successive steps by aspartic semialdehyde dehydrogenase, with $NADPH_2$ serving as the source of hydrogen for both the reactions.

There are three isofunctional forms or isozymes of aspartokinase. In one, the aspartokinase activity is inhibited by lysine; in the second, it is inhibited by threonine; and in the third, it is inhibited by methionine. The advantage of the three isofunctional forms of this enzyme is that when lysine is in excess, it will inhibit only its own production by affecting the activity of its sensitive aspartokinase. In this case, threonine and methionine biosynthesis will continue unchecked. Similarly, an excess of either of the other two amino acids will block its own continued production by feedback inhibition, without affecting the synthesis of the other end products of that branched synthetic pathway. Thus, multiplicity of isofunctional forms allows an efficient and fine regulation of amino acid synthesis in branched synthetic pathways (see Shannon, 1968).

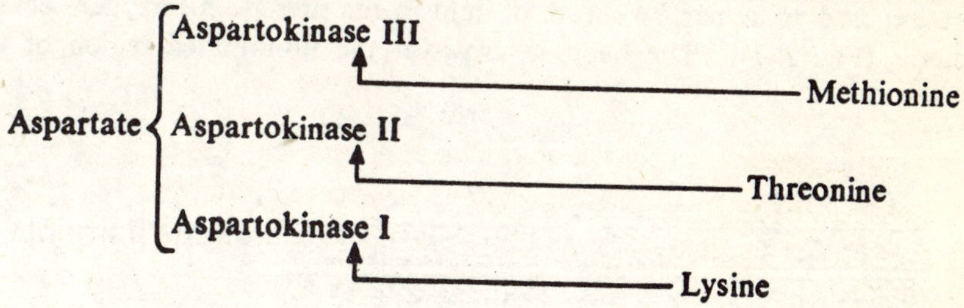

REFERENCES

Anonymous, Enzyme Nomenclature. Recommendations of the Internat. Union of Biochem. Elsevier, Amsterdam (1965).

Bernhard, S., The Structure and Function of Enzymes. Benjamin, New York (1968).

Huffaker, R. C., Peterson, L. W., Protein turnover in plants and possible means of its regulation. *Ann. Rev. Plant Physiol.*, **25**, 363–92 (1974).

Moss, D. W., Enzymes. Oliver and Boyd, Edinburgh (1968).

Phillips, D. C., The three-dimensional structure of an enzyme molecule. *Scientific American*, **215**, 78–93 (1966).

Shannon, L. M., Plant isoenzymes. *Ann. Rev. Plant Physiol.*, **19**, 187–210 (1968).

4: Photosynthetic Apparatus

The Chloroplast

INTRODUCTION

It has long been known that the chloroplast is the organelle for the absorption of light and the evolution of oxygen in photosynthesis. This was demonstrated as early as 1884 by Engelmann. He performed an elegant experiment in which different regions of a filament of the green alga *Spirogyra* were exposed to a narrow beam of light in the presence of oxygen-seeking bacteria (Fig. 4-1). The bacteria crowded the illuminated region of the

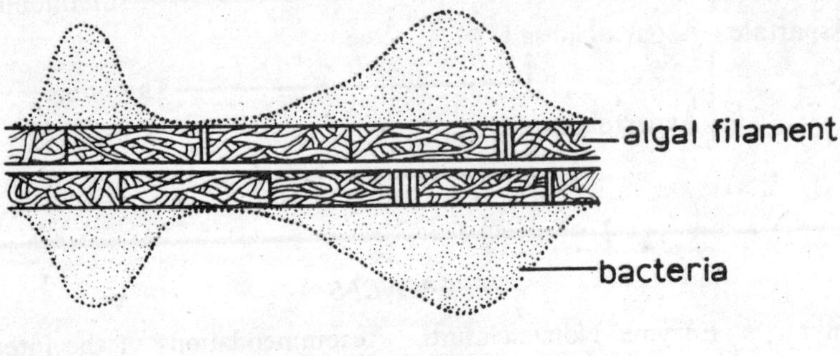

Fig. 4-1 Growth of aerobic bacteria along different regions of a *Spirogyra* filament exposed to different colours of light. (After Engelmann, 1884.)

chloroplast of the algal cell, as this was the region richest in oxygen (evolved photosynthetically). When similar experiments were later conducted by exposing the filament to the light of different colours, it was observed that maximum photosynthesis occurred in the orange-red part of the spectrum. A significant biochemical breakthrough was made by Hill (1937) who demonstrated that isolated chloroplasts, when exposed to light in the presence of water and a suitable electron acceptor, evolved oxygen.

Blackman (1905) successfully divided the entire process of photosynthesis into two component reactions, termed light and dark reactions. The ability of isolated chloroplasts to carry out both these reactions was shown by Arnon *et al.* (1954). However, their results did not indicate the sites of the two reactions within the chloroplast even though this was vaguely known. The problem was finally settled by Park and Pon (1963) who showed that the water-soluble proteinaceous nonpigmented stroma portion participated in dark reactions whereas the water-insoluble pigmented portion containing thylakoids carried out light reactions. Their results localized the Hill reaction and photophosphorylation in photosynthetic lamellae, and CO_2-fixation reactions in the stroma region.

ORGANIZATION

The double-membraned photosynthetic apparatus of plant cells is generally termed the chloroplast. In contrast, the photosynthetic apparatus of procaryotes is termed the chromatophore; it lies naked in the cell cytoplasm and is generally lamellar or vesicular in appearance. Chloroplasts may be discoid, ellipsoid, stellate, spiral, cup-shaped, reticulate, or dumb-bell-shaped. Their size and morphology are, however, extremely variable. In general, they are 4–15 μm long and 3–5 μm thick. Each bounding membrane of the chloroplast is nearly 70 Å thick, has a subunit structure, and the total thickness of the two membranes including the intermembranal space is approximately 300 Å. The internal region of the chloroplast consists of two portions, the granular matrix, called the stroma, and the complex internal membrane systems, called lamellae or thylakoids.

The seat of biogenesis of lamellar membranes is not yet defined. Some workers believe that the inner double-plastid membrane is involved in the evolution of lamellae through the process of budding and vesicle formation. This view has not yet been substantiated by autoradiographic findings which show no movement of label from plastid to lamellar membranes, or by chemical analysis of membrane lipid composition which shows significant qualitative difference between the lipid composition of the plastid envelope and lamellar membranes (Arntzen and Briantais, 1975). There is evidence to suggest the operation of a multistep assembly process which participates in the biogenesis of chloroplast membranes (see Arntzen and Briantais, 1975). A system of anastomosing tubules, termed the peripheral reticulum, has been observed in the vicinity of the inner chloroplast envelope in C_4 plants. Some C_3 plants are also reported to possess this reticulum in their chloroplasts under conditions of environmental stress. However, the functional significance of the peripheral reticulum is not known (see Laetsch, 1974).

The granular stroma matrix contains ribosomes, DNA fibrils, and osmiophilic globules, termed plastoglobuli, which are composed primarily of plastoquinones, vitamin K, α-tocopherolquinone, and α-tocopherylquinone. Plastoglobuli have been thought to function as extralamellar pools of membrane lipids which are utilized in the biosynthesis of membranes and accumulated during the biodegradation of membranes. A lamella or thylakoid consists of two membranes, each about 7.0 μm thick, and separated from each other by a space of about 4–70 μm. An isolated thylakoid appears under the electron microscope as a closed, flattened disc. The thylakoids traverse the stroma and at places get stacked up like piles of coins, thus forming grana. On the average, there may be 40–60 grana per chloroplast and

10–100 thylakoids per granum. Chloroplasts of red, brown, and golden-brown algae lack grana, and their thylakoids run singly or are stacked in bands. However, chloroplasts of many green algae have a distinct grana-like organization of their thylakoids.

Detergent treatment of the chromatophores of certain bacteria, e.g., *Rhodopseudomonas rubrum*, has resulted in the separation of two photochemically-active fragments, each containing a bacteriochlorophyll reaction centre (Vernon, 1968). In this connection, it is important to note that photosynthetic bacteria, such as *Chromatium* and *R. rubrum*, have been found to possess two photosystems operating in parallel, one driving cyclic photophosphorylation and the other mediating a noncyclic electron transport linked to the oxidation of electron donors. The noncyclic electron transport mediating photosystem is analogous to the PSII of higher plants in respect of oxidizing the external electron donor (see Olson, 1970).

ULTRASTRUCTURE OF THYLAKOIDS

As thylakoids are the sites of photophosphorylation, photoreduction, and photochemical evolution of oxygen, many attempts have been made to elucidate the pattern of the molecular organization of thylakoid membranes by using a variety of techniques, e.g., electron microscopy, X-ray diffraction, use of polarization optics, enzymatic digestion, immunology, and freeze-fractionation. The terminology commonly used to describe the chloroplast thylakoid system in higher plants (Fig. 4-2) is that of Weier *et al.* (1965). Each granum

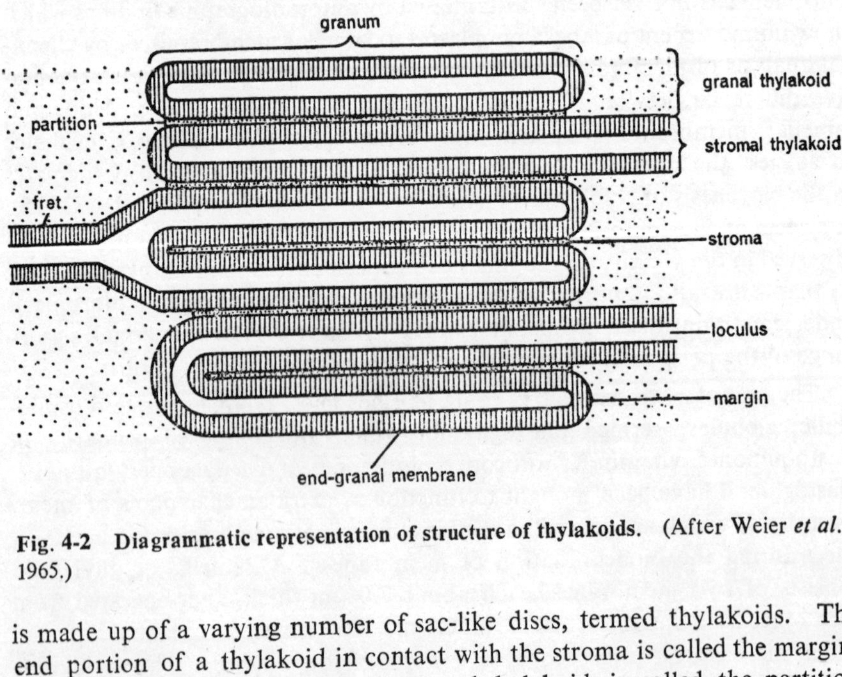

Fig. 4-2 Diagrammatic representation of structure of thylakoids. (After Weier *et al.*, 1965.)

is made up of a varying number of sac-like discs, termed thylakoids. The end portion of a thylakoid in contact with the stroma is called the margin. The region between two tightly appressed thylakoids is called the partition and the space within a tholakoid, the loculus. Thylakoids in the stroma are termed stromal thylakoids and their membrane is called the fret membrane. A thylakoid at the margin is known as the end-granal membrane.

Different interpretations of the membranous interconnections have given rise to various models for chloroplast-lamellar architecture. However, most workers agree that stromal and granal thylakoids form an interconnecting anastomosing system of membrane-bound compartment within the chloroplast, which is thought to facilitate rapid diffusion of various photosynthetic substrates. The isolation of chloroplasts in a low-salt medium causes a loss of grana, and continuous sheets of paired membranes are seen in their place. However, these membranes soon become restacked into grana when the chloroplast is again transferred from a low- to a high-salt medium. Obviously, grana formation is a reversible process, depending upon the ionic strength of the chloroplast-isolation medium. The tendency of the membranes to become reversibly organized into grana-like stacks seems to be a function of the distinctive area of membrane specialization and suggests that chloroplast-lamellar architecture is not as rigid as commonly believed.

Chloroplast lamellae contain two kinds of particles on the exterior surface: the larger particle (about 120 Å), easily removed from the lamellar surface by washing with water, shows a high specific activity for ribulose diphosphate carboxylase enzyme; the smaller particle (about 100 Å), released from the lamellar surface after washing with ethylenediaminetetraacetic acid (EDTA), possesses calcium-dependent ATPase activity. This particle functions in the terminal step of ATP formation during photophosphorylation; it is called the coupling factor because of its involvement in the coupling of photophosphorylation to photosynthetic electron transport in chloroplast membranes. The loss of the coupling factor from chloroplast lamellae results in a loss of the photophosphorylation ability of chloroplasts.

Another kind of particle sometimes bound to chloroplasts is the phycobilisome. It contains water-soluble pigments (phycoerythrin and phycocyanin) of red and blue-green algae.

The organization of thylakoid membranes is evidently particulate and involves two major classes of particles (Muhlethaler *et al.*, 1965; Branton and Park, 1967). These particles are different from surface-bound particles (just described) in respect of location, distribution, composition, and function; they occur within the membranes. The smaller-sized particles are more externally located than the larger-sized ones which are more internally located, i.e., the interior of the membranes shows an asymmetric distribution of these two classes of particles. Large freeze-etch particles are seen only in the stacked lamellar region, i.e., in granal and not in stromal lamellae. In contrast, the smaller particles are seen in stromal lamellae and end-granal membranes.

Several attempts have been made to understand the structure-function correlation of thylakoid membranes. It seems that lamellar fragments rich in small-sized particles are active in PSI whereas those rich in large freeze-etch particles are active predominantly in PSII (Arntzen *et al.*, 1972).

In a kinetic study of photosynthesis, Emerson and Arnold (1932) estimated about 2500 chlorophyll molecules cooperating in the fixation of one molecule of CO_2 or the evolution of one molecule of oxygen. Since eight quanta have been found to participate in the photosynthetic CO_2 fixation or oxygen evolution, subsequent research suggested a division of 2500 by 8 for a proper numerical estimate of chlorophyll molecules per reactive centre or per photosynthetic unit (PSU); this comes to about 250–300 chlorophyll molecules per PSU. Park and Biggins (1964) found regular repeating subunits within shadowed lamellar membranes and termed each subunit a quantasome on the basis of the presence of about 230 chlorophyll molecules per subunit. They considered quantasomes to be the morphological expression of the kinetically-determined PSU. Park and Sane (1971) equated the large freeze-etch particles with the 'quantasome core', and stated that a quantasome results from the association of these large particles with portions of the small freeze-etch particles within lamellar membranes.

The concept of the quantasome as a structural form of PSU runs into difficulty in view of the generally accepted mode of photosynthetic electron transport from water to $NADP^+$ as being a two-quanta process (being mediated by two photosystems, PSI and PSII) which has to operate four times to bring about the fixation of one molecule of CO_2. On this basis, a PSU would be expected to contain approximately 600 chlorophyll molecules. This PSU which contains both photosystems is composed of two reaction-centre complexes with about 200–300 chlorophyll molecules in each photosystem. The isolation of PSI from PSII and their functional characterizations are strongly suggestive of distinct and separate sites for the two photosystems as opposed to the same functional site which is implicit in the quantasome concept of Park and Biggins. Goodenough and Straehelin (1971) have reported two chloroplast mutants of *Chlamydomonas reinhardii* which have normal photosynthesis, but do not have stacked grana-like lamellar membranes and the large freeze-etch particles equivalent to the quantasome core.

Digitonin fractionation of lamellar membranes produces fragments rich in small freeze-etch particles which are active in PSI. Evidently, the functional components of PSI are not localized in the quantasome core which has been equated with the large freeze-etch particle found enriched in the detergent-derived PSII fraction. These and other investigations have also revealed that PSI activity is more externally localized and PSII is more internally localized within the lamellar membrane. However, it would be erroneous to conclude from these structure-function correlation studies, which merely indicate a probability, that the membrane subunits seen under the electron microscope are the actual site of photochemical activity; there is as yet no concrete evidence to confirm this possibility.

The formation of grana is not a prerequisite for the appearance of PSII activity in chloroplasts. The PSII activity precedes the stacking of lamellae into grana. It has been speculated that some components of PSII might be involved in the stacking, but there is no clue to the nature of these components. In this regard, studies on chloroplasts of C_4 plants have been very productive. All C_4 plants do not show a structural dimorphism of their chloroplasts, but some, e.g., sorghum and sugarcane, have typically dimorphic chloroplasts in which the mesophyll chloroplasts have well-developed grana whereas the bundle-sheath chloroplasts are agranal. It is important to note that bundle-sheath chloroplasts in young, developing leaves of C_4 plants do contain grana; these are lost in some species (e.g., sorghum and sugarcane) during chloroplast maturation (see Laetsch, 1974). Agranal bundle-sheath chloroplasts, on the other hand, show PSII activity, though at a lower efficiency than the PSII activity of grana-containing mesophyll chloroplasts (see Arntzen and Briantais, 1975). Some grana-deficient chloroplast mutants of *Chlamydomonas reinhardii*, pea, and soyabean, have been reported to have better PSII activity than the wild, parental strains; this further rules out any role of grana in PSII activity of photosynthesis (see Arntzen and Briantais, 1975). The chloroplasts of red and brown algae lack grana, but still carry out both PSII and PSI activities quite effectively; this also reinforces the foregoing conclusion. Thus, the precise physiological significance of granal organization in plant chloroplasts remains uncertain at present.

DEVELOPMENT

Shoot meristems of angiosperms contain minute disc-like bodies, called proplastids which originate from pre-existing proplastids. Their internal organization, as revealed under the electron microscope, looks homogeneous, except for a few vesicles which seem floating in a cytoplasm-like matrix. In dark-grown angiospermous plants, proplastids differentiate prolamellar bodies or semicrystalline bodies which are made up of a regular lattice-like network of tubules. A prolamellar body is a storage structure for membrane components. Sometimes a few thylakoids are seen to emerge from the prolamellar body and to join with the inner membrane of the chloroplast envelope. It is suggested that spaces within the interconnecting tubules are contiguous with, but isolated from, the stroma. Recently, Lang and Rae (1967) have observed thylakoid lattices in old vegetative cells of *Anabaena* which are very much like the prolamellar bodies in developing chloroplasts. The significance of this thylakoid-like lattice in the phylogeny of the chloroplast can be properly evaluated only after detailed studies of the biogenesis of thylakoids in several blue-green algae.

The internal structure and biochemistry of the differentiating proplastids change after a brief exposure to light. Illumination for less than five minutes breaks up the crystalline arrangement of the prolamellar body into an irregular network of interconnecting tubules. This step requires very little energy and is reversible. It is important to mention here that light plays a crucial role in the differentiation of proplastids into chloroplasts. First, light is required for the photoconversion of protochlorophyllide into chlorophyllide. The action spectrum for the conversion of protochlorophyllide into chlorophyllide is identical with the absorption spectrum of protochlorophyllide *in vivo*, with a maximum of about 650 nm. The chlorophyllide is further photoreduced to chlorophyll. The other important regulatory effect of light in chloroplast development from proplastids operates through the phytochrome system. The phytochrome-mediated control of chloroplast development is observed in leaf tissue; in other tissues, e.g., excised roots, potato tuber, and tobacco cell cultures derived from wound callus, chloroplast development depends on blue light and the photoreceptor in this control is believed to be either a carotenoid or a flavoprotein (Kirk, 1970).

The assembly of all the components of thylakoid membranes which is necessary for the transformation of proplastids into mature chloroplasts seems to occur through a multistep process. A number of greening experiments using intermittent illumination of etiolated leaves have provided useful information about the structure-function relationship in developing chloroplasts. The greening chloroplast membranes primarily accumulate chlorophyll-a; this is followed by the appearance of PSI activity, PSII activity, and, finally, the stacking of thylakoids into grana (see Arntzen and Briantais, 1975). However, the control mechanisms involved in the regulation and formation of the chloroplast constituents in their correct relative amounts along the time scale, during the transformation of proplastids into chloroplasts, are not yet known. The processes involved in the assembly of these constituents into thylakoid membranes are also open to conjecture. Nevertheless, some unicellular algae exhibit a kind of nutritional control in chloroplast development. Dark-grown algae are unable to form chloroplasts, whether in the dark or in the light, in a medium containing an assimilable organic carbon source. The absence of organic carbon from the culture medium induces chloroplast development after a certain lag-phase, thus suggesting a repressive effect of organic carbon on chloroplast formation. This repressive effect can be overcome by the addition of a nitrogen source to the culture medium either in the dark or in the light. Further investigation revealed that it is the carbon-nitrogen ratio, rather than the absolute level of either component, that controls chloroplast development in these algae (see Kirk, 1970). The entire sequence of chloroplast development is shown diagrammatically in Fig. 4-3.

Chemical analysis of mature chloroplasts shows that, on a dry weight basis, they are composed of 35–55% protein, 18–37% lipid, nearly 6% chlorophyll, 2% carotenoid, approximately 1–3% nucleic acid, and 5–8% inorganic matter. The major chloroplast lipids include mono- and digalactosyl diglycerides which are carriers of alpha-linolenic acid.

ORIGIN, PHYLOGENY, AND GENETICS

As early as 1905, Mereschkowsky had suggested that the chloroplasts of algae and higher plants are derived from ancient photosynthetic procaryotes, of the nature of blue-green algae, which entered the cells of primitive eucaryotes as intracellular symbionts and, during the course of evolution, became stabilized as permanent symbionts within eucaryotic cells. New findings on the pigmentation of blue-green algae and higher plants tended to

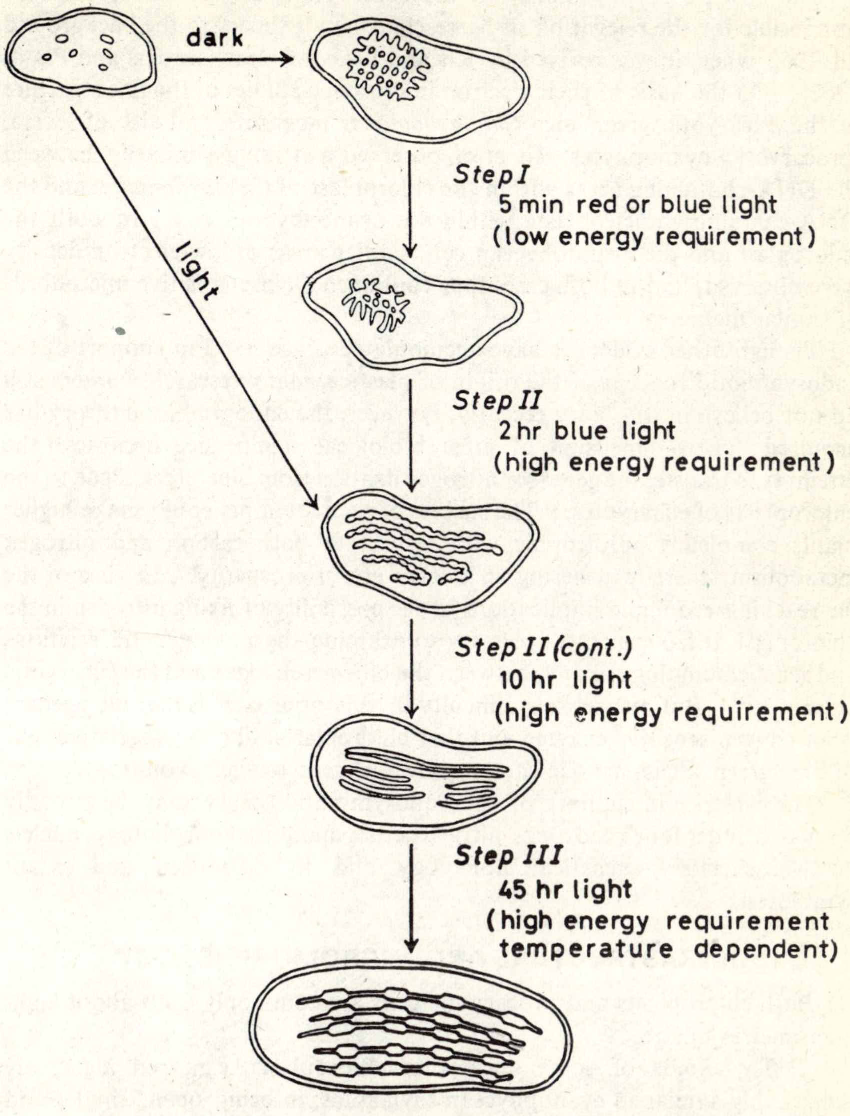

Fig. 4-3 Various stages in development of chloroplast. (After Devlin, R. M., Barker, A. V., Photosynthesis. Van Nostrand Reinhold, New York, 1971.)

discredit this theory. It was argued that if the chloroplasts had indeed descended from ancient cyanophytes, they would be expected to show some trace of the phycobilin pigments characteristic of cyanophytes. However, no such pigments are found in eucaryotes, except in Rhodophyta, Cryptophyta, and *Cyanidium* (the phytochrome system of green plants was not properly characterized; the phytochrome, which is widespread in many eucaryotic plants, is now known to be chemically very similar to the phycobilins of cyanophytes).

The absence of phycobilins in the chloroplasts of most eucaryotes was responsible for the relegation of Mereschkowsky's theory to the background till 1962 when it was revived by Ris and his coworkers (see Ris and Plaut, 1962). On the basis of their electron microscope studies of the fine structure of the eucaryotic green alga *Chlamydomonas moewusii*, and also of several (procaryotic) cyanophytes, Ris *et al.* observed a striking similarity between the DNA-containing areas within the chloroplast of *Chlamydomonas* and the DNA-containing nucleoplasm within the cyanophycean cell. In both the chloroplast and the cyanophycean cell, similar areas of low-electron density were observed, and in both cases they contained DNase-sensitive microfibrils of similar diameter.

Though other evidences have accumulated since 1962 in support of the endosymbiotic concept of the origin of plastids, many research workers still do not believe in it. Very recently, however, the endosymbiotic theory has acquired a new dimension of great biological significance because of the attempts to transfer the genes for nitrogen fixation from blue-green algae to the chloroplasts of eucaryotes. The success of such attempts could make higher plants completely autotrophic with respect to both carbon and nitrogen metabolism, thereby ushering in a new era of prosperity. In view of the far-reaching economic implications of the possibility of fixing nitrogen in the chloroplast, it has become necessary to examine the phylogenetic relations and genetic homology, if any, between the blue-green algae and the eucaryotic chloroplast. But one serious difficulty in this approach is that nitrogenase is an oxygen-sensitive enzyme and that chloroplasts, like the vegetative cells of blue-green algae, are the site of photosynthetic oxygen evolution.

The evidence in support of the endosymbiotic theory may be broadly discussed under four heads, viz., ultrastructure and micromorphology, nucleic acid biochemisty, genetical homology and hybridization, and extant symbioses.

ULTRASTRUCTURE AND MICROMORPHOLOGY

(1) Both chloroplasts and procaryotic cells are commonly upto about eight micrometres in size.

(2) Thylakoids of some eucaryotic chloroplasts, e.g., red algae, are remarkably similar to cyanophycean thylakoids, in being open, single, and disposed peripherally in sheets that are more or less parallel. In both blue-green algae and chloroplasts, the thylakoids originate from the bounding membrane or the inner layer of the chloroplast membrane.

Laetsch (1969) has observed a striking similarity between the thylakoid arrangement of the chloroplasts found in bundle-sheath cells of certain tropical grasses and that of some blue-green algae. A peripheral reticulum of anastomosing tubules, such as is commonly observed in ultrathin sections of blue-green algae, was seen also in the chloroplasts of bundle-sheath cells.

(3) The bounding layers of the endosymbiont cyanelles, *Cyanocyta korschikoffiana* (within *Cyanophora paradoxa*) and *Skujapelta nuda* (within *Glaucocystis nostochinearum*; see Fig. 4-4), appear more like those of chloro-

plasts than those of free-living blue-green algae.

Both plastids and procaryotes have the inner membrane highly infolded and variously convoluted to increase the surface area.

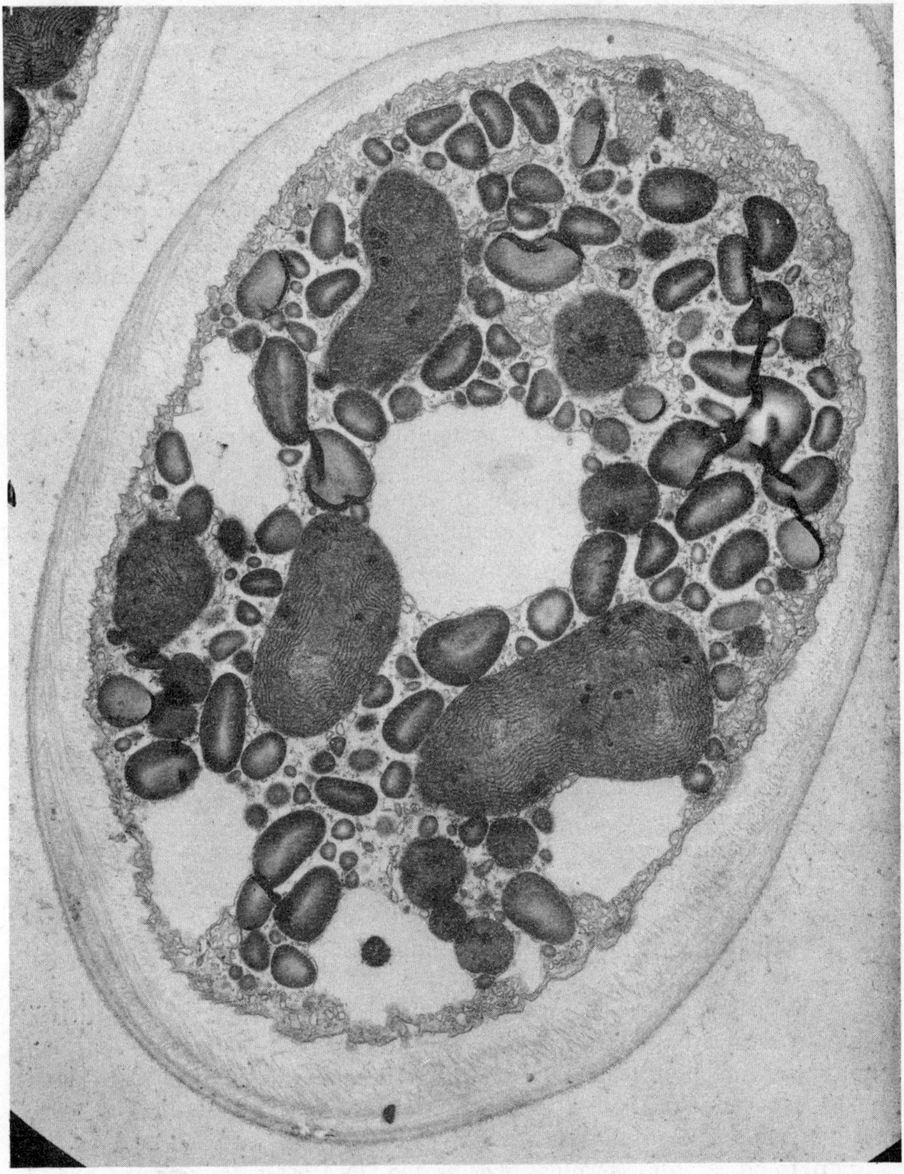

Fig. 4-4 Electron micrograph of *Glaucocystis nostochinearum*. (Courtesy: Dr. W. T. Hall, ×12,000.)

(4) DNA fibrils, ribosomes, and plastoglobuli are found in chloroplasts as well as cyanophytes. The dimensions of plastid plastoglobuli, though variable, are generally similar to those of the lipid granules or α-granules of

cyanophytes, and, in all probability, these three terms mean the same thing.

Fine granules, termed phycobilisomes, have been found disposed along the thylakoids of blue-green algae as well as the eucaryotic red algal chromatophores (Gantt and Conti, 1966). These phycobilisomes, which are simply phycobiliprotein aggregates, are attached on the stromal side of the thylakoids in both groups and are commonly located in rows, forming a two-dimensional crystalline array.

The resemblance between the fine structure of the chloroplast and that of intracellular blue-green algal symbionts lacking a cell wall is particularly striking. Thus, the cyanophyte symbiont of the flagellate *Cyanophora paradoxa* possesses photosynthetic thylakoids and a central area containing DNA, and is delimited from the host cytoplasm by two unit membranes. The entire structure looks remarkably like a chloroplast.

NUCLEIC ACID BIOCHEMISTRY

(1) A striking similarity has been observed between DNA-containing areas of blue-green algae and those of certain chloroplasts (Ris and Plaut, 1962; Ris and Singh, 1962).

(2) Chloroplast DNA as well as procaryotic DNA, but not nuclear DNA, can undergo extensive renaturation after heat denaturation at 60°C for two hours.

(3) Both plastids and procaryotes contain a unique base composition. The chloroplast DNA from both algae and higher plants very often does not contain 5-methylcytosine whereas the nuclear DNA frequently contains about 2–5% of 5-methylcytosine. The DNA from both cyanophytes and plastids is double-stranded.

(4) In both plastids and procaryotes (and also eucaryotic nuclei), DNA and RNA can be synthesized and replicated.

In vitro synthesis of DNA in isolated chloroplasts of *Euglena gracilis* has been demonstrated by Scott *et al.* (1968; see Tewari, 1971).

The enzyme DNA polymerase has been found to be associated with the chloroplast DNA along thylakoids. An RNA polymerase with the potential of transcribing chloroplast DNA into mRNA has also been detected within the chloroplast.

Thus, these enzymes, along with the 70S ribosomes found in chloroplasts, seem to meet all the requirements for the replication, transcription, and translation of chloroplast DNA. It has been estimated that chloroplast DNA can code for about a thousand different proteins containing as many as 200 amino acids in their primary structures (Kirk, 1966).

(5) In procaryotes as well as plastids, the DNA is histone-free and bound to membranes; in contrast, nuclear DNA of eucaryotes is associated with histones and not bound to membranes.

Tewari and Wildman (1969) have observed striking differences between DNA-dependent RNA polymerases of nuclei and those of chloroplasts. The chloroplast polymerase was found to be much more active than the nuclear polymerase and this difference in activity was attributed to the absence of histones in chloroplast DNA as opposed to their presence in nuclear DNA. However, it has also been shown that plastid DNA synthesis in tobacco is dependent to some extent on nuclear DNA synthesis. Photosynthetic organisms thus seem to constitute very interesting systems in which different control mechanisms operate in the transcription and translation of genes in nuclear DNA and of those in plastid DNA (Tewari, 1971).

(6) Quantitative estimates of DNA have demonstrated a striking resemblance between procaryotes and plastids, and distinct differences between these and nuclear DNA.

(7) There is a remarkable similarity between the ribosomes of chloroplasts and the ribosomes of procaryotes, but a sharp difference between these ribosomes and the cytoplasmic ribosomes of eucaryotes.

A few exceptions apart, the ribosomes of plastids and procaryotes with a sedimentation coefficient of 70S are smaller than 80S cytoplasmic ribosomes and have the 16S and 23S type of rRNA in contrast to the 18S and 25S type of cytoplasmic ribosomes. Both chloroplast ribosomes and plastid ribosomes seem to have a characteristic cleavage furrow which is apparently lacking in cytoplasmic ribosomes (Bruskov and Odintsova, 1968). Chloramphenicol inhibits the incorporation of amino acids into proteins in chloroplast ribosomes as well as procaryotic ribosomes, but does not do so in cytoplasmic ribosomes. On the other hand, actidione blocks protein synthesis on cytoplasmic ribosomes whereas chloroplast and procaryote ribosomes are quite insensitive to it.

(8) The protein synthesis mechanism in both procaryotes and plastids seems similar and requires formylmethionine tRNA.

GENETICAL HOMOLOGY AND HYBRIDIZATION

Analyses on different aspects of the biology of eucaryotic chloroplasts and on the genetics of blue-green algae strongly indicate that the genetic processes of both cyanophytes and chloroplasts may be very similar.

The study of cyanophycean genetics was pioneered by Kumar (1962) who succeeded in isolating mutant strains of the unicellular blue-green alga *Anacystis nidulans* and, having crossed them, obtained the first definite indication of gene recombination in these procaryotes. Since then, many papers have been published on various aspects of cyanophycean genetics, including mutagenesis and recombination (Houghton, 1972; Kumar, 1975).

Our knowledge of the occurrence of certain genetic processes in chloroplasts was for a long time based on indirect effects and observations of such phenomena as the division and segregation of chloroplasts in dividing cells, the existence of plastid clones whose identities are perpetuated through several cell divisions, and the maternal inheritance pattern of many plastid-associated traits (see Whitton *et al.*, 1971). It has also been acknowledged that plastids arise from only pre-existing plastids and not *de novo*. Experimental studies on the ultraviolet (UV) induction of bleached mutations in *Euglena* chloroplasts considerably strengthened the concept of genetic autonomy of the chloroplast, especially in view of the observation that such bleached mutations had an action spectrum peak at 260 nm (corresponding to a DNA target) and were found to be localized in the cytoplasm and not the nucleus. The theory that the chloroplast has at least some genetic autonomy has received strong and direct support from repeated demonstrations in recent years of DNA, RNA, ribosomes, and the protein-synthesizing enzymes within the chloroplasts of several algae and higher plants. Ridley and Leech (1970) have reported the *in vitro* division of chloroplasts isolated from young leaves of *Vicia faba*; should this involve real replication, rather than mere fragmentation of the plastid during sectioning, it would constitute direct evidence in support of the concept of genetic autonomy of the chloroplast.

PLASTID MUTATIONS

A good example of suspected mutations in chloroplast DNA has been described by Wildman (1971) in a cytoplasmic mutant of tobacco. This mutant exhibits leaf variegation which is transmitted maternally. The mesophyll cells from variegated leaves contain normal and defective chloroplasts in the same cells (no defective chloroplasts were detected in wild-type tobacco). The existence of such mixed populations of chloroplasts in the same cell ruled out a mechanism for DNA repression because it would be difficult to imagine how a repressor could switch off transcription in some chloroplasts and not in others. The only probable explanation is the occurrence of a mutation in chloroplast DNA which prevents transcription of the entire genome, thus leading to the establishment and perpetuation of defective chloroplast clones.

The *iojap* gene in maize has also been reported to induce plastid mutations (Rhoades, 1950). In fact, a gene-induced mutation of green plastids in barley to white plastids, producing leaf variegation, was reported as early as 1928 by Imai. Michaelis (1968) has examined gene-induced mutations of cytoplasmic genetic elements affecting the colour of the chloroplast in *Epilobium*.

In general, however, very few cases of chloroplast DNA mutation have been thoroughly studied. Surzycki *et al.* (1970) have explained the apparent paucity of such mutations on the basis that the chloroplast genome may be extensively polyploid or redundant and, hence, any single gene mutation would remain undetected for many generations.

Considerable research on the genetics of chloroplasts in *Euglena* has strengthened the theory that chloroplasts contain at least some of the information needed for their own maintenance.

Although the existence of genes within chloroplasts is now generally acknowledged, it remains to be seen whether all chloroplast DNAs contain similar or partially related genes. Suitable experiments involving hybridization between the chloroplast DNA of different kinds of higher plants and algae should provide valuable information on this point.

HYBRIDIZATION STUDIES

Chloroplast ribosomal RNA originates from chloroplast DNA and not from nuclear DNA; in eucaryotes, rRNA originates from nuclear DNA.

Surzycki and Gillham (1971) have used rifampicin to differentially inhibit the transcription of chloroplast DNA but not of nuclear DNA. At a certain concentration, rifampicin treatment led to phenotypic loss of resistance in certain antibiotic-resistant, uniparentally-inherited mutants of *Chlamydomonas reinhardii.* This observation provided strong indication of the mutations being localized in chloroplast DNA rather than nuclear DNA. According to Ruth Sager (see Preer, 1971), the uniparentally-inherited genes in *C. reinhardii* which she mapped most probably represent chloroplast DNA.

On the basis of their studies on *C. reinhardii,* Surzycki *et al.* (1970) have suggested that the genetic apparatus of the chloroplast is of hybrid origin, with the ribosomal RNA being derived from the chloroplast, and the DNA polymerase, the DNA-dependent RNA polymerase, and at least some of the ribosomal proteins being synthesized outside the chloroplast. According to this research team, the genetic system of the *C. reinhardii* chloroplast is fully integrated with the nuclear-cytoplasmic genetic system, and the final control over the structure and function of the chloroplast involves the participation of both these genetic systems.

Very little cross-hybridization occurs between the chloroplast DNA and the nuclear DNA extracted from tobacco leaves.

In *Euglena gracilis,* only about 15% of chloroplast genetic material was found to be homologous with that in the nucleus; on the other hand, substantial hybridization between rRNA from blue-green algae and the plastid DNA of *E. gracilis* has been observed. This indicates considerable genetic homology between the chloroplast of *Euglena* and the blue-green algal cell.

EXTANT SYMBIOSES

Symbioses between algae and other organisms (algae, protozoa through higher plants, and animals) are extremely widespread and diverse, involving thousands of different examples from various phyla of plants and animals.

Blue-green algae occur symbiotically (as cell components) in protozoa, amoebae, green algae (*Gloeochaete, Glaucocystis*; see Fig. 4-4), diatoms, fungi (*Geosiphon pyriforme*), etc. In such associations, blue-green algae play the role of chloroplasts.

Recent analyses of contemporary symbioses have substantially strengthened the very interesting possibility of intracellular endosymbionts and extrachromosomal genetic elements playing an important role as isolating mechanisms in evolution (see Preer, 1971).

The widespread existence of diverse symbiotic associations involving blue-green algae and many kinds of other organisms confirms the biological importance of such symbioses and supports the validity of the endosymbiotic theory. The artificial creation of symbiotic associations between blue-green algae and various other organisms (perhaps eucaryotic algae to begin with), by growing them in mixed cultures under suitable environmental conditions, should furnish more direct evidence in favour of the endosymbiotic theory.

COMMENTS

The several lines of evidence just summarized indicate that the chloroplasts of eucaryotes have most probably originated and evolved from ancient free-living photosynthetic procaryotes which became endosymbiotic within eucaryotic cells and, in course of time, were transformed into present-day chloroplasts which seem to resemble entire procaryotic cells morphologically, biochemically, and genetically. The observable differences between the properties of chloroplasts and cyanophytes do not seem very serious and could have well arisen as a direct consequence of the extended period of adaptation that chloroplasts must have undergone during the course of their endosymbiotic existence.

Raven (1970) has put forward certain views on the origin and evolution of chloroplasts in different phyla of algae and has speculated on possible phylogenetic connections among these ph la in relation to the origin of higher plants. His views are represented diagrammatically in Fig. 4-5. The similarities observed between such different phyla as Cyanophyta and Rhodophyta, Phaeophyta and Pyrrophyta, and Chlorophyta and Euglenophyta, may be very reasonably explained by postulating an independent, symbiotic origin of the chloroplast in each case. Raven has further postulated that endosymbionts within eucaryotic organisms have lost many of their functions to the nucleus during the course of their evolution.

The recent discovery (see Lewin, 1975) of a marine cyanophyte, *Synechocystis didemni* Lewin, which is exceptional among blue-green algae in possessing both chlorophyll-a and chlorophyll-b strongly supports Raven's scheme as well as Mereschkowsky's views on the symbiotic origin of green algal chloroplasts. The evidence presented in this chapter can provide at least a strong theoretical basis for the phylogenetic homology between cyanophytes and chloroplasts. When this is considered in conjunction with another argument (still largely speculative), viz., that the chloroplast may still harbour the *nif* gene (for nitrogenase enzyme) determining nitrogen fixation, albeit in an inactive or defunct form (perhaps because of the prevailing aerobic conditions), the case for some possibility of transferring the gene from blue-green algae to the chloroplasts of crop plants (or at least of green algae) acquires a certain credibility. Some likely approaches in this direction may include the joint culture of isolated chloroplasts (admittedly, a most formidable task yet) with blue-green algae under anaerobic or reducing conditions through several generations, and the introduction of nitrogen fixers into the roots of crop plants in an attempt to establish artificial symbioses comparable to those existing in, for instance, the roots of cycads.

The successful establishment of the *nif* gene within the chloroplast may be expected to require: (1) acceptance of the "foreign" DNA; (2) survival and replication of the DNA introduced in the cytoplasm or the nucleus; and

(3) expression of the foreign DNA in terms of proteins (enzymes) in the new cytoplasmic environment. The problem is further complicated by the ability of different cell types of higher plants to activate genes to differing extent. However, the recent suspicion of a strong phylogenetic connection between

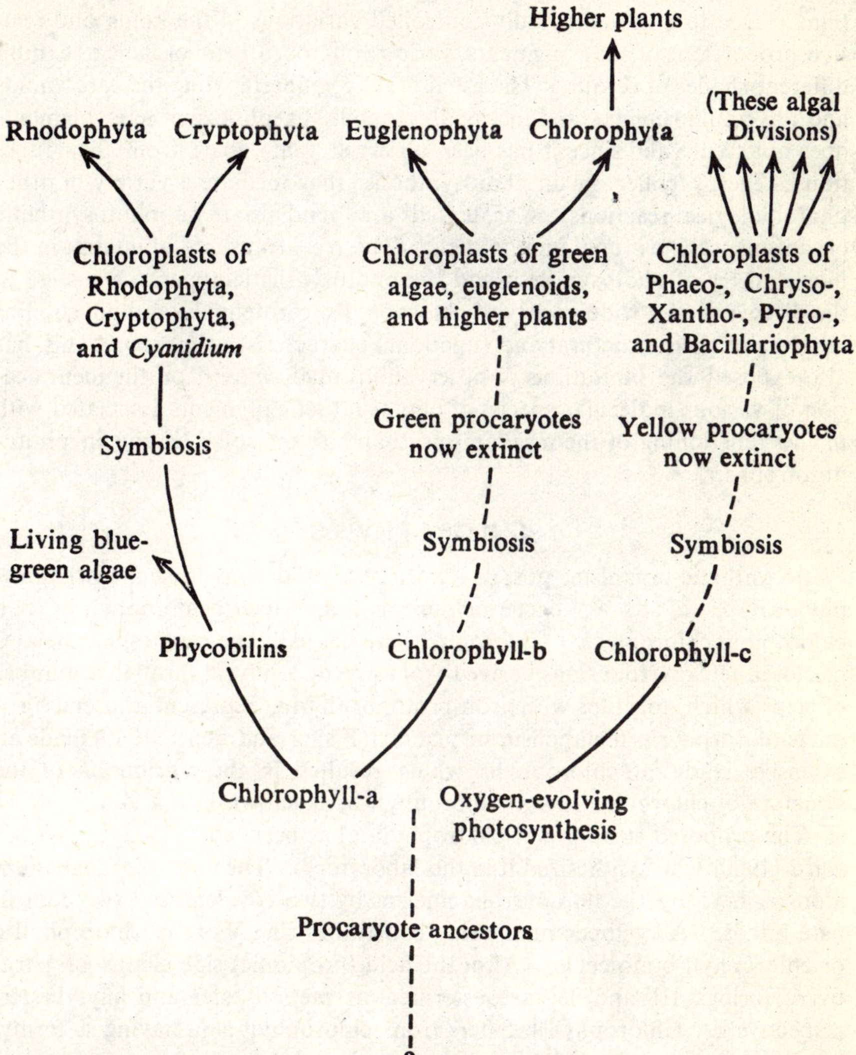

Fig. 4-5 Scheme for evolution and phylogeny of chloroplast. (After Raven, 1970.)

chloroplasts and cyanophytes and the emergence of the endosymbiotic theory as one of the exciting hotspots in modern biology encourage the hope that the problem of transference of the gene (for nitrogen fixation) from blue-green algae to the chloroplasts of crop plants might be solved in the not too distant future, probably within two decades.

Photosynthetic Pigments

INTRODUCTION

Photosynthetic organisms contain two principal kinds of pigments, the fat-soluble chlorophylls and carotenoids, and the water-soluble phycobiliproteins. Due to environmentally-controlled variations in the kinds and relative proportions of these pigments, various photosynthetic organisms exhibit different shades of colour. The usual practice of designating the carotenoids and phycobiliproteins, on functional grounds, as photosynthetic pigments does not seem valid since it has been observed that, apart from their function as energy collectors in photosynthesis, they mediate a variety of other photobiological reactions. Carotenoids are found also in nonphotosynthetic organisms, such as fungi and animals. The recent upsurge of interest in the biochemistry of photosynthesis and the photosynthetic apparatus has led to the discovery of a wide variety of chlorophylls, carotenoids, and phycobilins in terms of their structural and functional characteristics *in vivo*. This has also exposed the limitations of observations made *in vitro* on the identification of various molecular species of photosynthetic pigments associated with *in vivo* functioning of the mechanisms that harvest solar energy in photoautotrophs.

CHLOROPHYLLS

Photosynthetic organisms possess a variety of chlorophylls, such as chlorophyll-a, -b, -c, -d, -e, bacteriochlorophyll-a, bacteriochlorophyll-b, and chlorobium chlorophyll. Chlorophylls are basically the magnesium chelates of closed tetrapyrrole rings derived from protoporphyrin through a number of steps which, together with iron-protoporphyrin, represent the end products of porphyrin metabolism in plants. Fisher and Stern (1940) made an extensive study of chlorophylls which resulted in the elucidation of the structure of chlorophyll-a and chlorophyll-b, as shown in Fig. 4-6.

The proposed structure of chlorophyll-a has been confirmed by Woodward (1960) who synthesized it in the laboratory. The nonionic magnesium atom is held by the tetrapyrrolic nucleus by two covalent and two coordinate bonds. A cyclopentanone ring constitutes ring V of the chlorophyll-a or chlorophyll-b molecule. Also, the acid (propionic) side chains of tetrapyrrole rings III and IV are esterified as methyl ester and phytyl ester, respectively. Chlorophyll-b differs from chlorophyll-a in having a formyl group instead of a methyl group at carbon-3 of the tetrapyrrole ring. Chlorophyll-d contains a formyl group instead of the vinyl group at the carbon-2 position, and it is believed that this pigment results from the oxidation of chlorophyll-a *in vitro*. In fact, chlorophyll-d has never been detected *in vivo*, a fact that casts doubts on its very existence as a natural pigment of photosynthesis in the red algae in which it has been reported, on the basis of *in vitro* studies. Chlorophyll-c lacks phytyl esterification, has vinyl or ethyl group at carbon-4, and possesses an unreduced ring IV. In these respects, it differs from chlorophyll-a. In nature, it exists as a mixture of two pigments because some molecules contain the ethyl group whereas others contain the vinyl group at carbon-4. The structure of chlorophyll-e is not known. Bacteriochlorophyll-a is similar to chlorophyll-a in structure, except that its carbon-2 has an acetyl group and its

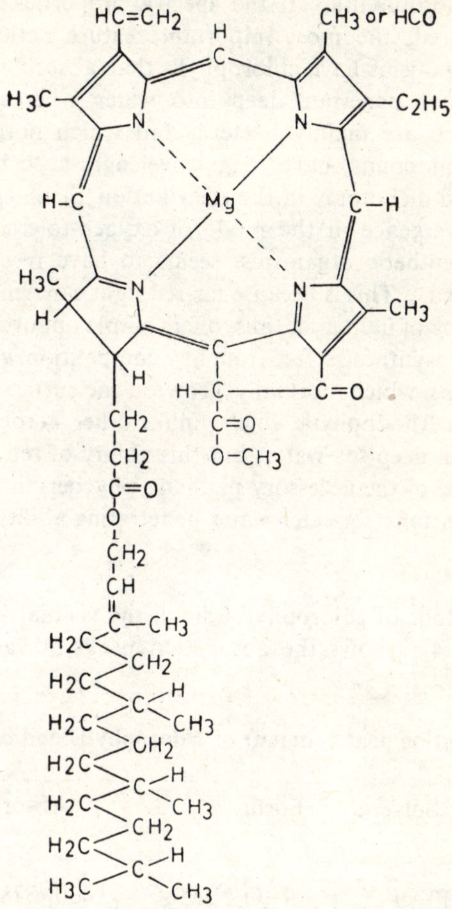

Fig. 4-6 Chemical structure of chlorophyll-a and chlorophyll-b.

carbon-3 and -4 are reduced. The structure of bacteriochlorophyll-b is not yet determined. The chlorobium chlorophylls which absorb at 650 nm and 660 nm seem to be mixtures of six pigments each; this view is based on the variation in groups attached to carbon-4, -5, and -8. Besides, it contains farnesyl ester in place of the phytyl ester found in other chlorophylls.

Chlorophyll-a is found in all oxygen-evolving photosynthetic organisms whereas chlorophyll-b is present in higher green plants as well as in two algal phyla, viz., Chlorophyta and Euglenophyta. Chlorophyll-c has been reported in algae alone and occurs in Phaeophyta, Pyrrophyta, Bacillariophyta, Cryptophyta, and some members of Chrysophyta. Chlorophyll-d and chlorophyll-e have been located in only Rhodophyta and Xanthophyta, respectively. All photosynthetic bacteria contain bacteriochlorophyll-a, except for *Rhodopseudomonas* which is stated to possess bacteriochlorophyll-b. It is noteworthy that each strain of green sulphur bacteria has either the 650 nm or the 660 nm form of chlorobium chlorophyll in addition to bacteriochlorophyll-a. If the spectral properties of various chlorophylls are compared, the most important feature noticed is that aerobic photosynthetic organisms have chlorophylls that absorb those wavelengths of light that cannot penetrate deep into water whereas their anaerobic counterparts, which are mainly bacteria and which normally live at the bottom of stagnant ponds, have long wavelength-absorbing chlorophylls. Such a pronounced dichotomy in the distribution of chlorophylls, coupled with a similar divergence in the mode of oxygen tolerance and metabolic activity in photosynthetic organisms, seems to have resulted mainly from their ecological need. This is because far-red light can penetrate deeper than shorter wavelengths of light and, thus, offers ample opportunity to anaerobic organisms to photosynthesize free from any competition with aerobic photosynthetic organisms which normally grow on the surface of water or land. Exceptions are the Rhodophyta which, unlike other aerobic photosynthetic organisms, grow in deep sea-water; but this ability of red algae seems to be due to the presence of the accessory pigment phycoerythrin which absorbs green light, known for its greater water-penetrating ability.

PROPERTIES

The absorption bands of chlorophyll-a lie in the visible region of the light spectrum. Table 4-1 shows the absorption peaks of various chlorophylls in solution *in vitro*.

Table 4-1 Absorption peaks (in nm) of chlorophylls and bacteriochlorophyll

Chlorophyll	Solvent	Formula	Absorption peaks (nm)
a	Ether	$C_{55}H_{72}O_5N_4Mg$	430, 578, 662
b	Ether	$C_{55}H_{70}O_6N_4Mg$	430, 595, 644
c	Ether	?	447, 579, 628 (weak peak)
d	Ether	?	447, 548, 688
e	Methanol	?	415 654
Bacterio-chlorophyll-a	Ether	$C_{55}H_{74}O_6N_4Mg$	358, 391, 577 773

In general, the *in vivo* absorption bands of chlorophylls lie at 10–20 nm higher wavelengths than their corresponding absorption bands *in vitro*. Besides, the kind of solvent used and the association of polar substances (e.g., water) with ether greatly affects the intensities and positions of the absorption bands. The dilution of chlorophyll solution in acetone or alcohol with water leads to colloidal aggregates of chlorophyll whose absorption maxima approach those observed *in vivo*. Thus, the broadening and shifting of chlorophyll absorption bands to longer wavelengths *in vivo* is attributed to the formation of chlorophyll aggregates of a dimeric and a polymeric nature (Brody and Brody, 1965). Both the study of the derivative spectra of chlorophyll-a *in vivo* and the absorption and curve analysis of model systems, contaiinng different concentrations of chlorophyll dissolved in nonpolar solvents and deposited in films, suggest four forms of chlorophyll-a; forms which absorb light of wavelengths 661.6 nm, 669.6 nm, 677.1 nm, and 683.7 nm (Brown, 1972). Forms of chlorophyll-a absorbing light at wavelengths longer than 685 nm have aroused considerable interest from the viewpoint of locating the real trap or site where solar energy is converted into chemical form. Recent studies indicate the existence of a form that absorbs 691 nm of chlorophyll-a in addition to the other special and rare form, called P700, which constitutes the reaction centre of PSI. Other far-red absorbing types of chlorophyll-a are also known, though with much less certainty (Brown, 1972). Substantial evidence indicates the existence of two spectral forms of chlorophyll-b *in vivo* absorbing light at 640 nm and 650 nm, respectively. Many workers believe that various *in vivo* spectral species of chlorophyll-a result from the aggregation of chlorophyll molecules with water. An alternate hypothesis is that these different spectral forms may be due to their differing patterns of complex formation with the proteins of membranes. This view is supported by the observation that a mutant of *Scenedesmus* lacks a functional PSI, the long wavelength-absorbing forms near 700 nm, and also a specific chlorophyll-protein complex.

Although the fact that chlorophyll is present in different forms (a, b, c, etc.) has been long established, the observation that these different kinds of chlorophyll occur in multiple molecular forms *in vivo* has not been widely appreciated. Well-planned experiments in the laboratory of Dr. C. S. French at Stanford University have revealed the existence of at least four multiple molecular forms of chlorophyll-a alone. The same is true of other chlorophylls, and also of various kinds of macromolecules found in the living cell in general.

PHYCOBILIPROTEINS

A phycobiliprotein consists of two parts: (1) the phycobilin or bile pigments of algae which are open-chain tetrapyrroles (unlike the cyclic type in chlorophyll) and constitute the chromophoric moiety of the biliproteins, and (2) the protein. The amalgam of the bile pigment with the protein of phycobiliprotein seems to involve two kinds of bonds: the first appears to be of the hydrogen type, since the treatment of the biliprotein with either heat or urea causes denaturation of the pigment leading to the quenching of its fluorescence property; the second kind is of a peptide nature and is therefore stronger, since the separation of bile pigment and protein components requires treatment with strong acids. The prefixes R- and C-, hitherto used to signify the source of phycobiliproteins from either Rhodophyta or Cyanophyta, are now employed to indicate the spectral characteristics of these pigments, irrespective of their source. Moreover, the biliproteins of a single spectral class differ in their pH-dependent dissociation and reaggregation, and amino acid composition and immunological-serological properties (Bogorad, 1967). The absorption maxima in the visible region of the aqueous solution of phycobiliproteins are given in Table 4-2.

Table 4-2 Absorption maxima (in nm) of phycobiliproteins

Phycobiliprotein	Absorption maxima
R-phycocyanin	614, 551
C-phycocyanin	615
R-phycoerythrin	565, 545, 500
C-phycoerythrin	565
Allophycocyanin	650, 600

BIOSYNTHESIS

Protoporphyrin IX is the precursor of the two most important biological macromolecules, viz., the hemes, which are iron tetrapyrroles, and the chlorophylls which are magnesium tetrapyrroles. The suffix IX denotes that of the 16 different isomers the ninth protoporphyrin is actually involved in the synthesis of hemes and chlorophylls. Protoporphyrin synthesis generally takes place by the condensation of succinyl CoA with glycine to form alpha-amino-beta-ketoadipic acid which, being highly unstable, immediately undergoes decarboxylation to yield delta-aminolevulinic acid (ALA). This enzymatic reaction is catalyzed by the enzyme ALA synthetase which has been located in all systems, except green plants where the transamination of gamma-, delta-dioxovaleric acid produced from α-oxoglutarate is believed to be the reaction leading to the formation of ALA in the presence of ALA transaminase. Pyridoxal phosphate is the prosthetic group of ALA synthetase:

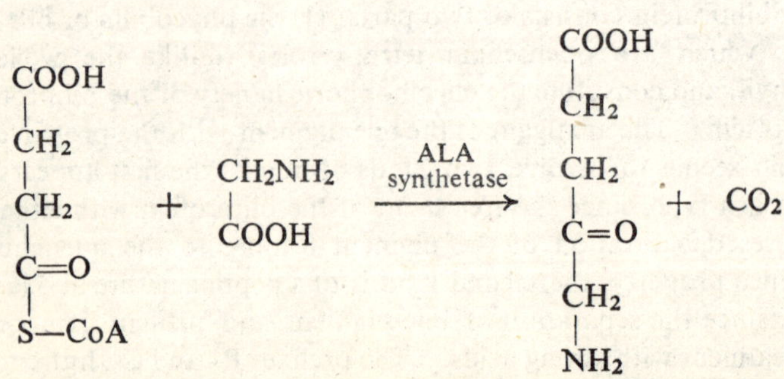

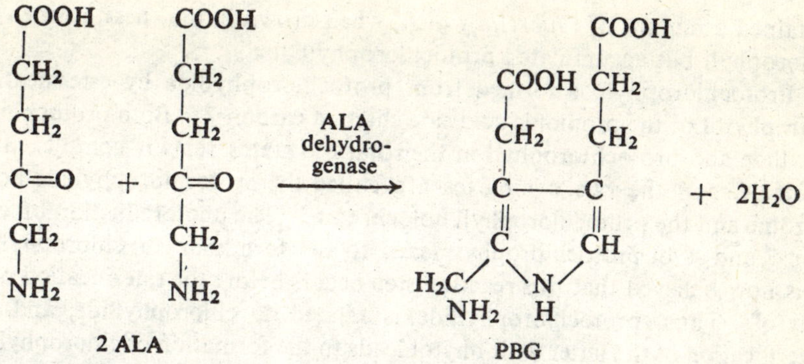

Two molecules of ALA are acted upon by ALA dehydrogenase, and a monopyrrole compound porphobilinogen (PBG) is formed by the elimination of two molecules of water. The presence of the enzyme ALA dehydrogenase has been demonstrated in a number of photosynthetic organisms. Four molecules of PBG are then consumed in the formation of the compound uroporphyrinogen III in a reaction catalyzed by two enzymes (synthetase and cosynthetase) which have previously been isolated from a variety of plant materials. However, the details of the mechanism of this reaction are still controversial.

Uroporphyrinogen III possesses four acetic acid (A) and propionic acid (P) side chains. The enzyme urogen decarboxylase decarboxylates the four A chains, thus producing the compound called coproporphyrinogen III. This compound undergoes oxidative decarboxylation in the presence of an oxidative decarboxylase to form protoporphyrinogen IX. In this reaction, the two P side chains of pyrrole rings I and II get converted into vinyl groups. Protoporphyrinogen IX is oxidized to form protoporphyrin IX by an unknown enzymatic reaction. Subsequent chelation of protoporphyrin IX with magnesium occurs in the presence of an unknown enzyme, thus leading to the formation of magnesium-protoporphyrin. The carboxylic group of the propionic acid located in ring III of Mg-protoporphyrin undergoes esterification with methanol, giving rise to Mg-protoporphyrin monomethyl ester; this ester, after the substitution of the vinyl group on carbon-4 of ring II by an ethyl group, followed by the formation of a cyclopentanone ring by the esterified P residue, forms protochlorophyllide-a or Mg-vinyl pheoporphyrin.

That such a sequence of chlorophyll biosynthesis may be correct is borne out by biochemical analysis of the chlorophyll mutants of *Chlorella* which are blocked at different steps along the biosynthetic chain of chlorophyll formation. Granick (1961) isolated a series of *Chlorella* mutants with blockages at various steps; one group of mutants accumulated protoporphyrin, another Mg-protoporphyrin, a third Mg-protoporphyrin monomethyl ester, and a fourth protochlorophyllide-a. Analyzing protoporphyrin metabolism in isolated chloroplasts of *Euglena*, Granick (1967) noted the formation of protoporphyrin from delta-aminolevulinic acid. Bryan and Bogorad (1963) obtained a mutant of *Chlorella* which, when grown in darkness, formed no chlorophyll but accumulated protochlorophyllide-a.

Protochlorophyll is formed from protochlorophyllide by esterification with phytol of the propionic acid side chain at carbon-7. Both protochlorophyllide and protochlorophyll in their natural states remain complexed to proteins, and the two complexes are called the protochlorophyllide holochrome and the protochlorophyll holochrome. The photoreduction of carbon-7 and -8 of protochlorophyll leads to the formation of chlorophyll-a. It is now believed that this reaction step occurs before the esterification with phytol. Thus, protochlorophyllide is related to chlorophyllide, and the esterification of the latter with phytol leads to the formation of chlorophyll-a. Observations of chlorophyll biosynthesis in pigment mutants of *Chlorella* and *Euglena* give ample support to the aforesaid biosynthetic sequence of chlorophyll formation in green plants.

There are no reports on the occurrence of protochlorophyllide-b or chlorophyllide-b. Chlorophyll-b in plants always appears after the development of chlorophyll-a. Nothing definite is known about chlorophyll-c. However, both chlorophyll-d and bacteriochlorophyll-a are believed to be derived from chlorophyll-a. In other words, all available evidence and fossil history (see Olson, 1970) indicate that chlorophyll-a was the first pigment of photosynthetic organisms to have developed during the evolution of photosynthetic pigments. An important difference between the biosynthetic patterns of chlorophyll-a and bacteriochlorophyll-a is that the latter is extremely sensitive to oxygen and is not synthesized by photosynthetic bacteria under aerobic conditions. Light, protein synthesis, ALA synthetase, and ATP levels seem to be some of the important regulatory controls of chlorophyll metabolism in photosynthetic organisms. However, further details about the mode of exercise of these controls have not yet been determined.

CAROTENOIDS

Carotenoids are unsaturated polyhydrocarbons of an aliphatic or alicyclic nature and are generally composed of eight isoprene units (each unit = C_5H_8). They fall into two classes, viz., carotenes and xanthophylls; the latter are oxygenated derivatives of the former. Carotenoids absorb light of wavelengths below 500 nm and the position of their long-wave absorption bands in the visible region, which are generally three, depends upon the number of conjugated double bonds (i.e., alternating double and single bonds) present. In other words, with the increase in the number of conjugated double bonds, the absorbance bands are displaced towards longer wavelengths (see Table 4-3).

Any of the following alterations in the geometrical configuration of carotenoids will lead to some shifts in the position of their absorption bands too:

(a) A shift in an ethylenic bond in a six-membered ring from its conjugated position to an isolated position leads to the displacement

Table 4-3 Relation between position of absorption bands and number of conjugated double bonds per carotenoid molecule, in petroleum ether

Pigment	Number of conjugated double bonds	Absorption bands		
		I	II	III
Phytoene	3	275	285	296
Phytofluene	5	331	348	367
β-carotene	11	427	450	485
Neurosporene	9	416	440	470
Lycopene	11	446	472	505

of absorption bands towards shorter wavelengths, e.g., β-carotene with one conjugated bond in each of its two six-membered rings has one of its prominent absorption bands at 425 nm whereas e-carotene with its isolated ethylenic bond has its corresponding absorption peak at 418 nm.

(b) The replacement of the terminal ethylenic bond by a 5,6-epoxide group lowers the absorption maxima, e.g., L-carotene has one of its absorption maxima at 445 nm whereas its epoxide derivative shows the corresponding peak at 442 nm.

(c) The organization of open-chain molecules into ring forms shifts the absorption bands towards shorter wavelengths, e.g., lycopene, the open-chain carotene, has its absorption peak of 446 nm shifted to 425 nm in β-carotene which has two closed rings.

(d) The addition of carbonyl groups along with the conjugated double bonds displaces the absorption peaks of carotenes towards higher wavelengths.

Apart from these four alterations, the kinds of isomerism and the nature of the solvents used for extracting carotenoids also change the positions of absorption bands. For instance, mesobiliviolin is the bile component of both phycocyanin and allophycocyanin and the bile pigment mesobilierythrin (mesobilirodin) of phycoerythrin. The incorporation of ^{14}C-delta-amino-levulinic acid into phycocyanin of *Cyanidium caldarium* provides indirect evidence for the existence of a porphyrin or heme precursor in the synthesis of bile pigments (see Bogorad, 1967).

The biosynthetic mechanisms of carotenoid formation also are not established, but it is believed that acetate is the ultimate source from which carotenoids derive their carbon atoms, involving hydroxymethyl glutaric, mevalonic, and dimethylacrylic acids as the intermediates. The active isoprene unit produced from mevalonic pyrophosphate is isopentenyl pyrophosphate which isomerizes into pyrophosphate. Both isopentenyl pyrophosphate and its isomer are 5-carbon compounds and these two react with each other to produce a 10-carbon compound, called geranyl pyrophosphate. Two molecules of geranyl pyrophosphate yield geranyl-geranyl-pyrophosphate, a 20-carbon compound. From two such 20-carbon molecules there arises a molecule of phytoene which is a 40-carbon compound.

Many kinds of carotenoid mutants of *Chlorella* were isolated by Claes (1967). One strain formed only phytoene whereas another formed phytoene, phytofluene, and e-carotene. The formation of xanthophyll was completely blocked in these strains. The mutant strain *9a* lacked both α- and β-carotene but had xanthophylls. However, the most interesting mutant was strain 5/520 which, when grown heterotrophically in darkness, did not form chlorophyll but produced xanthophylls along with small amounts of phytoene, phytofluene, e-carotene, neurosporene, and prolycopene. When the dark-grown cultures were exposed to illumination, β-carotenes together with an unidentified carotene, designated as X-carotene (possibly β-zeaxanthin), were formed at the cost of carotenoids produced under dark, heterotrophic conditions. These observations led to the formulation of the following scheme for carotenoid production in *Chlorella* (see Claes, 1967):

Phytoene Phytofluene e-carotene Proneurosporene Prolycopene	$\xrightarrow[N_2]{\text{Light}}$	α-carotene β-carotene β-zeaxanthin	$\xrightarrow[O_2]{\text{Dark}}$	Xanthophylls

Of these pigments, phytoene is the most saturated, and prolycopene the most unsaturated acyclic hydrocarbon. It has been suggested that the formation of cyclic carotenoids, such as carotenes, starts either at the level of proneurosporene or prolycopene. The foregoing scheme also suggests that xanthophyll formation occurs in the dark under aerobic conditions. The oxygen of xanthophylls seems to come not from water but from elementary (gaseous) oxygen. With the aid of $^{18}O_2$ and $H_2{}^{18}O$, it has been observed that the oxygen of lutein arises from molecular oxygen and not from water.

FACTORS AFFECTING PIGMENTATION

Studies on pigment mutants of algae have established the participation of genetic elements in the control of the quantity and quality of algal pigmentation. The effects on pigmentation of environmental factors, such as nutrition, light, temperature, and antibiotics, will now be described.

CHLOROPHYLLS

The essentiality of magnesium and nitrogen, but not of potassium or phosphorus, for the formation of chlorophyll in *Chlorella* was shown by Hase *et al.* (1957). When *Chlorella* cells were grown in a medium free of nitrogen and magnesium, they remained etiolated, but no etiolation occurred in a medium lacking phosphorus or potassium.

Allen (1958) noted rapid bleaching of chlorophyll-a in a mutant of *Chlorella* lacking chlorophyll-b, and from this observation she concluded that chlorophyll-b stabilizes and protects chlorophyll-a against bleaching by light. Intense illumination was observed to decrease the quantity of both chlorophyll-a and phycocyanin in *Cyanidium caldarium.*

Some naturally-occurring strains of *Chlorella* and *Chlamydomonas* can form chlorophyll from protochlorophyllide-a even in the dark whereas their mutants are unable to do so and require light for the conversion of protochlorophyllide-a into chlorophyll (see Granick, 1967). This suggests that the critical step in the conversion of protochlorophyllide-a into chlorophyll is the reduction of the pyrrole ring which seems to be genetically controlled and can proceed both in the light and the dark, depending on the nature of the genetic block. *Euglena* has an absolute requirement of light for the maintenance of its chlorophyll (Wolken, 1958). In an investigation of the simultaneous effects of both light and temperature on the ratio of chlorophyll-a 673 and chlorophyll-a 683 forms in *Anacystis nidulans*, Halldal (1958) found the ratio to be highest in cells grown at 4000 lux and at 30–38°C, and lowest in cells grown in dim light. This implies that chlorophyll-a 673 is less sensitive to bleaching than chlorophyll-a 683, or that its formation needs more intense light than that of chlorophyll-a 683. Brown and French (1959), from the results of their experiments on the effect of intense illumination by red light on the bleaching of various chlorophylls in colloidal-pigment suspensions of *Chlorella pyrenoidosa,* concluded that chlorophyll-b is the most stable and chlorophyll-a 695 the least stable pigment and that chlorophyll-a 673 is comparatively less susceptible than chlorophyll-a 685 to the bleaching action of light. Moreover, the relative proportions of pigments can vary considerably with age too (see Bogorad, 1967).

In *Euglena*, the formation of photosynthetic pigments is completely inhibited by treatment with such agents as streptomycin, chlorotetracycline, acriflavine, and heat. Such treatment also frequently leads to the production of albino mutants in this alga.

PHYCOBILIPROTEINS

Many workers have studied the effect of the quality and intensity of light on the production of phycobiliproteins in algae. Brody and Emerson (1959) noted enhanced formation of phycoerythrin in *Porphyridium cruentum* after the alga had been treated with low-intensity green light (but not blue light). *Anacystis nidulans* forms more phycocyanin in white light of low intensity whereas high-intensity light retards the formation of biliprotein (Myers and Kratz, 1955). Chlorophyll-deficient mutants of *Cyanidium caldarium* were reported to form phycocyanin and allophycocyanin in the light of wavelengths 450 nm or 600 nm whereas the wavelengths 550 nm, 600 nm, and 700 nm were observed to be ineffective in this respect; the synthesis of these pigments was completely blocked in the dark (see Bogorad, 1967). These observations suggest that in *C. caldarium* the photoreceptor of biliprotein formation is neither phycocyanin nor allophycocyanin and that the photoreceptor itself appears to be the precursor of the biliproteins.

Fujita and Hattori (1962) have investigated the effects of various factors on the formation of phycobiliprotein in *Tolypothrix tenuis*. This blue-green alga forms phycocyanin, phycoerythrin, as well as allophycocyanin. When grown in red light, it becomes rich in phycocyanin and allophycocyanin, but phycoerythrin predominates in green light. Intense illumination bleaches the alga and such cultures fail to form biliproteins in a medium lacking a nitrogen source, even when exposed to light of moderate intensity. However, if the alga is grown in darkness or in red light in a medium containing nitrate, phycocyanin and allophycocyanin is formed. If these cultures are kept in green light, instead of being exposed to red light or darkness, phycoerythrin is synthesized.

The foregoing observations suggesting the formation of phycocyanin and allophycocyanin in dark and in red light, and the production of phycoerythrin in green light, indicate the involvement of a common precursor. This is supported by the finding that the effect of green light can be reversed by red light.

Chloramphenicol inhibits the formation of biliprotein in *C. caldarium* and *T. tenuis*. This indicates the importance of protein synthesis for the formation of biliprotein.

CAROTENOIDS

Some general effects of environmental factors on carotenoid synthesis have already been dealt with. Though very few detailed studies of these aspects seem to have been made, it may be assumed that oxygen, temperature, light, metabolites, and metabolic inhibitors considerably influence the synthesis and accumulation of carotenoids.

FUNCTIONS

All photosynthetic pigments, except chlorophyll-a, act as accessory pigments in the sense that the light energy absorbed by them is ultimately transferred to a special form of chlorophyll-a, called P680 or P700, which initiates the process of conversion of solar energy into chemical form. This process is common to all photosynthetic organisms, but the carbon compounds eventually manufactured as a result of photosynthesis show considerable diversity in nature, quality, and quantity. This is especially true of algae in which different phyla vary in the composition of photosynthetic pigments as well as in the nature of the food reserves stored. Precise reasons for such variations in the nature of photosynthetic products remain unknown, but the idea that they may conceivably be an outcome of the differences among photosynthetic pigments, though highly conjectural, merits consideration.

At present, the only distinct function of chlorophylls is to drive the photochemical reactions of photosynthesis. Govindjee and Rabinowitch (1961; see Govindjee, 1975) discovered a class of pigments in certain unicellular algae whose absorption bands were in the far-red region between 720 nm and 800 nm. From a comparison of the absorption spectra of the newly-discovered pigments with the action spectrum (Fig. 4-7) of photosyn-

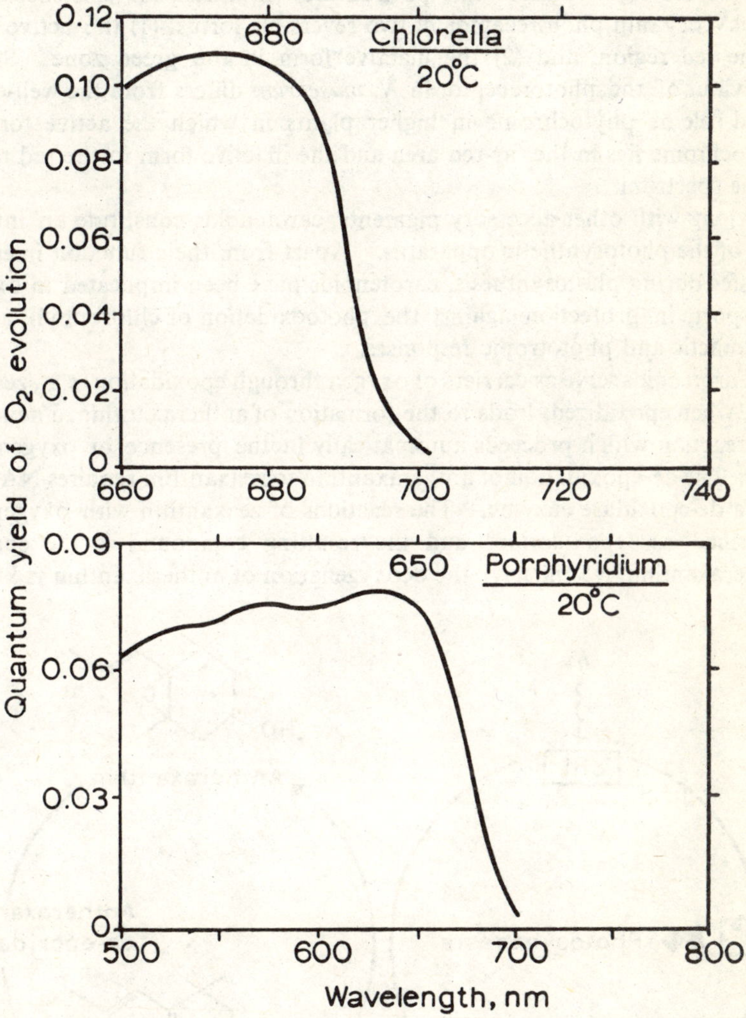

Fig. 4-7 Quantum yields of oxygen evolution by *Chlorella* and *Porphyridium* at different wavelengths of light.

thesis in *Chlorella pyrenoidosa* and *Porphyridium cruentum*, they concluded that the pigments are responsible for the observed photoinhibition of photosynthesis at wavelengths longer than 680 nm (approximately). But the manner in which these pigments inhibit photosynthesis is open to conjecture. Allophycocyanin, which is somewhat similar to phytochrome in chemical and spectral characteristics and absorbs light of 650 nm wavelength, has been shown to trigger the photomorphogenetic development of *Nostoc muscorum* (Lazaroff, 1973). This was concluded after comparing the absorption spectrum of allophycoçyanin with the action spectrum of morphogenesis which requires 650 nm light for the initiation of cellular differentiation in *N. muscorum*. However, the fact that green light inhibits morphogenesis whereas red light reverses this phenomenon indicates the existence of the allophycocyanin photoreceptor in two reversible forms: (1) the active form in the red region, and (2) the inactive form in the green zone. Such a behaviour of the photoreceptor in *N. muscorum* differs from the well-established role of phytochrome in higher plants in which the active form of phytochrome lies in the far-red area and the inactive form in the red region of the spectrum.

Along with other accessory pigments, carotenoids constitute an integral part of the photosynthetic apparatus. Apart from their function in energy transfer during photosynthesis, carotenoids have been implicated in oxygen transport, in protection against the photooxidation of chlorophyll, and in phototactic and phototropic responses.

Carotenoids serve as carriers of oxygen through epoxidation, e.g., zeaxanthin, when epoxidized, leads to the formation of antheraxanthin, a nonenzymic reaction which proceeds automatically in the presence of oxygen and light. The de-epoxidation of antheraxanthin to zeaxanthin requires NADPH and a de-epoxidase enzyme. The reactions of zeaxanthin with oxygen are described as 'epoxidation' and the resulting compound is the epoxide antheraxanthin. Similarly, the deoxygenation of antheraxanthin is known

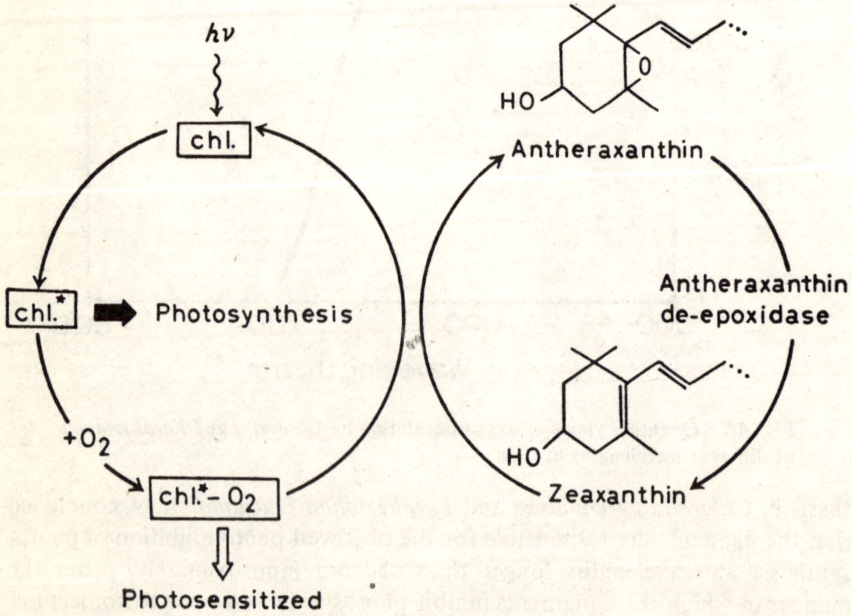

Fig. 4-8 Epoxidation, de-epoxidation cycle.

as de-epoxidation and the resulting de-epoxide is zeaxanthin. Thus, the two carotenoids seem to constitute a chemical buffer system which confers protection against lethal photosensitized oxidation of chlorophyll by scavenging unwanted oxygen and channelling it into epoxide formation (Fig. 4-8). In the presence of light and oxygen, chlorophyll-a molecules get excited by absorbing photons of light, and these excited molecules are deactivated during the subsequent steps of photosynthesis. Some of the excited molecules may react with molecular oxygen, thus entering into a peroxide configuration. If these oxygenated and excited chlorophyll molecules are not deoxygenated they lead to lethal photosensitized oxidation, a situation that may be avoided by the zeaxanthin-antheraxanthin system: zeaxanthin consumes the oxygen of the excited chlorophyll molecule and is changed into antheraxanthin which subsequently gets reduced by the photochemically-generated NADPH and regenerates zeaxanthin. This epoxide cycle thus antagonizes the lethal photosensitization of chlorophyll molecules produced by the joint action of light and oxygen. Coloured carotenoid-deficient mutants of *Chlorella*, *Chlamydomonas*, and *Anabaena* have been observed to undergo lethal photosensitization under aerobic conditions in the presence of light, indicating the significance of carotenoids in protecting the cells against lethal photodamage.

Carotenoids are also believed to participate in phototactic movements and responses of various algae, including their zoospores and gametes.

REFERENCES

Allen, M. B., Possible functions of chlorophyll-b, *Brookhaven Sympos. Biol.*, **11**, 339–42 (1958).

Arnon, D. I., Allen, M. B., Whatley, F. R., Photosynthesis by isolated chloroplasts. *Nature*, **174**, 394–96 (1954).

Arntzen, C. J., Briantais, J., Chloroplast structure and function. In Govindjee (ed.), Bioenergetics of Photosynthesis, 52–113, Academic Press, New York (1975).

Arntzen, C. J., Dilley, R. A., Peters, G. A., Sharp, E. R., Photochemical activity and structural studies of photosystems derived from chloroplast grana and stroma lamellae. *Biochim. Biophys. Acta*, **256**, 85–107 (1972).

Blackman, F. F., Optima and limiting factors. *Ann. Bot.*, **19**, 281–95 (1905).

Bogorad, L., Chloroplast structure and development. In San Pietro, A., Greer, F. A., Army, T. J. (eds.), Harvesting the Sun, 191–210, Academic Press, New York (1967).

Branton, D., Park, R. B., Subunits in chloroplast lamellae. *J. Ultrastruct. Res.*, **19**, 283–94 (1967).

Brody, M., Emerson, R., The effect of wavelength and intensity of light on the proportions of pigments in *Porphyridium cruentum*. *Amer. J. Bot.*, **46**, 433–40 (1959).

Brody, S. S., Brody, M., An experiment showing that P700 can be an aggregated form of chlorophyll-a. *Arch. Biochem. Biophys.*, **110**, 583–85 (1965).

Brown, J. S., Forms of chlorophyll *in vivo*. *Ann. Rev. Plant Physiol.*, **23**, 73–86 (1972).

Brown, J. S., French, C. S., Absorption spectra and relative photostability of different forms of chlorophyll in *Chlorella*. *Plant Physiol.*, **34**, 305–9 (1959).

Bruskov, V. I., Odintsova, M. S., Comparative electron microscope studies of chloroplast and cytoplasmic ribosomes. *J. Mol. Biol.*, **32**, 471–73 (1968).

Bryan, G. W., Bogorad, L., Protochlorophyllide and chlorophyll formation in response to iron nutrition in a *Chlorella* mutant. In *Microalgae and Photosynthetic Bacteria, Jap. Soc. Plant Physiol.*, Tokyo Univ. Press (1963).

Claes, H., Action spectrum of light dependent carotenoid synthesis in *Chlorella vulgaris*. In Goodwin, T. W. (ed.), Biochemistry of Chloroplasts, Vol. 1, 441–44, Academic Press, New York (1967).

Emerson, R., Arnold, W., Photochemical reaction in photosynthesis. *J. Gen. Physiol.*, **16**, 191–205 (1932).

Engelmann, T. W., Farbe und Assimilation. *Bot. Zeit.*, **42**, 81–83, 97–105 (1884).

Fisher, H., Stern, A., Die Chemie des Pyrrols, Vol. 2. Akademische Verlagsges, Leipzig (1940).

Fujita, Y., Hattori, A., Changes in composition of cellular material during formation of phycobilin chromoproteins in *Tolypothrix tenuis*. *J. Biochem.* (Tokyo), **52**, 38–42 (1962).

Gantt, E., Conti, S. F., Granules associated with the chloroplast lamellae of *Porphyridium cruentum*. *J. Cell Biol.*, **29**, 423–34 (1966).

Goodenough, U. W., Straehelin, L. A., Structural differentiation of stacked and unstacked chloroplast membranes. Freeze etch electron microscopy of wild type and mutant strains of *Chlamydomonas*. *J. Cell Biol.*, **48**, 594–619 (1971).

Govindjee (ed.), Bioenergetics of Photosynthesis. Academic Press, New York (1975).

Granick, S., Magnesium protoporphyrin monoester and protoporphyrin monomethyl ester in chlorophyll biosynthesis. *J. Biol. Chem.*, **236**, 1168–72 (1961).

Granick, S., Heme and chlorophyll biosynthetic chain. In Goodwin, T. W. (ed.), Biochemistry of Chloroplasts, Vol. 2, 373–410, Academic Press, New York (1967).

Halldal, P., Pigment formation and growth of blue-green algae in crossed gradients of light intensity and temperature. *Physiol. Plantarum*, **11**, 401–20 (1958).

Hase, E., Morimura, Y., Tamiya, H., Some data on the growth physiology of *Chlorella* studied by the technique of *synchronous culture*. *Arch. Biochem. Biophys.*, **69**, 149–65 (1957).

Hill, R., Oxygen evolved by isolated chloroplasts. *Nature*, **139**, 881–83 (1937).

Houghton, J. A., The genetics of blue-green algae. *Sci. Prog.*, Oxford, **60**, 407–12 (1973).

Imai, Y., A consideration of variegation. *Genetics*, **13**, 544–62 (1928).

Kirk, J. T. O., Nature and function of chloroplast DNA. In Goodwin, T. W. (ed.), Biochemistry of Chloroplasts, Vol. 1, 319–40, Academic Press, New York (1966).

Kirk, J. T. O., Biochemical aspects of chloroplast development. *Ann. Rev. Plant Physiol.*, **21**, 11–42 (1970).

Kumar, H. D., Apparent genetic recombination in a blue-green alga. *Nature*, **196**, 1121–22 (1962).

Kumar, H. D., Genetics of blue-green algae. In Mohan Ram, H. Y., Shah, J. J., Shah, C. K. (eds.), Form, Structure and Function in Plants, 69–80, Sarita Prakashan, Meerut (1975).

Laetsch, W. M., Relationship between chloroplast structure and photosynthetic carbon fixation pathway. *Sci. Prog.*, Oxford, **57**, 323–51 (1969).

Laetsch, W. M., The C_4 syndrome: A structural analysis. *Ann. Rev. Plant Physiol.*, **25**, 27–52 (1974).

Lang, N. H., Rae, P. M. M., Structures in a blue-green alga resembling prolamellar bodies. *Protoplasma*, **64**, 67–74 (1967).

Lazaroff, N., Photomorphogenesis and Nostocacean development. In Carr N. G., Whitton, B. A. (eds.), The Biology of Blue-green Algae, 279–319 Blackwell, Oxford (1973).

Lewin, R. A., A marine *Synechocystis* (Cyanophyta, Chroococcales) epizoic on ascidians. *Phycologia*, **14**, 153–60 (1975).

Mereschkowsky, C., Über Natur und Ursprung der Chromatophoren im Pflanzenreiche. *Biol. Zentralbl.*, **25**, 593–604 (1905).

Michaelis, P., Beitrage zum Problem der Plastiden-Abanderung. IV. Über das Plasma und Plastidenabanderungen auslasende, isotopen-(^{32}P)-induzierte Kerngen von *Epilobium*. *Mol. Gen. Genet.*, **101**, 257–306 (1968).

Mühlethaler, K., Moor, H., Szarkowski, J. W., Ultrastructure of the chloroplast lamellae. *Planta*, **67**, 305–16 (1965).

Myers, J., Kratz, W. A., Relation between pigment content and photosynthetic characteristics in a blue-green alga. *J. Gen. Physiol.*, **39**, 11–22 (1955).

Olson, J. M., The evolution of photosynthesis. *Science*, **168**, 438–46 (1970).

Park, R. B., Biggins, J., Quantasome: size and composition. *Science*, **144**, 1009–11 (1964).

Park, R. B., Pon, N. G., Chemical composition and substructure of lamellae isolated from *Spinacea oleracea* chloroplasts. *J. Mol. Biol.*, **6**, 105–14 (1963).

Park, R. B., Sane, P. V., Distribution of structure and function in chloroplast lamellae. *Ann. Rev. Plant Physiol.*, **22**, 395–430 (1971).

Preer, J. R., Jr., Extrachromosomal inheritance: hereditary symbionts, mitochondria, chloroplasts. *Ann. Rev. Genet.*, **5**, 361–406 (1971).

Raven, P. H., A multiple origin for plastids and mitochondria. *Science*, **169**, 641–46 (1970).

Rhoades, M. M., Gene induced mutation of a heritable cytoplasmic factor producing male sterility in maize. *Proc. Natl. Acad. Sci.*, USA, **36**, 634–35 (1950).

Ridley, S. M., Leech, R. M., Division of chloroplasts in an artificial environment. *Nature*, **227**, 463–65 (1970).

Ris, H., Plaut, W., Ultrastructure of DNA-containing areas in the chloroplast of *Chlamydomonas*. *J. Cell Biol.*, **13**, 383–91 (1962).

Ris, H., Singh, R. N., Electron microscope studies on blue-green algae. *J. biophys. biochem. Cytol.*, **9**, 63–80 (1962).

Surzycki, S. J., Goodenough, U. W., Levine, R. P., Armstrong, J. J., Nuclear and chloroplast control of chloroplast structure and function in *Chlamydomonas reinhardi*. In Miller, P. L. (ed.), Control of Organelle Development, 13–37, Cambridge Univ. Press, London (1970).

Tewari, K. K., Genetic autonomy of extracellular organelles. *Ann. Rev. Plant Physiol.*, **22**, 141–68 (1971).

Tewari, K. K., Wildman, S. G., Function of chloroplast DNA. II. Studies on DNA-dependent RNA polymerase activity in tobacco chloroplasts. *Biochim. Biophys. Acta*, **186**, 358–72 (1969).

Vernon, L. P., Photochemical and electron transport reactions of bacterial photosynthesis, *Bacteriol. Rev.*, **32**, 243–61 (1968).

Weier, T. E., Stocking, C. R., Bracker, C. E., Risley, E. B., Structural relationships of the internal membrane system of *in situ* and isolated chloroplasts of *Hordeum vulgare*. *Amer. J. Bot.*, **52**, 339–47 (1965).

Whitton, B. A., Carr, N. G., Craig, I. W., A comparison of the fine structure and nucleic acid biochemistry of chloroplasts and blue-green algae. *Protoplasma*, **72**, 325–57 (1971).

Wildman, S. G., An approach towards defining the role of chloroplast DNA in the reproduction and differentiation of chloroplasts in higher plants. In Harris, P. J. (ed.), Biological Ultrastructure: The Origin of Cell Organelles, 91–105, State Univ. Press, Oregon (1971).

Wolken, J. J., The chloroplast structure, pigments and pigment-protein complexes. *Brookhaven Sympos. Biol.*, **11**, 87–100 (1958).

Woodward, R. B., Total synthesis of chlorophyll. *J. Amer. Chem. Soc.*, **82**, 3800–3802 (1960).

5: Photosynthesis: Light Reactions

INTRODUCTION

In photosynthesis, light is essential for the formation of organic matter and oxygen from water and carbon dioxide:

$$CO_2 + 2H_2O + \text{Light} \xrightarrow{\text{Chlorophyll}} (CH_2O) + H_2O + O_2 \qquad (1)$$

The whole process is completed in two stages, the light and the dark. The first stage starts with the dehydrogenation of water and ends with the production of a reduced coenzyme, called nicotinamide adenine dinucleotide phosphate, and of adenosine triphosphate. The reduced coenzyme, usually abbreviated as $NADPH_2$ (its correct representation should be $NADPH + H^+$, with the oxidized form being written as $NADP^+$), is the natural reductant of phosphoglyceric acid (PGA) which results from the fixation of CO_2 in ribulose-1,5-diphosphate. ATP is used up in activating the CO_2-acceptor molecule and in driving the highly endergonic reaction of PGA reduction.

Thus, the function of photosynthetic light is to generate two chemically-stable, high-energy compounds, viz., $NADPH + H^+$ and ATP, which are consumed in the reduction of PGA resulting from CO_2 fixation in the ATP-activated molecule of ribulose diphosphate. The further fate of reduced PGA may be the synthesis of carbohydrate, lipid, or protein. The CO_2 fixation, the PGA reduction, and the formation of carbohydrate, lipid, or protein may occur in the light or in the dark. The second stage, i.e., the dark stage, involves several chemical reactions, the fixation of CO_2, and the synthesis of a carbohydrate. Both the light and dark phases involve a series of complex reactions, and although the path of CO_2 in photosynthesis is almost clear, that of light energy is still not definite. Since isolated chloroplasts can incorporate CO_2 into various cellular substances in the dark in the presence of ATP and $NADPH + H^+$, the formation of these two compounds, rather than CO_2 assimilation, is the key role of photosynthesis. These two photosynthetic products are also collectively referred to as the 'assimilatory power' of plants as their availability determines the overall efficiency of CO_2 assimilation.

It is now evident that photosynthesis is a process in which two light reactions operate in series through an interconnecting electron transport chain. $NADPH + H^+$ and ATP are the products of their cooperative action. The two light reactions are activated by two pigment systems or photosystems, PSI and PSII. The product of the photochemical reaction of PSI is a strong reductant (X) and a weak oxidant (the reaction centre of

PSI), and the product of the photochemical reaction of PSII is a weak reductant (Y) and a strong oxidant (the reaction centre of PSII). The strong reductant (X) reduces $NADP^+$ with the help of ferredoxin and the ferredoxin $NADP^+$ reductase; the strong oxidant oxidizes water, thus leading to the release of oxygen. The weak reductant (Y) and the weak oxidant interact; the electron of the reduced Y is passed on to the oxidized PSI, thus restoring each to its normal state. The net result of the two photoacts is to reduce $NADP^+$ with electrons from water. This is, in essence, plant photosynthesis, and it obviously represents a case of oxidation-reduction reaction between an oxidant $NADP^+$ of the redox potential of nearly -0.4 volt and a reductant, water, of the redox potential of nearly 0.8 volt. The path of an electron from a substance of positive potential to that of negative potential is always an uphill path and can operate only at the expense of energy, the magnitude of which will depend upon the potential difference between the two substances. Its converse is the downhill path which always operates with the liberation of energy from the electrons moving downhill. The potential difference between water and $NADP^+$ is nearly $[0.8 - (-0.4) =]$ 1.2 volts. The free energy changes in biological reactions are better understood and represented in terms of a redox potential scale expressed in volts. One electron volt is the energy consumed or liberated during the uphill or downhill transfer of a single electron across a potential difference of one volt, and is equal to 23 kcal/mole.

From water to $NADP^+$, an electron has to overcome a potential barrier of 1.2 volts, or an energy barrier of 27 kcal. The required energy is provided from the light absorbed by both PSI and PSII. The mechanism by which light energy sends the electrons across a potential barrier of 1.2 volts is the same as that which transforms the mechanism of light energy in photosynthesis. A major part of the absorbed light energy finds its way into $NADPH + H^+$ and a minor part of it gets stored in the form of ATP through pseudocyclic or cyclic phosphorylation.

As we have seen, the formation of $NADPH + H^+$ involves an uphill journey of photoelectrons, but that of ATP is intimately associated with their downhill movement. In fact, the light-induced formation of ATP, termed photophosphorylation, is very much analogous to oxidative phosphorylation in mitochondria in that the formation of ATP in both is coupled to the downhill path of electrons.

The two light reactions consist of two separate phases, namely, the photophysical phase which is insensitive to temperature and includes events such as the absorption of light quanta and their migration to the reaction centres, and the photochemical phase which is sensitive to temperature, starts subsequent to the migration of light quanta to the reaction centres, and terminates with the formation of $NADPH + H^+$, ATP, and oxygen.

PHOTOPHYSICAL PHASE

After absorbing a photon, a pigment molecule, say chlorophyll-a, becomes excited; this state lasts nearly 5×10^{-9} sec. The cells, if they have to photosynthesize, must somehow catch this shortlived energy and convert it into a more stable and manageable form. Their failure to do so results in the loss of absorbed energy through several ways. The collectors of light energy are the photosynthetic pigments which include chlorophylls, carotenoids, and phycobilins. They are found in both photosystems PSI and PSII, but the precise proportion is still uncertain. The reaction centre of PSI is a special form of chlorophyll-a which absorbs light at 700 nm and is thus termed P700; the reaction centre of PSII has likewise been identified as P680. In fact, these two reaction centres are the main photosynthetic pigments for the conversion of light energy into chemical form. The function of the remaining pigments seems to be to transfer their absorbed light quanta to P680 and P700.

It is now believed that inductive resonance (defined as the light-induced, electronically-excited state of the first molecule, the donor molecule, whose fluorescence frequencies correspond with the absorption frequencies of the second molecule, the acceptor molecule) is the probable mechanism by which a pigment molecule transfers its light quantum to its neighbour. The absorbed light, if not used in some chemical reaction, is emitted in the form of fluorescence. The wavelength of the emitted photon is always either equal to or longer than that of the absorbed photon. In terms of energy, the emitted light has a lower energy content than the absorbed photon. According to the resonance mechanism, a photon absorbed by one molecule can move to neighbouring molecule, provided that the latter's absorption spectrum overlaps the former's emission spectrum. If this is so, a photon or light quantum may visit several such pigment molecules before being used up in fluorescence or some other process. This resonance transfer occurs not only between like molecules but also between unlike molecules, provided that the receiver absorbs at the same or a slightly longer wavelength as does the emitter. For example, if a solution containing a mixture of chlorophyll-a (660 nm) and chlorophyll-b (644 nm) is exposed to a monochromatic light (644 nm) absorbed by chlorophyll-b alone, the fluorescence observed is not that of chlorophyll-b but of chlorophyll-a. Conversely, if the mixture is irradiated with monochromatic light absorbed by chlorophyll-a alone, the fluorescence here too is always that of chlorophyll-a and not of chlorophyll-b. This is because chlorophyll-a can accept photons from chlorophyll-b, but the latter cannot accept photons from chlorophyll-a. Chlorophyll-b absorbs photons of lower wavelength (=higher energy content) than chlorophyll-a. The natural direction of quantum flow is from smaller wavelength or higher energy content to longer wavelength or lower energy content. Thus, the photons cannot travel from chlorophyll-a to chlorophyll-b, but can always move from chlorophyll-b to chlorophyll-a (Fig. 5-1).

The relation between P680 (or P700) and other pigments is the same as that between chlorophyll-a and chlorophyll-b. Accessory pigments gather light quanta which automatically migrate towards reaction centres. But once the light quanta reach the centres, they cannot flow back because of the wavelength or energy barrier. In other words, P680 and P700 act as a trap or sink for light quanta.

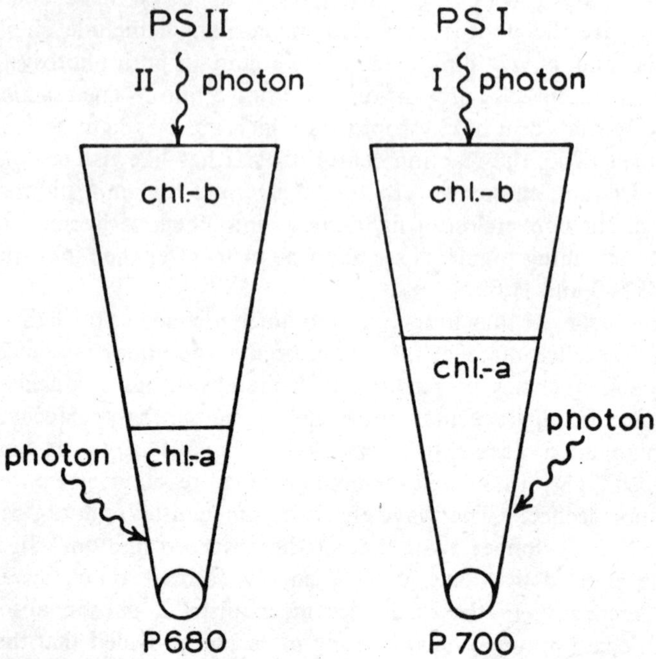

Fig. 5-1 Relative proportions of chlorophyll-a and -b in pigment systems I and II.

In such a model of quantum flow, the relative life of the excited accessory pigment and of the reaction centre is of crucial importance. If the two have the same excited life, most light quanta from accessory pigments may not find a place in reaction centres and will thus go to waste. This wastage can be prevented if the excited life of accessory pigments is much longer than that of reaction centres. In this case, even if both become excited simultaneously, reaction centres will dispose of their excitation energy quicker than will accessory pigments, thus making themselves available for accepting the excitation energy from the accessory pigments. The observed excited life of chlorophyll-a is nearly 1000 times longer than that of P700.

The entry of a light quantum into the reaction centre marks the end of the photophysical phase and the start of the photochemical phase.

PHOTOCHEMICAL PHASE

The reaction centre of each photosystem becomes excited by an incoming quantum. The net result is that both P680 and P700 are oxidized by losing an electron. The electron ejected from P700 may be passed on to $NADP^+$ or returned to it through a number of electron carriers. If the lost electron goes to $NADP^+$, P700 is left oxidized. To make the oxidized P700 work further, it must be reduced or returned to its normal state by some means. The electron from P680 finds its way to the oxidized P700 through an electron transport chain that interconnects the two photosystems. This leaves P680 in an oxidized state. Another electron transfer chain operates between water and P680 and the electron of water moves through this chain to bring about the reduction of the oxidized P680. The net effect of the whole operation is the light-induced transport of electrons from water to $NADP^+$ (Fig. 5-2). However, when the electrons of water are passing through the interconnecting electron transport chain that operates between PSI and PSII, they release a part of their energy which is used up in phosphorylating ADP to ATP (noncyclic photophosphorylation).

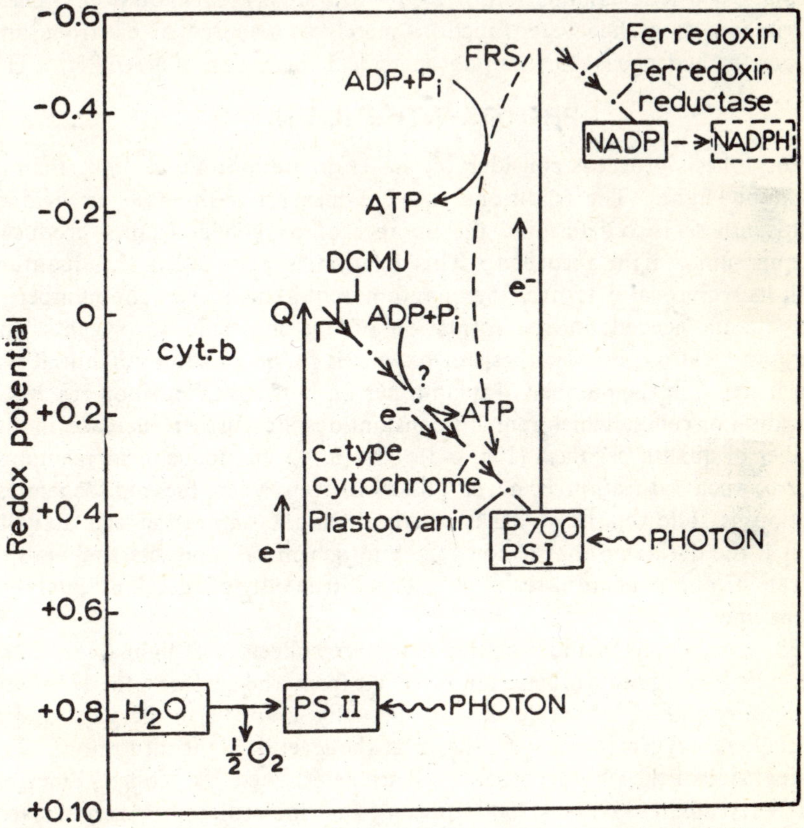

Fig. 5-2 Series model or Z-scheme of photosynthesis. (Modified after Hill and Bendall, 1960.)

As indicated earlier, the electron lost by P700 may return to it through a series of electron carriers. It is during such a cyclic path of the electrons that a substantial amount of energy also becomes available for the formation of ATP (cyclic phosphorylation as in Fig. 5-2). This is how the two photochemical reactions bring about the formation of NADPH + H^+, ATP, and oxygen in plant photosynthesis. It should be noted, however, that when PSI alone is operating *in vivo*, then the only product formed is ATP.

As shown in Fig. 5-2, the source of oxygen in photosynthesis is water. Since, according to photosynthetic equation (1), one molecule of oxygen is evolved per molecule of CO_2 utilized, what is the proof that the oxygen evolved comes from water and not from CO_2? When algae are allowed to photosynthesize in water which contains the heavy isotope of oxygen (^{18}O) and CO_2 which contains the light isotope of oxygen (^{16}O), the oxygen produced is always found to be of the ^{18}O kind. This indicates that water is the source of photosynthetic oxygen. Under certain conditions, algae and higher plant chloroplasts can use hydrogen gas for the reduction of $NADP^+$. It has always been observed that hydrogen-adapted algae carry out photosynthetic assimilation of CO_2 without oxygen evolution. These experiments show that water functions merely as a source of electrons and protons (or hydrogens) for the photochemical formation of $NADPH + H^+$.

PHOTOSYNTHETIC UNIT

Photosynthesis proceeds considerably faster in discontinuous light than in continuous light. The traditional way of measuring the rate or yield of photosynthesis is to determine the number of oxygen molecules produced per quantum of light absorbed. This relation is also called the quantum yield; its reciprocal is termed the quantum requirement, i.e., the number of light quanta needed for the formation of one molecule of oxygen. The quantum yield of photosynthesis or oxygen is found to show an initial rise with increase in the number of quanta per flash of light, but soon reaches a saturation or ceiling which remains constant despite further increase in the number of quanta per flash (Fig. 5-3). In purely photochemical reactions, *in vitro*, such saturation is never observed even when the photochemical reaction is followed by a dark reaction. Light saturation can actually occur if the dark step that follows the photochemical step sets the limit to the rate of photochemical reaction. This is true only of reactions catalyzed by enzymes.

Since chlorophylls and other pigments are collectors of light quanta, the concept of the photosynthetic unit evolved from the attempt to determine how many chlorophyll molecules are actually involved in the yield of one molecule of oxygen in photosynthesis, at the level of light saturation. The original size of the photosynthetic unit as determined by Robert Emerson and W. Arnold (1932) was 2500 molecules per molecule of oxygen evolved. That is, the photochemical reaction reaches saturation when only a few chlorophyll molecules out of a population of 2500 have individually received a quantum during the flash. If each chlorophyll molecule were to act as a photocatalyst converting light energy into chemical form, light saturation should not have occurred until every chlorophyll molecule had received a quantum from the flash. This suggested that chlorophylls are simply collectors of light quanta and that conversion of light energy into chemical form is mediated by molecules of enzymatic nature which are presumed to occur in the ratio of one molecule per 2500 chlorophyll molecules. Thus, the redefined photosynthetic unit was thought to consist of 2500 chlorophyll molecules associated with a single enzyme molecule.

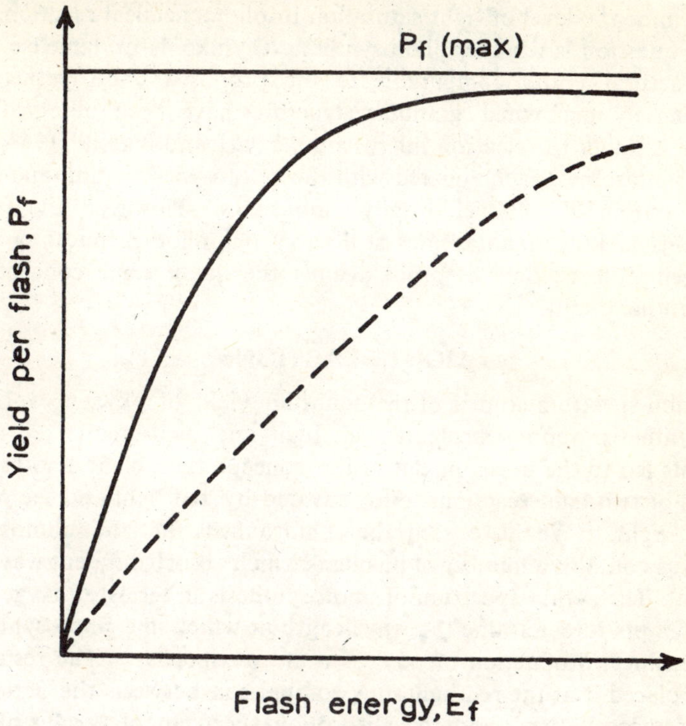

Fig. 5-3 Light saturation of photosynthesis at shorter (broken line) **and longer** (solid line) **dark intervals.** (After Emerson and Arnold, 1932.)

With advances in the knowledge of photochemical reactions, the redefined photosynthetic unit underwent major revision. The two photochemical reactions, each activated by a quantum of red or far-red light, cooperate in splitting water into electrons, protons, and oxygen, and transporting the liberated electrons to $NADP^+$; the reduction of one molecule of $NADP^+$ needs two electrons (or hydrogens) of water or liberates half a molecule of oxygen. Since the two photosystems transfer one electron from water to $NADP^+$ in one operation, they must operate twice to completely reduce a molecule of $NADP^+$, or four times to evolve a molecule of oxygen. As a single operation of the two photochemical reactions occurs at the cost of two light quanta, an operation repeated four times over must have at least eight quanta. Since the evolution of a molecule of oxygen in a saturating flash is known to involve 2500 chlorophyll molecules, there must be one enzymatic centre per (2500/8) nearly 300 chlorophyll molecules.

The concentration of the reaction centre P680 or P700 has been estimated in relation to chlorophyll molecules and the estimate comes to about one molecule of P700 per 300–400 chlorophyll molecules. Most of the electron carriers of the chloroplast also occur in the ratio of one carrier molecule to 300–400 chlorophyll molecules. It should be noted that the hypothetical enzymatic centre just referred to is now tentatively identified with the reaction centre P680 or P700 whose concentration or availability is known to determine the level of light saturation in photochemical reaction.

The question is whether photosynthetic thylakoids or lamellae contain any structural entity comparable to the proposed photosynthetic unit. Some densely pigmented granular structures have been observed in the thylakoids under the electron microscope. Park and Biggins (1964) termed them quantasomes, each equated with the photosynthetic unit and estimated to contain 300–400 chlorophyll molecules. However, the fact that thylakoids lacking quantasomes still carry out photochemical production of oxygen (Hill reaction) is proof against the quantasome concept of the photosynthetic unit.

LIGHT REACTIONS

The action spectrum studies of the quantum yield of oxygen evolution in photosynthesis in monochromatic red light and dichromatic red and far-red lights led to the development of the concept that plant photosynthesis consists of two light reactions, one activated by red light and the other by far-red light. We have seen that chloroplasts or the photosynthetic apparatus contains a number of pigments which absorb different wavelengths of light. The action spectrum of photosynthesis in terms of oxygen evolution attempts to determine the wavelength at which the photosynthesis or photochemical production of oxygen is most vigorous. The results have always placed it at the red light. A comparison between the action spectrum of photosynthesis and the absorption spectrum of various photosynthetic pigments has always suggested that chlorophyll-a is the most effi-

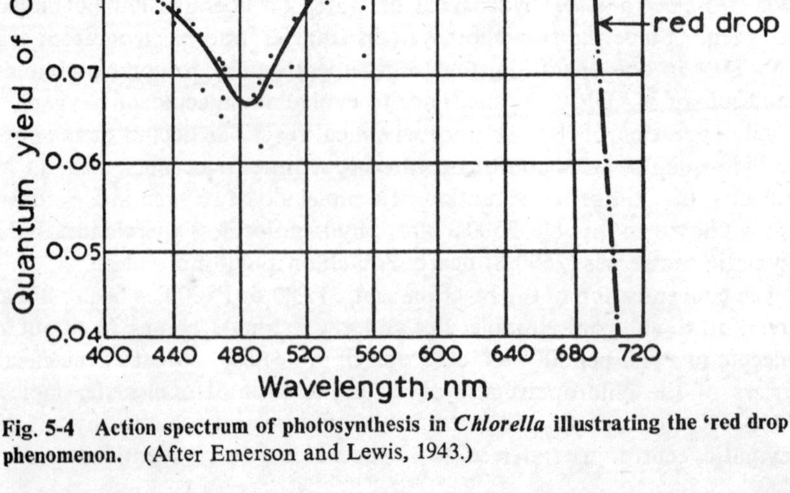

Fig. 5-4 Action spectrum of photosynthesis in *Chlorella* illustrating the 'red drop' phenomenon. (After Emerson and Lewis, 1943.)

cient pigment of photosynthesis and that chlorophyll-b or phycobilins take second place in efficiency. The carotenoids, except fucoxanthin, are extremely poor supporters or sensitizers of photosynthesis.

The action spectrum of photosynthesis appears as illustrated in Fig. 5-4. The quantum yield curve shows a dip in the region 440–520 nm which is exclusively absorbed by carotenoids, remains constant in the region 600–680 nm, and suddenly drops in the region above 680 nm. Since carotenoids contribute very little of their absorbed light towards photosynthesis, the dip observed between 440 nm and 520 nm is quite expected. As the wavelength zone 600–680 nm is absorbed by chlorophyll-a and chlorophyll-b, it is quite understandable that the maximum quantum yield occurs throughout this region. The surprising feature is the sudden drop in quantum yield at wavelengths longer than 680 nm, the region absorbed by chlorophyll-a alone. This phenomenon, known as the 'red drop', was first discovered by Emerson and Lewis (1943).

In 1957, Emerson and his coworkers showed that far-red light that is photosynthetically inefficient can be made fully efficient by combining it with another monochromatic beam of shorter wavelength absorbed by chlorophyll-b or phycobilins. The quantum yield from the two combined beams, red and far red, was found to be much greater than the sum of the yields from the two separate beams (Fig. 5-5). The enhancement in yield (or oxygen production) due to the combination of supplementary red light with far-red light is known as the Emerson effect (E) and is defined as

$$E = \frac{\text{rate of oxygen evolution in combined beams} - \text{rate of oxygen evolution in red beam}}{\text{rate of oxygen evolution in far-red beam}}$$

Emerson concluded that photosynthesis involves two light reactions, one

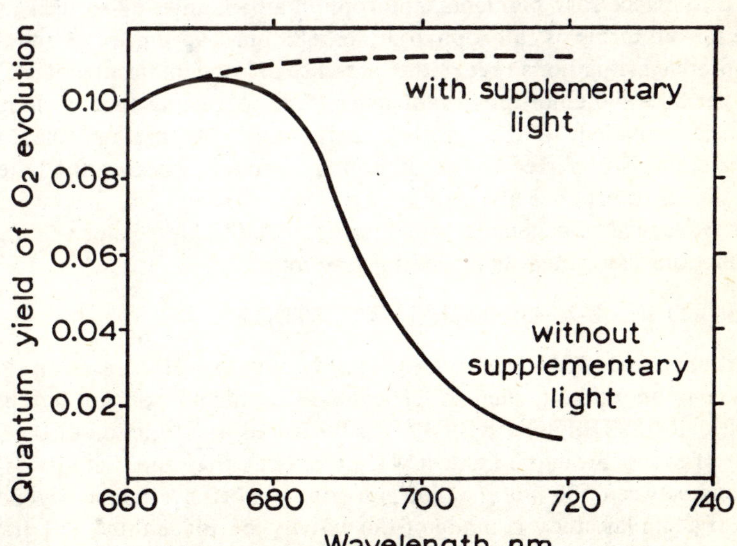

Fig. 5-5 Emerson effect as observed in *Chlorella*.

driven by chlorophyll-a and the other by the accessory pigment chlorophyll-b or phycobilins.

The operation of two photochemical reactions in photosynthesis is now undisputed, but Emerson's suggestion that one photochemical reaction is sensitized by chlorophyll-a and the other by the accessory pigment is untenable. The reasons are: (1) light quanta absorbed by accessory pigments chlorophyll-b and phycobilins are transferred very efficiently to chlorophyll-a, causing chlorophyll-a to fluoresce; if the accessory pigment were engaged in a separate photochemical reaction as stipulated by Emerson, there would be no cause for it to transfer its quanta to chlorophyll-a; and (2) chloroplast mutants of *Hordeum vulgare* lacking chlorophyll-b still show the Emerson effect; this too is evidence against the Emerson hypothesis that one photochemical reaction is sensitized by the accessory pigment.

Recent studies on chloroplast fractions have revealed the existence of two functionally and physically distinct photochemical systems, PSI and PSII. PSI corresponds to the photochemical reaction driven by far-red light and PSII to that driven by red light. The red drop phenomenon is now shown to be the photochemical activity of PSI alone. The two photosystems have been well-characterized physically, biochemically, and functionally (Table 5-1). Their physical separation has been achieved by treating the chloroplasts with the detergent digitonin or Triton which fragments them into two fractions of different density. The two particulate fractions can be easily separated in a high-speed ultracentrifuge. The heavier fraction includes PSII, and the lighter fraction includes PSI. PSII is believed to be located in the stromal thylakoid membrane and includes most of the accessory pigments, chlorophyll-a 670, and P680. PSI has the entire far-red forms of chlorophyll-a (i.e., 683 nm, 695 nm, etc.) and P700. Quantitative estimations reveal the presence of one molecule of P680 or P700 per 300–400 chlorophyll molecules, thus suggesting that each photosystem is equivalent to the photosynthetic unit. At present, the French Press method is preferred to the digitonin treatment because the latter has been shown to produce artefacts.

Physiological, biochemical, and genetical studies have helped to analyze the functions associated with the two fractions.

PHYSIOLOGICAL AND BIOCHEMICAL STUDIES

The illuminated PSII fraction can carry out the Hill reaction in the presence of an oxidant, such as ferricyanide or dichlorophenol indophenol (DCIP). If the Hill activity of PSII is measured as a function of the wavelength of monochromatic light, it will be noticed that this activity is most vigorous between 600 nm and 700 nm, but almost negligible beyond 680 nm. If a similar study is made of the activity of the lighter PSI fraction in monochromatic light, using oxygen evolution as a measure of its photochemical activity, PSI will appear photochemically inactive. But when it is supplemented with certain cofactors, such as ferredoxin, ferredoxin $NADP^+$

reductase, $NADP^+$, and the ascorbate-DCIP couple, and when its activity is estimated in terms of the photoreduction of $NADP^+$, it is photochemically very active in the far-red region (670–700 nm) and catalyzes the reduction of $NADP^+$ as long as the electron donor is available. A PSI preparation containing the electron acceptor pyocyanin or phenazine methosulphate (PMS) and lacking the ascorbate-DCIP couple and ferredoxin $NADP^+$ reductase carries out cyclic photophosphorylation instead of reducing $NADP^+$. This, with the information that PSII participates in the Hill reaction, and PSI in $NADP^+$ reduction or cyclic phosphorylation (depending upon the nature of the added electron carriers), there is no further mystery about the red drop phenomenon. It is now known that photosynthesis occurs throughout the red and far-red regions. If oxygen evolution is used as the measure, the red drop occurs, since PSII becomes inactive at or above 680 nm. PSI is active in both red and far-red light, but its activity is always greater in the latter than in the former. This is because PSI has both red and far-red absorbing accessory pigments whereas PSII has only the red absorbing pigments and, hence, cannot be activated by the far-red quanta. If pyocyanin- or PMS-catalyzed photophosphorylation is the parameter, the quantum yield of photosynthesis will show a rise rather than a decline in far-red light as this process is mediated by the photochemical activity of PSI alone.

The results of *in vivo* studies also suggest that the mode of action of two photosystems is nearly the same in both fractionated and whole chloroplast preparations. These analyses have been made by measuring the quantum yield of $NADP^+$ reduction in illuminated chloroplasts which had been treated (or in controls, untreated) with DCMU or chlorophenyl dimethyl urea (CMU) (see Fig. 5-2). The DCMU or CMU is a strong photosynthetic inhibitor which blocks photosynthesis by inactivating PSII or inhibiting the Hill reaction. The DCMU- or CMU-treated chloroplast would be practically devoid of PSII activity in comparison with untreated chloroplast. As shown in Fig. 5-4, untreated chloroplasts show near-constant quantum yield between 650 nm and 680 nm which drops severely above 680 nm. In contrast, the quantum yield of treated chloroplasts containing ascorbate-DCIP or DCIPH shows a rise above 680 nm, and the total yield at 700–710 nm is nearly double that of untreated chloroplasts at 680 nm.

A curious observation is the severe drop in the yield of reduced $NADP^+$ by untreated chloroplasts in the far-red region despite their having active PSI. This decline is very similar to the red drop observed when oxygen yield is the criterion *in vivo*. Why this is so, when both *in vitro* and DCMU findings clearly show that photoreduction of $NADP^+$ is catalyzed by PSI and not by PSII, is clear: PSII becomes inactive in far-red light and, then, water is no longer the source of electrons for the PSI-catalyzed photoreduction of $NADP^+$. According to this interpretation, if PSI is supplied with a suitable electron donor which does not need an active PSII for transporting its electrons to PSI, there should be a rise rather than a drop in the yield of reduced $NADP^+$ in the far-red region. This actually happens when the untreated or DCMU-treated chloroplasts are supplemented with ascorbate-DCIP or DCIPH and incubated in far-red light. Here the function of DCMU is to inactivate the PSII in the same way as does far-red light, thus preventing the further use of water as an electron source for $NADP^+$ reduction. The ascorbate-DCIP or DCIPH acts as an effective substitute for water and PSII in that it can directly meet the need of electrons for the photoreduction of $NADP^+$ catalyzed by PSI.

Why do broken chloroplasts need artificial electron carriers for their photochemical activity whereas intact chloroplasts do not need them? It is believed that some essential components of the electron transport chain are lost during chloroplast fractionation. Such fractions cannot be fully active unless supplied with artificial electron carriers which can suitably replace the lost components. DCIP is one of the most widely used artificial carriers. It is coloured when oxidized and colourless when reduced. This makes it a useful reagent for classroom demonstration of the Hill reaction in which the oxidized form of the dye serves as the acceptor of electrons from water. In PSI-catalyzed photoreduction of $NADP^+$, the reduced form of dye (DCIPH), rather than water, is the source of electrons for $NADP^+$ reduction. Pyocyanin or PMS substitutes for the lost component of the cyclic electron transport chain, and unless one of the two is added there will be no cyclic electron transport and, hence, no cyclic phosphorylation. The function of the ascorbate in the ascorbate-DCIP couple is to keep the DCIP in reduced form, i.e., the couple and the DCIPH are equivalent functionally.

Thus, the photoreduction of $NADP^+$ with the ascorbate-DCIP couple or DCIPH as the electron source depends upon the operation of PSI, but that with water as the electron source requires the cooperative action of both PSII and PSI.

GENETICAL STUDIES

Photosynthetic mutants of the green algae *Chlamydomonas reinhardii* and *Scenedesmus obliquus* have greatly contributed to an understanding of the mechanism and function of the two light reactions in photosynthesis (see Levine, 1969). A mutant of *Scenedesmus* lacking PSII activity does not show the Hill reaction and fails to use water as an electron source for the photochemical reduction of $NADP^+$. Other mutants of *S. obliquus* and *C. reinhardii* which show neither PSI activity nor $NADP^+$ reduction nor cyclic photophosphorylation have been isolated. These results strongly suggest that $NADP^+$ reduction is dependent on the activity of PSI which is also required for cyclic photophosphorylation (see Chapter 2).

SEQUENCE OF PHOTOSYNTHETIC ELECTRON TRANSPORT CHAIN

Photosynthetic electron transport is defined in terms of the capacity of chloroplast fragments to photoreduce $NADP^+$ in the presence of ferredoxin

$NADP^+$ reductase with water as the electron donor. The only known component of the electron transport chain between water and PSII is manganese. The manganese involvement in the photooxidation of water is inferred from the inability of Mn-deficient chloroplasts to carry out the Hill reaction or oxygen evolution. The electron transport chain between PSII and PSI includes plastoquinone (Y), cytochrome-f, plastocyanin, and that between PSI and $NADP^+$ includes X, ferredoxin and ferredoxin $NADP^+$ reductase. The path of electrons from water to $NADP^+$ is known as the noncyclic path because the electrons do not return to water. In this process, a close stoichiometry is observed between water oxidation and $NADP^+$ reduction, i.e., one molecule of $NADP^+$ is reduced per molecule of water oxidized. This is true also for ascorbate-DCIP- or DCIPH-supported reduction of $NADP^+$ by PSI. In contrast, the cyclic path catalyzed by PSI alone and in which the electron of PSI returns to it after a cyclic journey through two-to-three unknown electron carriers, involves no net loss or gain of electrons (see Fig. 5-2). Some characteristics of identified electron carriers are now summarized.

ELECTRON CARRIERS

The hypothetical electron carriers Y and X are believed to be equivalent to plastoquinone and ferredoxin reducing substance (FRS), respectively. In general, most electron carriers occur in chloroplasts at a concentration of one electron carrier per 300–400 chlorophyll molecules.

Plastoquinone Plastoquinone (Y), the oxidant of P680 (PSII), is a derivative of benzoquinone and has a redox potential of nearly 0.0 volt. Chloroplasts lacking plastoquinone show no oxygen evolution or Hill reaction. Photochemically-active PSII shows no fluorescence with it and this is why it is also known as the quencher of PSII fluorescence. Fluorescence is a measure of the wasteful light energy, and its application in understanding the relation between photophysical and photochemical events of photosynthesis has proved very helpful. Without plastoquinone, PSII gives strong fluorescence.

Cytochromes These are proteins, each containing an iron-chelated porphyrin group, and are constituents of the electron transport chain of both chloroplasts and mitochondria. Their iron moiety can alternate between ferrous and ferric form, depending upon whether they receive or lose electrons. The ease with which they gain or lose electrons makes them the most important biocatalysts of oxidation-reduction reactions in living cells.

Cytochrome-b_6 and cytochrome-f found in chloroplasts have a redox potential of 0.03 volt and 0.36 volt, respectively. Red light reduces cytochrome-b_6 and far-red light oxidizes cytochrome-f. The two cytochromes are, however, restored to their normal state in the dark.

Plastocyanin This is a copper containing protein which is blue when oxidized and colourless when reduced. Like cytochrome, it is oxidized by far-red light and reduced by red light. Its redox potential (0.37 volt) lies between cytochrome-f and P700 (0.42 volt).

The reaction centres P700 and P680 (having a redox potential close to that of oxygen) are also included in the electron transport chain as without them water cannot donate its electron to $NADP^+$. During photosynthesis, both the pigments become oxidized, the P700 giving its electron to FRS, and P680 to plastoquinone. The oxidized P700 is reduced by the electron from cytochrome-f, and P690 is reduced by the electron from water. The reduced plastoquinone reduces the oxidized cytochrome-f with the help of cytochrome-b_6. Thus, if the path of the electron from water to $NADP^+$ is followed, it will be noticed that the known sequence is P680, plastoquinone, cytochrome-b, cytochrome-f, plastocyanin, P700, FRS, ferredoxin, and ferredoxin $NADP^+$ reductase. This also explains how the weak reductant of PSII and the weak oxidant of PSI interact during photosynthesis. The primary acceptor X is considered identical to the bound form of ferredoxin. P450 is also considered identical to X.

The FRS, not yet well-characterized, has a more negative potential (−0.6 volt) than that of hydrogen or ferredoxin (−0.42 volt). Its involvement in the photosynthetic electron transport chain is inferred from the observed failure of the photoreduction of ferredoxin in its absence. Reduced ferredoxin cannot reduce $NADP^+$ without the flavoprotein enzyme, termed ferredoxin $NADP^+$ reductase. Ferredoxin is a nonheme iron protein and mediates a number of oxidation-reduction reactions. The reduced ferredoxin can be generated in photosynthesis (with water or hydrogen) and from respiration. It can in turn pass on its electrons to NAD^+, $NADP^+$, nitrate, or hydrogenase, through specific enzyme-mediated systems (Fig. 5-6).

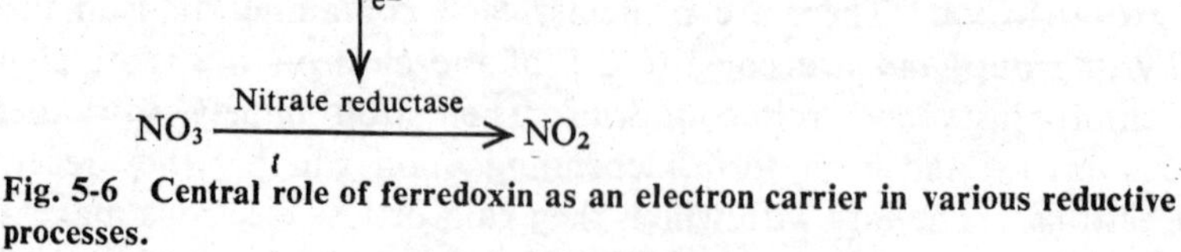

Fig. 5-6 Central role of ferredoxin as an electron carrier in various reductive processes.

For instance, reduced ferredoxin can pass on its electrons to hydrogenase which combines electrons and protons to form molecular hydrogen. This reaction is reversible since the enzyme also attacks molecular hydrogen, breaking it into electrons and protons and transferring the released electrons to ferredoxin. Another important reaction catalyzed by ferredoxin is nitrogen fixation in which the reduced carrier transfers its electrons to the nitrogensae enzyme; the latter reduces nitrogen gas to ammonia.

The proposed working sequence of the electron carriers in photosynthesis, commonly known as the Series model or Z-scheme, is based on three kinds of evidence: (1) the redox potential of the known electron carriers; (2) the antagonistic effect of red and far-red light on the two cytochromes (b and f); and (3) the photosynthetic behaviour of algal mutants lacking one or the other component of the photosynthetic electron transport chain.

Figure 5-7 gives the redox potentials of various natural and artificial photosynthetic electron carriers.

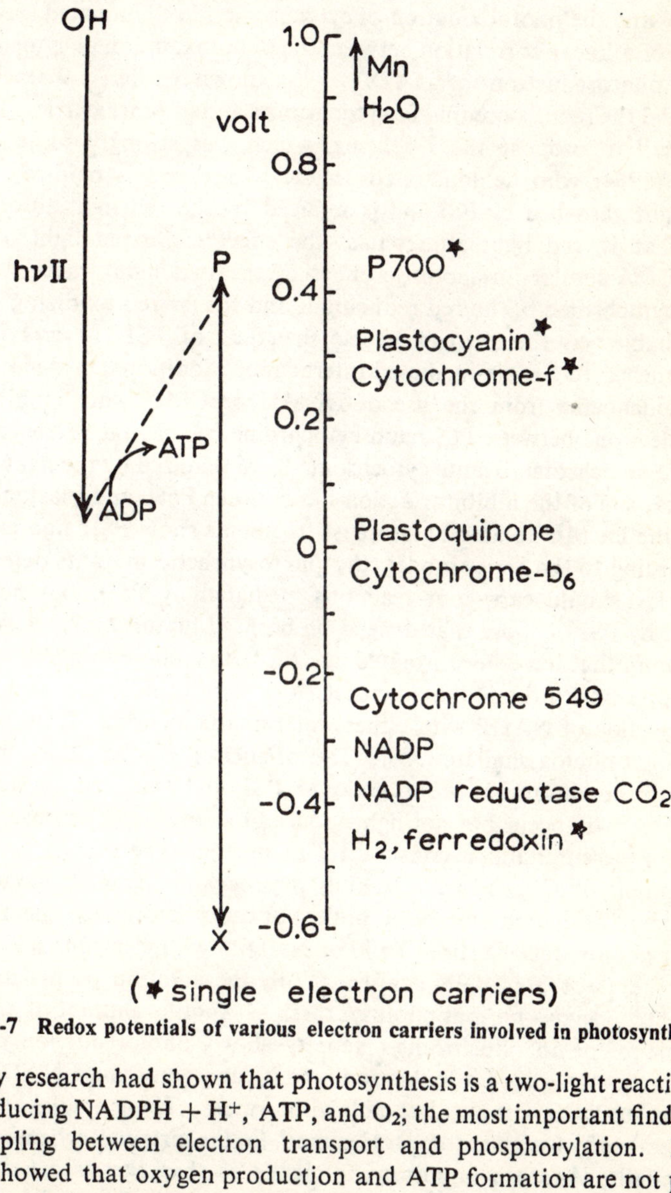

Fig. 5-7 Redox potentials of various electron carriers involved in photosynthesis.

Early research had shown that photosynthesis is a two-light reaction process producing NADPH + H^+, ATP, and O_2; the most important finding was the coupling between electron transport and phosphorylation. Further studies showed that oxygen production and ATP formation are not possible unless the weak reductant (Y) and the weak oxidant (X) interact. Later, this interaction was found to be mediated by cytochrome-b and cytochrome-f. The fact that, in oxidative phosphorylation, ATP formation is always linked with the downhill migration of electrons along the respiratory cytochromes led Hill and Bendall (1960) to conclude that photophosphorylation may also be coupled to the downhill path of electrons between PSII and PSI. The observed redox potential gap of nearly 0.33 volt between cytochrome-b_6 (0.03) and cytochrome-f (0.36) was mainly responsible for this conclusion. Meanwhile, Hill had observed the photoreduction of cytochrome-b_6 and the photooxidation of cytochrome-f in illuminated leaves; the existence of a linear correlation between the photooxidation of cytochrome-f and the photoreduction of $NADP^+$ was known earlier. These observations and the redox potential of cytochromes and of other carriers led Hill and Bendall to propose the Z-scheme, which was strongly supported by Duysens (1964) who demonstrated that cytochrome-f is oxidized by the far-red light absorbed by PSI and is reduced by the red light absorbed by PSII. That is, red light antagonizes the effect of far-red light on cytochrome-f. A similar antagonistic effect of the two light waves was also seen on cytochrome-b, the red reducing it and the far-red oxidizing it. The only probable way the PSI can reverse the effect of PSII or *vice versa* is through their cytochrome-mediated interaction. Additional evidence for the Series model came from the use of DCMU or CMU which inhibits the flow of electrons between P690 and cytochrome-b_6. In the presence of this inhibitor, cytochrome-b and cytochrome-f are oxidized by the light of PSI. The actual site of the inhibitor action lies between P680 and plastoquinone, because the DCMU-treated chloroplast fragments show PSII fluorescence.

According to the Series model, the photosynthetic mutants defective in P700 or PSI should carry out reactions mediated by PSII, but not those mediated by PSI or those that depend on both. Mutants of *S. obliquus* and *C. reinhardii* that have been isolated are those that show the Hill reaction when supplied with the Hill oxidant DCIP or ferricyanide, but do not show photoreduction of NADP with either water or the ascorbate-DCIP couple. They do not photoassimilate CO_2. The prediction of the Series model is that under conditions of nonfunctional PSI, light-induced reduction of cytochromes will occur but not light-induced oxidation. The observed behaviour of these mutants is consistent with such an expectation.

A mutant of *C. reinhardii* lacking plastocyanin cannot photooxidize cytochrome-f or cytochrome-b_6 or both, but can photoreduce them. Also it cannot photoreduce oxidized P700 or $NADP^+$ with electrons from water or from the ascorbate-DCIP couple. With the addition of plastocyanin, the mutant resumes normal photosynthesis. Another mutant of this alga with impaired or no cytochrome-f activity shows photoreduction of cytochrome-b, but not its photooxidation, and it can photoreduce $NADP^+$ with the ascorbate-DCIP couple, but not with water. These findings support the place already assigned to cytochrome-f in the original scheme of Hill and Bendall. Another important inference is that the ascorbate-DCIP couple donates its electron to PSI at the plastocyanin level. This shows also how it can bypass the requirement of PSII for the photoreduction of $NADP^+$.

Some mutants of the alga lack cytochrome-b_6 or plastoquinone. In the former, the photoreduction of cytochrome-f does not occur whereas its photooxidation occurs readily. In the latter, PSII activity is completely missing, but PSI activity is normal. If the missing components are added to the reaction mixture, normal photosynthesis is restored.

PHOTOPHOSPHORYLATION

The nomenclature cyclic and noncyclic photophosphorylation is based on the path of electron transport. Although the operation of noncyclic phosphorylation in photosynthesis is now well documented, cyclic phosphorylation is known to occur only in isolated chloroplasts. For instance, it is not yet resolved whether both kinds of photophosphorylation operate simultaneously or one at a time *in vivo*. Many workers have doubts about the importance of cyclic phosphorylation in normal photosynthesis. Similarly, nothing is known concerning the nature of the high energy intermediate which is produced in light and consumed in the dark in phosphorylating ADP to ATP. Also the quantum yield of photophosphorylation (i.e., the molecules of inorganic phosphate phosphorylated per quantum) is far from clear. Controversy still exists regarding the precise site of conversion of light energy into ATP.

Photophosphorylation apparently occurs in two steps, viz., the photochemical step, and the dark step in which the high energy intermediate produced in the light is used to form ATP from ADP and inorganic phosphate. This concept of two-stage phosphorylation was formulated by Jagendorf (1962) on the basis of results obtained from an experiment: since photophosphorylation needs ADP, inorganic phosphate (P_i), and magnesium ions in addition to the photosynthetic apparatus and light, chloroplast preparations lacking ADP, P_i, and Mg^{++} were first illuminated with photosynthetic light for a very brief period (a fraction of a second) and then the whole preparation was quickly transferred to another reaction vessel containing the required cofactors of photophosphorylation. It was seen that for every disappearing molecule of ADP and P a molecule of ATP was formed. The preillumination must have initiated the formation and accumulation of high energy intermediates which were used up in the formation of ATP during the ensuing dark period.

Photophosphorylation, like oxidative phosphorylation, is coupled to the downhill photosynthetic electron transport. The evidence in support of this view is: (1) ADP, P_i, and Mg^{++} are known to cause a threefold to fourfold increase in the rate of photosynthetic electron transport; (2) inhibitors of photosynthetic electron transport, such as DCMU and CMU, restrain photophosphorylation; and (3) photosynthetic mutants with a defective electron transport chain lack photophosphorylation.

The mechanism proposed for coupling between photophosphorylation and electron transport is generally interpreted by assuming a reaction step common to both the processes. The reaction step is believed to involve a coupling factor of proteinaceous nature which is thought to block the passage of electrons under conditions of a limiting supply of ADP and P_i, and to allow their normal course under nonlimiting conditions. This is how the uncouplers of photophosphorylation, such as ammonium ion, are thought to permit the normal flow of electrons with no associated phosphorylation.

CYCLIC PHOTOPHOSPHORYLATION

Arnon *et al.* (1954) discovered that illuminated chloroplasts, deprived of CO_2 and $NADP^+$, can produce ATP and water from ADP and P_i, respectively, when supplied with a catalytic amount of FMN or vitamin K and oxygen. This process involved no net chemical change other than the formation of ATP and water:

$$ADP + P_i \rightarrow ATP + H_2O$$

When FMN or vitamin K from a photochemically-active chloroplast preparation was replaced by pyocyanin or PMS, the net chemical change was again the formation of ATP and water from ADP and P_i, respectively. However, the difference between FMN- and PMS-catalyzed photophosphorylation was that the former needed oxygen to occur whereas the latter showed no involvement of oxygen. Arnon termed the oxygen-dependent, FMN-catalyzed photophosphorylation as pseudocyclic (Fig. 5-8) and the oxygen-independent, PMS-catalyzed process as cyclic photophosphorylation. The pseudocyclic process involves the reduction of FMN with the

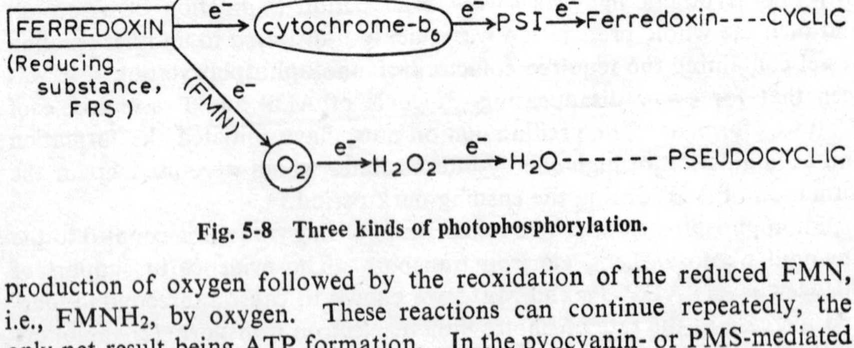

Fig. 5-8 Three kinds of photophosphorylation.

production of oxygen followed by the reoxidation of the reduced FMN, i.e., $FMNH_2$, by oxygen. These reactions can continue repeatedly, the only net result being ATP formation. In the pyocyanin- or PMS-mediated cyclic phosphorylation, the added cofactor (pyocyanin or PMS) is reduced when there is no production of oxygen and the reduced cofactor is then reoxidized by an unknown component of the electron transport chain and not by oxygen. The overall net effect is again the formation of ATP.

The operation of pseudocyclic photophosphorylation depends upon the ability of chloroplast preparations to produce molecular oxygen; the DCMU-treated chloroplasts do not show FMN-catalyzed photophosphorylation although the PMS-catalyzed process remains fully effective. This means that the source both of hydrogen (electrons) for the reduction of FMN and of oxygen that parallels the FMN reduction must be water. In fact, noncyclic photophosphorylation which shows a stoichiometric relation between the oxidant O_2 and the reductant $NADPH + H^+$ can become pseudocyclic if FMN, which is an autooxidizable hydrogen acceptor, is available in catalytic amounts. The FMN, with the help of light and chlorophyll, breaks and rejoins the water molecule, and the single water molecule should be sufficient for meeting the requirement of pseudocyclic photophosphorylation as long as the chloroplast survives (Fig. 5-9). The formation of ATP during the pseudocyclic process depends upon the reduction of FMN. The reaction between $FMNH_2$ and oxygen is believed to yield insufficient energy for any significant ATP synthesis.

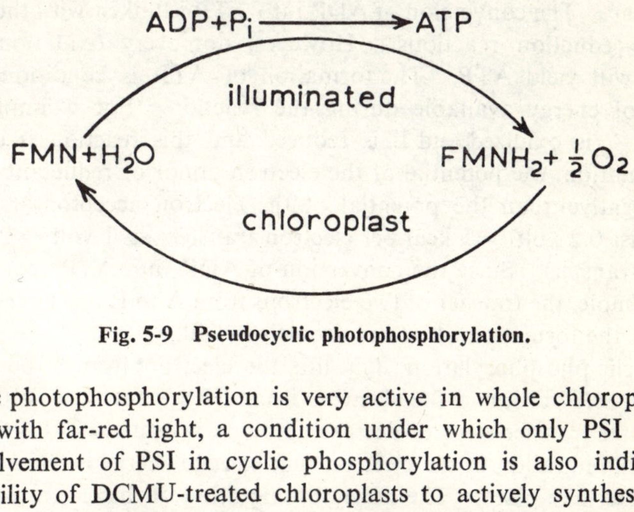

Fig. 5-9 Pseudocyclic photophosphorylation.

Cyclic photophosphorylation is very active in whole chloroplasts illuminated with far-red light, a condition under which only PSI is active. The involvement of PSI in cyclic phosphorylation is also indicated by (1) the ability of DCMU-treated chloroplasts to actively synthesize ATP; (2) the inability of P700-lacking algal mutants to show far-red-sensitized phosphorylation; (3) the lack of the enhancement effect of red light on the far-red-sensitized phosphorylation; and (4) the ability of the detergent-treated light chloroplast fraction (PSI) to synthesize only ATP in the absence of the ascorbate-DCIP couple and $NADP^+$. The operation of cyclic phosphorylation is also inferred from some *in vivo* experiments. For instance, algae that grow in a hydrogen atmosphere in the complete absence of oxygen develop the ability to use molecular hydrogen instead of water to reduce CO_2 in the light; this process is very efficient in the light zone absorbed by PSI. This sort of photosynthesis which involves no evolution of oxygen is termed anaerobic photosynthesis and its mechanism is explained as follows: the hydrogen-adapted algae develop the enzyme hydrogenase which allows hydrogen to react with $NADP^+$ via ferredoxin and ferredoxin $NADP^+$ reductase; the reduced $NADP^+$ reduces CO_2 to organic carbon;

the function of the light is simply to activate PSI for the generation of ATP through a cyclic process, and the ATP so formed helps both CO_2 fixation and its reduction by NADPH + H^+. In anaerobic photosynthesis, the reduction of $NADP^+$ does not apparently need photosynthetic light and so is unlike that occurring during aerobic photosynthesis in which $NADP^+$ reduction is a light-dependent process. That PSI alone is truly involved in anaerobic photoassimilation of CO_2 is further shown by the absence of the Emerson enhancement effect on this process. Another example is the photoassimilation by algae of organic substrates, such as glucose and acetate. When algae photoassimilate glucose or acetate, they require photosynthetic ATP and are known to suppress the photoreduction of $NADP^+$ and the photochemical evolution of oxygen. In such cases, the process is anaerobic and is driven solely by PSI-mediated cyclic photophosphorylation. This is further supported by the inability of a mutant of *C. reinhardii* lacking P700 to show anaerobic assimilation of glucose.

The site, rate, and mechanism of cyclic photophosphorylation are not altogether clear. The conversion of ADP into ATP is linked with the exergonic oxidation-reduction reactions. However, not every oxidation-reduction reaction will yield ATP. The formation of ATP is conditioned by the amount of energy available during the reaction. For example, if, in a reaction, A is oxidized and B is reduced and this reaction is coupled to ATP formation, the potential of the electron donor or reductant A must be lesser (negative) than the potential of the electron acceptor or oxidant B by at least 0.2 volt (=5 kcal per electron transfer, as 1 volt = 23 kcal per electron transfer). Since the conversion of ADP into ATP requires nearly 7–9 kcal/mole, the transfer of two electrons from A to B must occur in order to permit the formation of one molecule of ATP.

In cyclic phosphorylation, light lifts the electron from P700 (0.42 volt) to FRS (−0.6 volt), and the electron from FRS returns to P700 through two-to-three transfer steps of decreasing redox potentials. It is during such a downhill migration of the electron that enough energy is released for ATP synthesis to take place. The downhill migration of the electron from FRS to P700 occurs through a potential gap of nearly one volt which is sufficient to produce at least two ATP molecules per electron transfer, or four molecules per two electrons transferred.

The cyclic electron path includes the part of the electron transport chain that connects the two photosystems. The precise point at which the cyclic electron enters the transport chain is debatable. It may enter at the plastoquinone, cytochrome-b or cytochrome-f, or plastocyanin level. The reason is that PSI-dependent phosphorylation in some cases requires plastoquinone, in others cytochrome-b or cytochrome-f, and, in at least one reported case, plastocyanin. The last observation is based on the behaviour of a mutant of *C. reinhardii* with no plastocyanin activity. This mutant lacks cyclic phosphorylation which can, however, be resumed normally after the mutant is supplied with exogenous plastocyanin. The general belief is that the site lies between FRS and the oxidizing side of PSI.

NONCYCLIC PHOTOPHOSPHORYLATION

In cyclic phosphorylation, no net electron transport from a donor to an acceptor is involved whereas noncyclic phosphorylation always proceeds with a net transport of electrons from a donor to an acceptor. When both the photosystems are functioning, the electron acceptor is $NADP^+$; when PSII alone is operative, the acceptor molecule must be an added carrier, such as ferricyanide or DCIP (Hill oxidant). However, in both the conditions, the electron donor is water. The photoreduction of $NADP^+$ by ascorbate DCIP or DCIPH is also a process of noncyclic electron transport, but it does not support any phosphorylation, although the potential gap of almost more than 0.2 volt between the ferredoxin-reducing substance and $NADP^+$ is reasonably sufficient to generate a molecule of ATP, provided the energy of the moving electron is not dissipated in the three transfer steps which lie between FRS and $NADP^+$.

Many evidences are available to show that noncyclic photophosphorylation includes active PSII, the entire electron transport chain lying on the oxidizing side of PSII, and a major portion of the chain, probably upto cytochrome-f, lying on the PSII-reducing side. DCMU inhibition of noncyclic photophosphorylation and its absence from the algal mutant with inactive or missing PSII, both indicate its dependence on the photochemical activity of PSII. The illuminated heavy chloroplast fractions with added DCIP or ferricyanide carry out both the Hill reaction and ATP synthesis.

In photosynthesis by illuminated chloroplasts or algae, two molecules of ATP are formed per two molecules of $NADP^+$ reduced or one molecule of oxygen evolved or two molecules of water oxidized:

$$ADP + P + NADP^+ + H_2O \rightarrow ATP + NADPH + H^+ + \tfrac{1}{2}O_2$$

In photosynthesis by illuminated PSII containing the Hill oxidant DCIP or ferricyanide, two molecules of ATP are also synthesized per four molecules of oxidant reduced or one molecule of oxygen produced or two molecules of water oxidized:

$$H_2O + ADP + P_i + 2DCIP \rightarrow ATP + 2DCIPH + \tfrac{1}{2}O_2 .$$

It is evident that the stoichiometric relation is the same as that between light-induced ATP synthesis and water oxidation whether the electron acceptor is $NADP^+$ or DCIP or ferricyanide.

The consideration of the potential gap and other evidences suggests that the site of noncyclic photophosphorylation lies between cytochrome-b_6 and cytochrome-f. There is genetical evidence that mutants of *Chlamydomonas* that have no cytochrome-f do not carry out noncyclic photophosphorylation. The other evidence is that illuminated chloroplasts show no significant photoreduction of cytochrome-f under a limiting rate of electron transport (chloroplasts devoid of ADP and P_i). When the rate-limiting conditions are relieved by the addition of phosphorylating reagents or an uncoupler of photophosphorylation, the photoreduction of cytochrome-f is resumed at a substantial rate. This suggests that a site of photophosphorylation must precede cytochrome-f in the electron transport chain.

The most reasonable hypothesis at present is the following: there are two sites for photophosphorylation, one for cyclic photophosphorylation and the other for noncyclic photophosphorylation; the ratio P : 2e is nearly one for both cyclic and noncyclic photophosphorylation; photophosphorylation mediated by both PSI and PSII (noncyclic) is a 4-quanta process, that mediated by PSI (cyclic) is a 1-quantum process, and that mediated by PSII (noncyclic) with added DCIP or ferricyanide is also a 2-quanta process, per molecule of ATP formed.

The various physiological activities of isolated chloroplasts are summarized in Table 5-1.

Table 5-1 Physiological activities of illuminated chloroplasts isolated in aqueous suspensions

Components present or supplied	Products formed	Process
Some hydrogen acceptor (A), e.g., benzoquinone, dichlorophenol indophenol, NAD^+, and $NADP^+$	O_2 + reduced form of hydrogen acceptor (AH_2)	Hill reaction
ADP, inorganic phosphate	ATP	Cyclic photophosphorylation
ADP, inorganic phosphate, $NADP^+$	ATP, NADPH, H^+, O_2	Noncyclic photophosphorylation
ADP, inorganic phosphate, $NADP^+$, CO_2	ATP, NADPH, H^+, O_2, carbohydrates	Complete photosynthesis

ENERGETIC EFFICIENCY OF PHOTOSYNTHESIS

To reduce one molecule of CO_2 to the level of carbohydrate in the Benson-Calvin cycle, the chloroplast needs two molecules of NADPH + H^+ and three molecules of ATP. But chloroplasts assimilating CO_2 in the C_4 cycle do so at the cost of two molecules of NADPH + H^+ and five molecules of ATP per CO_2 molecule. As explained earlier, the two photosystems harvest the light energy and store it in the form of ATP and NADPH + H^+. A molecule of NADPH + H^+ stores 52 kcal of light energy, and a molecule of ATP stores nearly 7–9 kcal of light energy. Four light quanta (two per photosystem), each of 40 kcal, are used up in the formation of a molecule of ATP and NADPH + H^+. Thus, if we measure the efficiency of photosynthetic conversion of light energy in terms of ATP and NADPH + H^+, it will come to [(52 + 7)100/160=] nearly 37%. This means that a major portion of the absorbed light energy is wasted.

Another way of measuring efficiency is in terms of CO_2 assimilation. We know that a molecule of oxygen is evolved for every molecule of CO_2

fixed and we know also that nearly 112 kcal of light energy is conserved in converting a molecule of CO_2 into carbohydrate. By the same token, nearly 673 kcal of light energy will convert six molecules of CO_2 into a hexose. Since photoassimilation of a molecule of CO_2 consumes eight light quanta, the total quantum consumption will rise to 48 per hexose molecule formed. The energetic efficiency of the process is:

$$\frac{672 \times 100}{48 \times 40} = \frac{67200}{1920} = 35\%$$

This figure is nearly two per cent less than the previous figure. The simple reason is that a part of the light energy stored in NADPH + H^+ and ATP may be lost during their participation in the energy-dependent process of CO_2 assimilation. No data are available regarding the energetic efficiency of C_4 plants. It is, however, suspected that such plants may be more efficient in storing solar energy than those having the Benson-Calvin cycle.

It is not yet clear why only two molecules of ATP are produced per molecule of oxygen evolved or CO_2 fixed, especially when the latter needs three, not two, molecules of ATP. From where does the extra molecule of ATP come? Does it come from cyclic photophosphorylation or does noncyclic photophosphorylation form three, not two, molecules of ATP per molecule of ATP evolved? Only future research may provide the answers to these questions.

COMPARISON OF HIGHER PLANT AND BACTERIAL PHOTOSYNTHESIS

Photosynthetic bacteria fall into three distinct groups: the green bacteria, e.g., *Chlorobium*; the purple sulphur bacteria, e.g., *Chromatium*; and the purple nonsulphur bacteria, e.g., *Rhodospirillum* and *Rhodopseudomonas*. Both green and purple sulphur bacteria are primarily autotrophic and use inorganic sulphur compounds (H_2S) as electron donors for the photosynthetic reduction of CO_2, but the purple nonsulphur bacteria are photoheterotrophic and utilize succinate or malate instead of H_2S. The pigments involved in bacterial photochemical reactions are carotenoids, bacteriochlorophyll, and Chlorobium chlorophyll. Of these, the forms containing bacteriochlorophyll have been more extensively studied and photosynthetically understood. The common bacterial photosynthetic equation is:

$$2H_2S + CO_2 \rightarrow 2S + H_2O + (CH_2O)$$

where the electron donor H_2S may sometimes be replaced by a molecular hydrogen. This equation is similar to that for higher plant photosynthesis except that in the latter H_2O, and not H_2S or H_2, is the source of electrons. As early as 1935, Van Niel pointed out the similarity between the two types of photosynthesis and concluded that the function of photochemical reactions in both plant and bacterial systems is the transfer of electrons from the donor to the acceptor molecule. Thus, the reason why no oxygen is produced in bacterial photosynthesis is that bacteria do not use water (see Table 5-2) and, hence, instead of oxidizing water to oxygen, oxidize H_2S to sulphur, succinate to fumarate, or malate to oxaloacetate. When Gaffron

(1940) showed that certain algae could be adapted to use hydrogen for the photoreduction of CO_2 with no production of oxygen, and when it was subsequently established that such photoreduction is mediated by only one photosystem (PSI), the reason for the failure of photosynthetic bacteria to oxidize water was linked with the operation of a single photosystem in them, i.e., the bacterial photosynthesis is a one-light reaction process.

Table 5-2 Comparison between green plant photosynthesis and bacterial photosynthesis

Green plant photosynthesis	Bacterial photosynthesis
Consists of two photosystems connected in series	Consists of two photosystems connected in parallel
H_2O is the only electron donor in photoreduction	H_2S, malate, and succinate are the usual electron donors in photoreduction
Oxygen is evolved	No oxygen is evolved
The reductant of CO_2 is NADPH + H^+	The reductant is NADH + H^+
The electron transport chain includes b-type cytochromes	The known cytochromes are all c type
Plastocyanin is an essential component of the electron transport chain	No plastocyanin is found
Occurs at wavelengths between 400 nm and 700 nm	Occurs at wavelengths longer than 700 nm
P700 and P680 are the reaction centres of PSI and PSII, respectively	One of the reaction centres is P890; the other, if any, remains unknown
Emerson enhancement effect occurs	No Emerson enhancement effect has been reported, probably because the two photosystems operate in parallel and not in series
Can be made bacterial type by allowing the plants to photosynthesize in an atmosphere of H_2 gas	Cannot be converted into green plant type photosynthesis

Bacteriochlorophyll occurs in three spectral forms absorbing light at 800 nm, 850 nm, and 890 nm. The 890 nm form is the reaction or energy-trapping centre and is abbreviated as P890. The reaction centre occurs in a concentration of 1 per 50 molecules of bacteriochlorophyll. Like P700, its redox potential is nearly 0.43 volt and its primary electron acceptor is believed to have a redox potential comparable to that of X (FRS). The electron transport chain consists of known and unknown electron carriers; the known ones include three c-type cytochromes, viz., ubiquinone, ferredoxin, and NAD^+. All the cytochromes show well-defined and characteristic absorption spectra which are specific for oxidized (ferric) and reduced (ferrous) states. The three c-type cytochromes found in bacteria are distinguished by the sharp absorption bands of their reduced states. Of the three reduced states, one absorbs at 552 nm, the other at 555 nm, and the third at 430 nm. Their redox values are 0.06 volt, 0.18 volt, and 0.32 volt, respectively. The NAD^+ has the same potential as $NADP^+$.

Until recently, bacterial photosynthesis was regarded as a photochemical reaction analogous to PSI of green plants. The relation between the photoreduction of CO_2 and photophosphorylation, which are two separate functions of PSI in plant photosynthesis, was presumed to be linked in bacterial photosynthesis, i.e., the single photosystem was supposed to drive both cyclic photophosphorylation and photoreduction simultaneously. The existence of a cyclic electron transfer chain capable of generating ATP is well-established for photosynthetic bacteria. However, noncyclic electron transport from exogenous donors (H_2S) to endogenous acceptors (ferredoxin and $NADP^+$) can be conceived of in two ways: (1) high energy intermediates produced in the cyclic pathway perhaps reverse the flow of electrons and transport them against the thermochemical gradient to the level of ferredoxin which, in turn, produces $NADH + H^+$, the natural reductant of CO_2 in bacterial photosynthesis; and (2) probably a distinct photochemical step or reaction essentially similar to the one operating in the cyclic system transports electrons in the same manner as in noncyclic phosphorylation (i.e., there are two independent steps involved here). The former possibility is based on the concept of one photochemical reaction whereas the latter assumes that bacterial photosynthesis consists of two light reactions operating in parallel, and not in series (as in green plants). The findings of Sybesma and Flower (Olson, 1970) have strengthened the second possibility, and the recent concept of bacterial photosynthesis as consisting of two parallel light reactions is generally acknowledged; of the two, one is engaged in cyclic photophosphorylation and the other in noncyclic electron transport from the donor to the acceptor molecule. The consideration of light-induced absorption changes in cytochromes as a function of light intensity and the effect of inhibitors on the photochemical activity of cytochromes are some of the evidences on which the concept of two-light reactions is based.

Figures 5-10 and 5-11 depicting the current models of photosynthetic mechanism in bacteria also indicate the possibility of how the two parallel photosystems could interact through a flavoprotein which can mediate the reverse flow of electron from the cyclic pathway to NAD^+, or from succinate to NAD^+, at the cost of high energy intermediates of cyclic photophosphorylation. The photoreduction of NAD^+ and bacterial photophosphorylation are 2-quanta processes. Much of the detail of bacterial photosynthesis is still being worked out and the proposed model can be

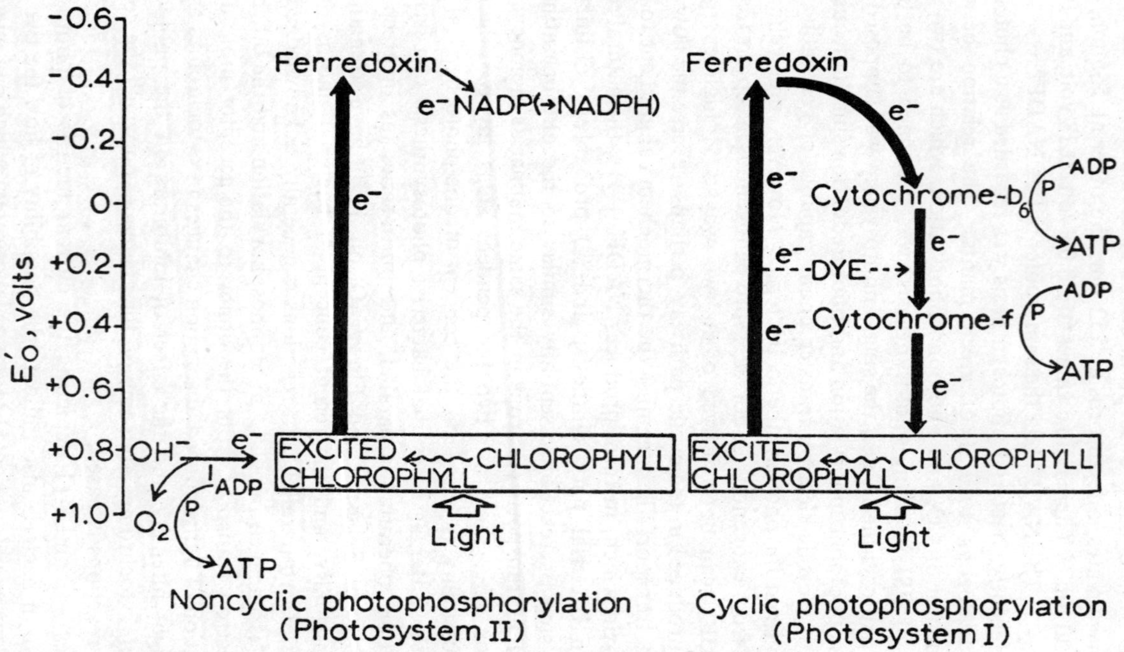

Fig. 5-10 Operation of cyclic and noncyclic photophosphorylation in parallel. (After Arnon *et al.*, 1967.)

finally accepted only after the missing components of photochemical reactions are identified and characterized.

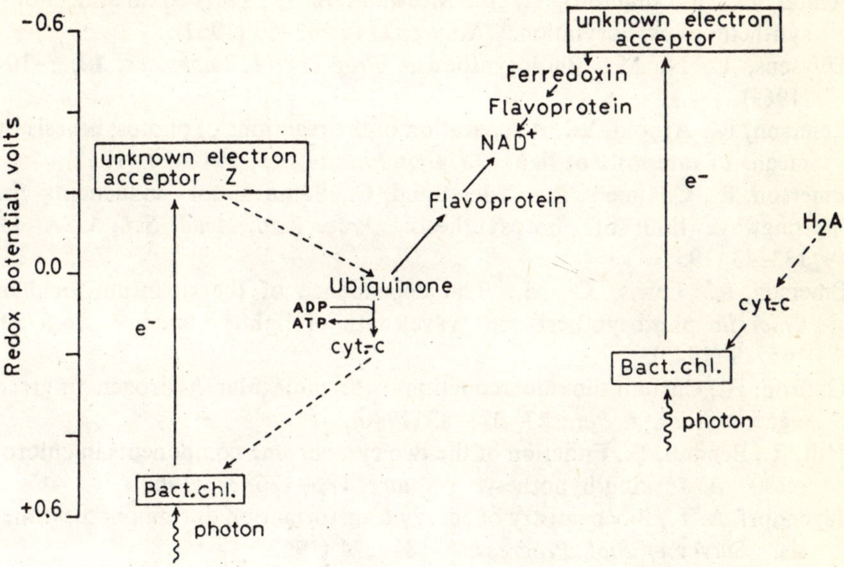

Fig. 5-11 Two-photosystem scheme of bacterial photosynthesis operating in parallel. (After Olson, 1970.)

Arnon *et al.* (1967) proposed a scheme for green plant photosynthesis in which the two photosystems were thought to act in parallel (Fig. 5-10) and not in series. PSII was thought to be associated with noncyclic electron transport from water to $NADP^+$ through the chain consisting of PSII, ferredoxin, and ferredoxin $NADP^+$ reductase; PSI was assumed to be connected exclusively with the cyclic pathway concerned with ATP synthesis. From the energetic viewpoint, the parallel scheme did not seem entirely satisfactory because it required that only one photon of red light was available to raise an electron from water to $NADP^+$. It is also believed that PSII includes two photoreactions, PSIIa and PSIIb, operating in series through a chain of electron carriers including cytochrome-b and plastocyanin. This suggestion is based on chloroplast fractionation studies which yielded two kinds of particles, PII and PI. PII particles were able to photooxidize cytochrome-b and could, in the presence of plastocyanin, evolve oxygen and photoreduce ferredoxin and, hence, also $NADP^+$. In contrast, PI particles were assigned the sole role of ATP synthesis as no evidence in support of the PSI participation in the photoreduction of $NADP^+$ with water was available. The most serious objection to this scheme seems to be the antagonistic effect of two wavelengths of light on cytochrome-f.

REFERENCES

Arnon, D. I., Allen, M. B., Whatley, F. R., Photosynthesis by isolated chloroplasts. *Nature*, **174**, 399 (1954).

Arnon, D. I., Tsujimoto, H. Y., Mcswain, B. D., Ferredoxin and photosynthetic phosphorylation. *Nature*, **214**, 562–66 (1967).

Duysens, L. N. M., Photosynthesis. *Progress in Biophysics*, **14**, 3–104 (1964).

Emerson, R., Arnold, W., A separation of the reactions of photosynthesis by means of intermittent light. *J. Gen. Physiol.*, **15**, 391–420 (1932).

Emerson, R., Chalmers, R., Cederstrand, C., Some factors influencing the longwave limit of photosynthesis. *Proc. Natl. Acad. Sci.*, USA, **43**, 133–43 (1957).

Emerson, R., Lewis, C. M., The dependence of the quantum yield of *Chlorella* photosynthesis on wavelength of light. *Amer. J. Bot.*, **30**, 165–78 (1943).

Gaffron, H., Carbon dioxide reduction with molecular hydrogen in green algae. *Amer. J. Bot.*, **27**, 273–83 (1940).

Hill, R., Bendall, F., Function of the two cytochrome components in chloroplasts: A working hypothesis. *Nature*, **186**, 136–37 (1960).

Jagendorf, A. T., Biochemistry of energy transformation during photosynthesis. *Survey of Biol. Progress*, **4**, 181–224 (1962).

Levine, R. P., Analysis of photosynthesis using mutants of algae and higher plants. *Ann. Rev. Plant Physiol.*, **20**, 523–40 (1969).

Olson, J. M., The evolution of photosynthesis. *Science*, **168**, 438–46 (1970).

Park, R. B., Biggins, J., Quantasome: size and composition. *Science*, **144**, 1009–11 (1964).

Van Niel, C. B., Photosynthesis of bacteria. *Cold Spring Harbor Sympos. Quant. Biol.*, **3**, 138–50 (1935).

6: Genetics of Photosynthesis in Algae

INTRODUCTION

For more than four decades, microalgae have played an important role in the study of photosynthesis. First selected as an experimental organism for photosynthesis studies by Otto Warburg in the 1920s, the unicellular green alga *Chlorella* has since come to occupy a central place in photosynthesis research, especially for the analysis of the dark reactions. The path of carbon in photosynthesis, first studied by Benson and Calvin in the 1940s and since known as the Benson-Calvin cycle, was mostly traced in cultures of this alga.

Initially confined to examination of the carbon pathway, algal studies have considerably diversified during the past two decades, yielding useful new information on the role of pigments in photosynthesis and related phenomena (Fork and Amesz, 1969) and the mechanisms of electron transport (Levine, 1969) and CO_2 fixation (Bassham and Jensen, 1967). In most of these analyses, microscopic algae, such as *Chlorella*, *Chlamydomonas*, *Scenedesmus*, *Anacystis*, and *Porphyridium*, have been used as the experimental subjects. These forms can be cultured fairly easily and their growth conditions can be reproduced. Their relatively short life cycles, coupled with fast growth rates and short generation durations, make them very suitable as experimental subjects. The scientific value of some of these forms has been greatly enhanced by the application of the technique of synchronous culture which can yield a cell population that has most of its cells at the same developmental and physiological state at a given time.

The exploitation of mutant strains of certain algae in delineating various aspects of the photosynthetic electron transport chain has introduced a new dimension in the study of photosynthesis in recent years. Just as the development of suitable techniques involving the use of radioactive isotopes and chromatography was instrumental in determining the path of carbon in the dark reactions of photosynthesis, so too three very important techniques involving the use of mutant strains, specific metabolic inhibitors, and difference spectroscopy have elucidated the light reactions and the path of the electron in photosynthesis. Of these three techniques, the use of mutants has received considerable attention in recent years; for instance, Levine and his coworkers (1968) have successfully used a number of gene mutants of the sexually-reproducing green flagellate *Chlamydomonas rein-*

hardii in a 'genetic' dissection of the process of photosynthesis into its various components.

METHODOLOGY AND TECHNIQUES

A number of acetate-requiring mutant strains of *Chlamydomonas reinhardii* have been produced. Many of these are either unable to carry out photosynthesis or do so at much slower rates than the wild type, even though their chlorophyll content is no less than that of the wild cells. Most of these mutations are in single genes, are inherited in Mendelian fashion, and seem to lie in nuclear rather than in chloroplast DNA (Levine and Goodenough, 1970). *C. reinhardii* has 16 linkage groups, as revealed by genetic analysis, and eight chromosomes, as actually observed cytologically. A scrutiny of the distribution pattern of Mendelian mutations affecting photosynthesis has revealed that 16 such gene mutations map on eight of the 16 linkage groups, and nine gene mutations affecting pigment synthesis belong to four of the remaining eight linkage groups.

A priori, the mutants of *C. reinhardii* that are defective in photosynthetic CO_2 fixation may be expected to fall in any of the following categories:

(a) those that are unable to synthesize some component of the photosynthetic electron transport chain;
(b) those in which the chloroplast structure and integrity have been damaged;
(c) those in which the capacity for photophosphorylation (ATP synthesis) has been lost; and
(d) those in which there is a loss of capacity to synthesize some enzyme concerned with the dark reactions of CO_2 fixation.

Actually, however, the vast majority of mutant strains have been shown to fall in category (a) and a few in category (d). No mutants of the type in (b) or (c) seem to have been reported.

Most of the studies on the genetic dissection of photosynthesis have been carried out on mutants that are blocked at some stage of the photosynthetic electron transport chain. In different mutants, genes have been revealed to govern the synthesis of as many as six components of the chain. These components are (1) Q, the electron acceptor of PSII; (2) cytochrome 559; (3) cytochrome 553; (4) plastocyanin; (5) 'M', an unidentified component; and (6) P700. None of these mutants is capable of photoreducing $NADP^+$ since they are blocked at some stage along the chain and, hence, the electron cannot migrate to $NADP^+$. However, if the affected or blocked component in a particular mutant is supplied exogenously, the capacity for the photoreduction of $NADP^+$ should be restored. This has been established in the case of the mutant strain blocked at the plastocyanin level; the addition of purified plastocyanin extracted from the wild cells was shown to restore the capacity for $NADP^+$ photoreduction in the mutant.

Very elegant techniques have been employed to decipher the sequence of movement of electrons along the photosynthetic electron transport chain in these six mutant strains. The basis of these techniques is the oxidation-reduction concept. When a molecule acquires an electron from another molecule, then the receiving molecule is reduced (since the acquisition of the electron is generally accompanied by the acquisition of hydrogen) and the molecule from which the electron is lost is oxidized (since the loss of an electron is generally accompanied by the loss of hydrogen).

In photosynthesis, the precise sequence in which the components of the electron transport chain act has been followed by comparing the mutant strains with the nonmutant controls with respect to changes in light-induced absorbance; these changes are an index of the redox status of the system. Thus, when cells or chloroplast fragments are illuminated with far-red light (about 710 nm) which is absorbed by PSI, the cytochromes (especially cytochrome 553) *lose* an electron which migrates towards the P700 trap, with the result that the cytochromes get *oxidized*. On the other hand, when the cells or chloroplasts are exposed to shorter wavelengths (about 660 nm) which activate PSII, the cytochromes *accept* an electron (from Q) and thus get reduced. The oxidized and *reduced* forms of cytochromes exhibit characteristically different absorption peaks and, hence, can be easily detected by means of a spectrophotometer.

When such light-induced changes in absorption characteristics were examined in the mutant strains *ac-115* and *ac-141* which lack cytochrome 559, it was observed that cytochrome 553 could be successfully oxidized by light but could not be reduced. On the other hand, in the mutant that lacks cytochrome 553, it was found that cytochrome 559 could be effectively reduced by light but could not be oxidized. These experiments elegantly demonstrated that cytochrome 559 lies on the PSII side of cytochrome 553 and that cytochrome 553 lies on the PSI side of cytochrome 559. Ebersold (1967) has further characterized the mutant strains *ac-115* and *ac-141* lacking cytochrome 559 by carrying out tests for dominance. In these tests, vegetative diploids carrying both mutant and wild-type genes were formed and it was observed that in both cases the mutations were recessive.

The technique used for inferring the path of movement of the electron from observations of light-induced changes in absorption characteristics has been used also for locating the precise site of the other components of the electron transport chain.

STUDIES OF PHOTOSYNTHESIS IN ALGAL MUTANTS

(1) Considerable support for the unique role of P700 in photosynthesis has resulted from investigations of mutants of *Scenedesmus obliquus* (mutant *8*) and *Chlamydomonas reinhardii* (mutant *ac-80a*). Papers on the light-induced absorption change at 700 nm and on the light-induced EPR (Electron Paramagnetic Resonance) signal in these mutants have shown that the mutations in both cases affect P700 (Givan and Levine, 1967). Bishop (1971) and Levine (1969) have rightly stressed that the mutants lacking P700 provide strong support for the Series model (Z-scheme) of photosynthesis.

(2) After suitable treatment of chloroplasts with certain detergents, particles active in PSI activity can be isolated. Laboratory experiments on

these particles have established that $NADP^+$ photoreduction depends largely on the presence of a functionally-active plastocyanin (see Bishop, 1971). Gorman and Levine (1966), on the basis of their work on the photochemical reactions of chloroplasts isolated from mutants of *C. reinhardii*, have furnished strong support for these findings. The mutant *ac-208* possesses both cytochromes 553 and 559 but lacks plastocyanin; another mutant *ac-206* has plastocyanin but lacks cytochrome 553. Studies of light-induced oxidation-reduction changes in these two mutants have substantiated the Series model of photosynthesis.

(3) Research on mutant strains of *Scenedesmus* and *Chlamydomonas* strongly supports the location of cytochrome 553 between M and plastocyanin (see Bishop, 1971). On the basis of their findings on the simultaneous generation of proton pump activity and cytochrome 553 photooxidation during the greening of mutant *y-1* of *Chlamydomonas*, Schuldiner and Ohad (1969) have proved that cytochrome 553 joins together the two photosystems, PSI and PSII; they have also suggested that the micromorphological changes in chloroplast structure occurring during the greening process may be closely related to the appearance of the proton pump and the coupling of the two photosystems.

(4) Boardman (1968), working on mutant strains *ac-115* and *ac-141* of *C. reinhardii*, has found that the loss of PSII activity is attended by a concomitant reduction in their cytochrome 559 content. This demonstrates the close association of cytochrome 559 with PSII.

(5) Some mutant strains of *Chlamydomonas*, unable to form chlorophyll in the dark, can be propagated as 'yellow strains' after growing in the dark for a few generations; these are devoid of chlorophyll and do not photosynthesize. When re-exposed to light, however, the cells become green and photosynthetic activity is restored (in view of this, such strains cannot be regarded as mutant in a genetical sense). The addition of chloramphenicol to such regreening cells does not block the recovery of chlorophyll and of PSI activity but does block the PSII-dependent Hill reaction. On the other hand, treatment with actidione inhibits the synthesis of chlorophyll but does not block the PSII activity (see Hoober *et al.*, 1969).

(6) Experiments on the *ac-20* mutant of *C. reinhardii*, a gene mutation mapped on a nuclear-linkage group, have yielded useful information about the site of synthesis of chloroplast components. In this mutant, the ability to synthesize chloroplast ribosomes is much less than that of wild-type cells. Mutant cells deficient in chloroplast ribosomes exhibit a greatly-reduced Hill reaction activity and ribulose diphosphate carboxylase reaction, but their chlorophyll content and PSI activities remain fairly normal. In ribosome-deficient cells, the levels of components associated with PSI are not significantly affected but certain components associated with PSII (e.g., cytochrome 559 and Q) and ribulose diphosphate carboxylase are significantly reduced or inhibited.

(7) Perhaps one of the most important contributions to the understanding of the photosynthetic electron transport chain comes from research on

algal mutants. Until recently, the chain was thought to include P680, Q, cytochrome 559, cytochrome 553, plastocyanin, P700, and, finally, FRS. Experiments on the *Chlamydomonas* mutant *ac-21* have uncovered the existence of an unidentified component (termed M) lying between cytochrome 559 and cytochrome 553. In this mutant, cytochrome 559 can be reduced by light but cannot be oxidized whereas cytochrome 553 can be oxidized but not reduced. The chemical nature of this component is not yet known.

Such studies of photosynthesis with algal mutants help in understanding the process in three significant ways:

(a) They seem to establish that the two photochemical systems PSI and PSII act in series and are connected (between Q and P700) through cytochromes in the form of a Z, in accordance with the Z-scheme or Series model as proposed by Hill and Bendall (1960).

(b) They have elucidated the relative roles of the chloroplast genome and the nuclear genome in the control of chloroplast structure and function. They have also indicated that the bulk of chloroplast components may be synthesized outside the chloroplast whereas a few may be synthesized within the chloroplast (Levine and Goodenough, 1970). According to Surzycki *et al.* (1970), the genetic system of the *C. reinhardii* chloroplast is concerned with many specific activities affecting the morphology and physiology of the chloroplast. This system seems to be well-integrated with the nuclear-cytoplasmic genetic system. Possibly, the overall control of plastid morphology and physiology involves the co-operative participation of both PSI and PSII.

(c) They have helped uncover the existence of unknown components in the photosynthetic electron transport chain. The unidentified component M, detected in mutant *ac-21*, is a good example. Another similar component, hitherto unknown and still unidentified, has recently been reported (see Bishop, 1971) and is thought to lie somewhere between P700 and ferredoxin.

Future investigations on different kinds of algal mutants, involving additional markers as well as other genera, should not only expose the existence of unknown reactions, processes, or components, but also go a long way in providing answers to many unsolved questions confronting present-day researchers on photosynthesis.

REFERENCES

Bassham, J. A., Jensen, R. G., Photosynthesis of carbon compounds. In San Pietro, A., Greer, F. A., Army, T. J. (eds.), Harvesting the Sun, 79–110, Academic Press, New York (1967).

Bishop, N. I., Photosynthesis: the electron transport system of green plants. *A. Rev. Biochem.*, **40**, 197–226 (1971).

Boardman, N. K., Cytochromes of developing and mutant chloroplasts. In Shibata, K., Takamiya, A., Jagendorf, A. T., Fuller, R. C. (eds.),

Comparative Biochemistry and Biophysics and Photosynthesis, 206–13, Tokyo Univ. Press (1968).

Ebersold, W. T., *Chlamydomonas reinhardi*: heterozygous diploid strains. *Science*, **157**, 447–49 (1967).

Fork, D. C., Amesz, J., Action spectra and energy transfer in photosynthesis. *Ann. Rev. Plant Physiol.*, **20**, 305–28 (1969).

Givan, A. L., Levine, R. P., The photosynthetic electron transport chain of *Chlamydomonas reinhardi*. VII. Photosynthetic phosphorylation by a mutant strain of *Chlamydomonas reinhardi* deficient in active P700. *Plant Physiol.*, **42**, 1264–68 (1967).

Gorman, D. S., Levine, R. P., Photosynthetic electron transport chain of *Chlamydomonas reinhardi*. VI. Electron transport in mutant strains lacking either cytochrome 553 or plastocyanin. *Plant Physiol.*, **41**, 1648–56 (1966).

Hill, R., Bendall, F., Function of two cytochrome components in chloroplasts: a working hypothesis. *Nature*, **186**, 136–37 (1960).

Hoober, J., Siekevitz, P., Palade, G., Formation of chloroplast membranes in *Chlamydomonas reinhardi y-1*: effects of inhibitors of protein synthesis. *J. Biol. Chem.*, **244**, 2621–31 (1969).

Levine, R. P., Genetic dissection of photosynthesis. *Science*, **162**, 768–71 (1968).

Levine, R. P., The analysis of photosynthesis using mutant strains of algae and higher plants. *Ann. Rev. Plant Physiol.*, **20**, 523–40 (1969).

Levine, R. P., Goodenough, U. W., The genetics of photosynthesis and of the chloroplast in *Chlamydomonas reinhardi*. *Ann. Rev. Genet.*, **4**, 397–408 (1970).

Schuldiner, S., Ohad, I., Biogenesis of chloroplast membranes. III. Light-dependent induction of proton pump activity in whole cells and its correlation to cytochrome-f photooxidation during greening of *Chlamydomonas reinhardi* mutant (*y-1*). *Biochem. Biophys. Acta*, **180**, 165–77 (1969).

Surzycki, S. J., Goodenough, U. W., Levine, R. P., Armstrong, J. J., Nuclear and chloroplast control of chloroplast structure and function in *Chlamydomonas reinhardi*. In Miller, P. L. (ed.), Control of Organelle Development, *S. E. B. Sympos.*, No. 24, 13–37, Cambridge Univ. Press, London (1970).

7: Photosynthesis: Carbon Fixation

INTRODUCTION

Plants obtain all their carbon and more than 90% of their dry weight from CO_2 in the atmosphere. They fix CO_2 in the light as well as in the dark, but about 99% of their carbon is derived from light fixation or carboxylation. The dark carboxylation is so low that it is outweighed by the losses incurred in respiration and most plants lose a tenth or so of the daytime gains during the night. However, some plants, e.g., members of the Crassulaceae, are unique because they fix significant amounts of CO_2 at night. It is widely believed that photosynthesis is a process in which CO_2 is converted into carbohydrate. This notion is inadequate since plants or even chloroplasts can, by a manipulation of environmental conditions, be made to synthesize mostly carbohydrates, amino acids (proteins) or fatty acids (fats), or even vitamins, coenzymes, and pigment molecules. Nonphotosynthetic organisms, such as animals, fungi, and bacteria, all fix CO_2 without the involvement of light. Photosynthetic organisms can also assimilate CO_2 in the dark, provided they have a sufficient quantity of ATP and NADPH + H^+ available. Thus, the mechanism of carboxylation in plants appears, on the whole, similar to that in animals and it is only the photochemical generation of the substances powering CO_2 assimilation by chloroplasts that makes green plants basically different from animals and other nonphotosynthetic organisms.

The first relatively stable product of CO_2 fixation in plants may be PGA, the unstable oxaloacetic acid which is rapidly converted into malic acid and/or aspartic acid, and glycolic acid or carbamyl phosphate. The route to PGA involves the CO_2-acceptor molecule ribulose-1,5-diphosphate, and the enzyme ribulose diphosphate carboxylase; the route to oxaloacetic acid depends upon phosphoenolpyruvate (PEP) and the associated enzyme PEP carboxylase; and the route to carbamyl phosphate is mediated by the enzyme carbamyl phosphate synthetase. Although the initial reactions of the glycollate pathway are not quite clear, one opinion states that a secondary branch of the carbon reduction cycle forms PGA. The mechanisms of the various pathways have been elucidated by feeding photosynthesizing plants with radioactive $^{14}CO_2$ and following the passage of radioactive carbon through different cellular or chloroplast fractions in relation to time. The isolation of various enzyme systems connected with the respective pathways has led to further evidence in support of the proposed mechanisms.

If plants are allowed to fix labelled CO_2 in the light and in the dark, the distribution pattern of radioactive carbon is found to be different under given conditions, thereby suggesting that light and dark carboxylations follow different paths. The usefulness of radioactive substances in tracking down biochemical routes is limited to short-term experiments. This is because of the rapid exchange of radioactive substance, say ^{14}C, among a variety of substances, some of which may be quite unrelated to the actual process under investigation. The usual procedure is to make a suspension of *Chlorella* (a unicellular green alga) or chloroplasts photoassimilate $^{14}CO_2$ in the form of bicarbonate for a few minutes and then to stop the activity of the photosynthetic enzymes by running a part of the suspension into boiling alcohol. The alcohol-soluble substances are extracted from the cells or chloroplasts and a two-dimensional chromatogram of the extract is then prepared and compared with a similarly prepared chromatogram of known substances. The location of radioactive spots on the chromatogram is facilitated by leaving the paper in contact with a photographic film which, when processed, will develop spots at places corresponding to those on the chromatogram, or by means of a Geiger-Muller Counter. The position of radioactive carbon within the molecule has been determined by chemically degrading the substance into smaller molecules, each containing a single carbon.

The Benson-Calvin, C_4, and glycollate cycles have been deciphered in the manner just outlined and are known to be operative in many plants. The carbamyl phosphate pathway is not so widespread, but has been demonstrated in some members of Cyanophyta (blue-green algae).

BENSON-CALVIN (C_3) CYCLE

The Benson-Calvin cycle, named after its discoverers and known also as the pentose phosphate reduction cycle, starts with the carboxylation of ribulose-1,5-diphosphate with CO_2 at carbon-2 (Fig. 7-1). The resulting 6-carbon carboxylated intermediate immediately splits into two molecules of PGA. The latter is reduced to triosephosphate at the expense of photochemically-generated assimilatory power (NADPH + H^+ and ATP):

$$\text{Ribulose-1,5-diphosphate} + CO_2 \xrightarrow{\text{Carboxylase}} \text{2 Phosphoglyceric acid}$$

$$\text{2 Phosphoglyceric acid} \xrightarrow[\text{Kinase + Dehydrogenase}]{\text{ATP + NADPH + H}^+} \text{2 Triosephosphate}$$

The triosephosphate regenerates the CO_2-acceptor molecule through a series of reactions. Recently, another possibility has been raised, viz., that the carboxylated intermediate may, through a reductive carboxylation, give rise to one molecule of triosephosphate and one molecule of PGA, instead of two molecules of PGA. This could be achieved if the newly-formed carboxyl group, bonded to carbon-2, were to be phosphorylated and reduced before liberation from the enzyme.

The Benson-Calvin cycle was originally proposed on the basis of two

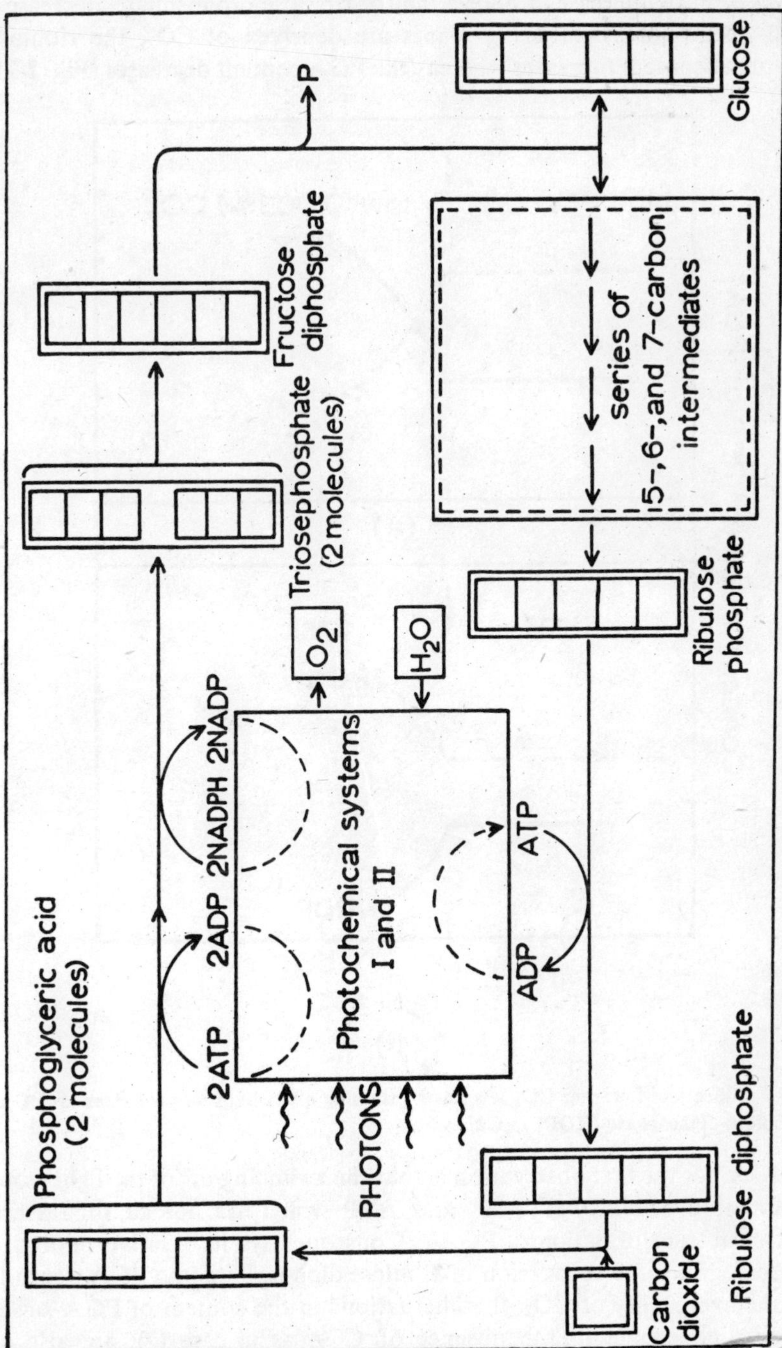

Fig. 7-1 Photosynthetic carbon reduction cycle. (After Levine, 1968.)

key observations: (1) when photosynthesizing plants are deprived of light, the level of PGA increases whereas that of ribulose diphosphate decreases, and (2) when photosynthesizing plants are deprived of CO_2, the ribulose diphosphate content increases whereas the PGA content decreases (Fig. 7-2).

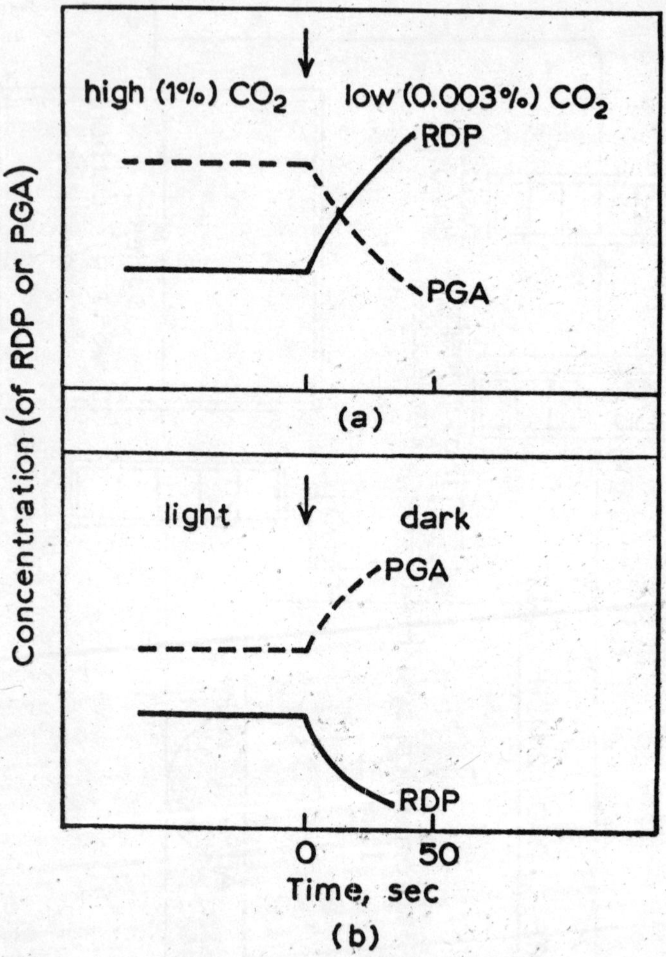

Fig. 7-2 Effects of lowering CO_2 level and switching off of light on pool sizes of PGA and ribulose diphosphate (RDP) in *Chlorella*.

The reason for the first observation is that the switching off of the light stops the formation of NADPH + H^+ and ATP which are needed for further biochemical transformation of PGA. Consequently, its relative pool size increases. Since the conversion of ribulose diphosphate into PGA depends upon the availability of CO_2, the fluctuations in the content of PGA or of ribulose diphosphate in the absence of CO_2, as in case (2), are also as expected.

In the light, plants incorporate most of the $^{14}CO_2$ into PGA and sugar phosphates; if they photosynthesize for five seconds, 80% of the ^{14}C is found in PGA, thus indicating that PGA is the first stable product fixation in the light. Further, the carbon of the carboxylic acid alone contains radioactivity, thus suggesting that PGA is formed by the addition of CO_2 by a carboxylic reaction to an acceptor substance which may be a 2-carbon molecule or a molecule with five carbon atoms. Radioactive carbon also appears in a number of pentose phosphate molecules, but its intensity is always higher in ribulose diphosphate. Further, the enzyme ribulose diphosphate carboxylase is found to catalyze *in vitro* the conversion of pentose diphosphate into PGA in the presence of CO_2. These findings support the concept of ribulose diphosphate as the acceptor molecule of CO_2 in the Benson-Calvin cycle. It was noticed that within five seconds of photosynthesis, only the carboxylic group carbon of PGA incorporated radioactivity, whereas two-to-three minutes of photosynthesis led to the labelling of all three carbons of PGA. This observation initiated the idea of the quantitative nature of the carbon reduction cycle in which three molecules of ribulose diphosphate react with three molecules of CO_2 to produce a net gain of one molecule of PGA.

The first reaction in the photosynthetic carbon reduction cycle is the activation of ribulose monophosphate with ATP mediated by ribulose diphosphokinase. Ribulose diphosphate changes from keto to enol form, a condition necessary for the carboxylation reaction to take place at carbon-2. The reduction of PGA to phosphoglyceraldehyde by NADPH $+ H^+$ is an endergonic reaction; hence, a molecule of PGA reacts with a molecule of ATP and the resulting 1,3-diphosphoglyceric acid then gets reduced to 3-phosphoglyceraldehyde by NADPH $+ H^+$ (Fig. 7-3). Thus, a molecule of CO_2 and ribulose phosphate gives rise to two PGA molecules, each of which utilizes one molecule of ATP and one molecule of NADPH $+ H^+$ to form a molecule of phosphoglyceraldehyde. Since the formation of ribulose diphosphate also requires ATP, the total ATP and NADPH $+ H^+$ involved in the conversion of a molecule of CO_2 into a carbohydrate intermediate are three and two molecules, respectively. In the same way, the net formation of a molecule of PGA from three molecules of CO_2 will need nine molecules of ATP and six molecules of NADPH $+ H^+$. Since hexose phosphate and triosephosphate are also predominant among the substances labelled after a brief period of photosynthesis with $^{14}CO_2$, and since PGA and triosephosphates are the intermediates identified in glycolysis in which NADPH $+ H^+$ (or NADH $+ H^+$) and ATP are generated, the working hypothesis proposed was that sugar might be formed in photosynthesis by a reversal of this process with photochemically-produced ATP and NADPH $+ H^+$. This supposition was strengthened by the demonstration of the required enzymes of the glycollate pathway in chloroplasts and the order of appearance and positioning of tracer carbon in triosephosphates and hexose phosphates.

Triosephosphate isomerase converts a part of the phosphoglyceraldehyde pool into dihydroxyacetone phosphate. The two triosephosphates condense in the presence of aldolase enzyme to produce a molecule of fructose-1,6-diphosphate. Carbon-3 and carbon-4 of the resulting hexose diphosphate showed labelling because carbon-1, -2, and -3 of dihydroxyacetone phosphate constitute carbon-1, -2, and -3 of fructose diphosphate, and carbon-1, -2, and -3 of phosphoglyceraldehyde constitute carbon-6, -5, and -4 of hexose diphosphate. A 4-carbon sugar phosphate molecule, called

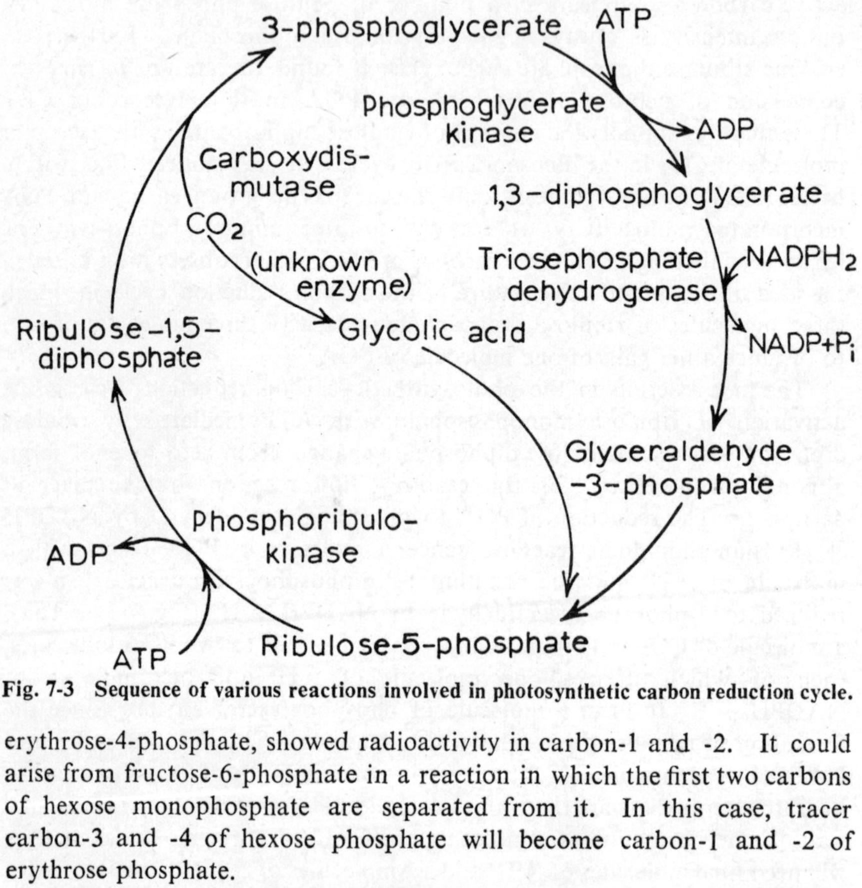

Fig. 7-3 Sequence of various reactions involved in photosynthetic carbon reduction cycle.

erythrose-4-phosphate, showed radioactivity in carbon-1 and -2. It could arise from fructose-6-phosphate in a reaction in which the first two carbons of hexose monophosphate are separated from it. In this case, tracer carbon-3 and -4 of hexose phosphate will become carbon-1 and -2 of erythrose phosphate.

A 7-carbon sugar, called sedoheptulose, is labelled in the same way as other intermediates and is said to result from erythrose phosphate and dihydroxyacetone phosphate in a reaction mediated by another aldolase enzyme. Both the presence of such an enzyme in chloroplasts and the position of radioactive carbons in the 7-carbon sugar diphosphate agree well with the proposed pathway. Similar enzymatic studies coupled with the path of ^{14}C distribution indicate the formation of a molecule of ribose phosphate and of xylulose phosphate from the reaction between sedoheptulose phosphate and 3-glyceraldehyde. A pentose isomerase enzyme isomerizes ribose-5-phosphate to ribulose-5-phosphate. Two kinds of phosphatases participate in the cycle: one converts fructose diphosphate into monophosphate, and the other converts sedoheptulose diphosphate into its monophosphate.

The transketolase reaction involves thiamine pyrophosphate (TPP) as a coenzyme and the 2-carbon fragment that it separates is known as phosphate glycolaldehyde which reacts with phosphoglyceraldehyde to form xylulose-5-phosphate. Ribulose-5-phosphate, coming as the end product of the cycle, gets phosphorylated in the first position with ATP to form its active stage.

Three molecules of ribulose diphosphate and three molecules of CO_2 interact to give six molecules of PGA, of which five will be consumed in driving the cycle; thus, the net gain is one molecule of PGA per three molecules of CO_2 assimilated or one molecule of hexose phosphate per six molecules of CO_2 consumed. Two molecules of triosephosphate are used in the formation of fructose diphosphate, and the third and fourth triose molecules are applied in the two transketolase reactions producing two molecules of xylulose phosphate. The fifth molecule reacts with erythrose phosphate to form sedoheptulose from which is derived a molecule of ribose phosphate. Two molecules of xylulose phosphate and one molecule of ribose phosphate give rise to three molecules of ribulose phosphate. Thus, five molecules of a 3-carbon compound result in three molecules of 5-carbon compound through a chain of well-understood reactions.

Of the many reactions in the C_3 cycle, two are very important:

(a) the reaction of CO_2 with the acceptor molecule ribulose diphosphate to form two molecules of phosphoglycerate; and

(b) the reduction of the phosphoglycerate produced to triosephosphate; this is just the reverse of the oxidation step (triosephosphate to phosphoglycerate) in dark respiration. The major difference is that the coenzyme involved in respiration is NAD^+ whereas that in the reductive step (photosynthesis) is $NADP^+$.

C_4 PATHWAY

The C_4 cycle is also sometimes known as the dicarboxylic acid cycle or the β-carboxylation pathway. As early as 1954, Kortschak and his associates (Kortschak *et al.*, 1965) reported the formation of C_4 dicarboxylic acids as the primary products of photosynthesis in sugarcane leaves. A few years later, it became known that, in these leaves, the label from $^{14}CO_2$ became incorporated primarily in malic and aspartic acids (in short-term fixation) and not in PGA, as proposed for other plants by Calvin and his group. These initial observations on sugarcane were further extended by Kortschak and his coworkers (1965) and confirmed by Hatch and Slack (1966) who subsequently extended this type of experimental finding to include specific tropical grasses and certain dicotyledonous plants.

Hatch and Slack found the initial labelling in oxaloacetic acid followed by malic and aspartic acids. This led to the hypothesis of the photosynthetic dicarboxylic acid β-carboxylation cycle in which phosphoenolpyruvic acid was shown to be the CO_2-acceptor molecule. The key feature of this C_4 cycle is that the primary carboxylation reaction is catalyzed by PEP carboxylase so that the resulting product is oxaloacetate. The malate is derived from oxaloacetate by reduction with NADPH + H^+ (produced in photosynthesis) in the presence of malic dehydrogenase. Aspartate can also be derived from oxaloacetate by a transamination reaction which is mediated by glutamate-aspartate transaminase. The operation of this cycle is now widely accepted. Oxaloacetate donates CO_2 to an appropriate receptor, such as ribulose diphosphate. This will produce pyruvate from which PEP arises at the expense of photosynthetically-generated ATP. The conversion of pyruvate into PEP is catalyzed by the enzyme pyruvate orthophosphate dikinase which is found in leaves and whose activity is light-dependent:

$$\text{Ribulose diphosphate} + \text{Oxaloacetate} + H_2O \xrightarrow{\text{Transcarboxylase}} 2\ \text{Phosphoglycerate} + \text{Pyruvate}$$

$$\text{Pyruvate} + \text{Orthophosphate} + \text{ATP} \xrightarrow{\text{Pyruvate dikinase}} \text{PEP} + \text{Pyrophosphate} + \text{AMP}$$

The most distinguishing anatomical feature of the leaf of C_4 plants is the radial arrangement of chlorenchyma cells around vascular bundles and its differentiation into an inner circle of large, thick-walled, cylindrical cells, called bundle-sheath cells, and outer layers of palisade cells, called mesophyll cells (Fig. 7-4). Bundle-sheath cell chloroplasts may have rudimentary grana (Fig. 7-5) as in *Zea mays* and sugarcane, or well-developed grana as in Bermuda grass (Fig. 7-6). Mesophyll cell chloroplasts of all C_4 plants have well-developed grana. It is important to note, however, that the chloroplasts in mesophyll and bundle-sheath cells grade all the way from no structural dimorphism to extreme dimorphism. No causal relationship between chloroplast structural dimorphism and C_4 photosynthesis has so far been established (see Laetsch, 1974).

The two kinds of cells, viz., bundle-sheath and mesophyll, and their chloroplasts have been the subject of several recent analyses (Laetsch, 1969) using the electron microscope (see Figs. 7-5 to 7-8). Very interesting ultrastructural differences have been observed, though dimorphic chloroplasts by no means occur in all C_4 plants. There is also evidence that dimorphic chloroplasts differ in respect of the quality and quantity of the enzymes involved in CO_2 assimilation. For instance, mesophyll cells seem to have almost all the PEP carboxylase whereas most of the ribulose diphosphate carboxylase and the associated enzymes are located in bundle-sheath cells. It is suggested that mesophyll cells carry out the dicarboxylic acid cycle and the oxaloacetate or malate so produced is transported to bundle-sheath cells and their chloroplasts which contain the enzyme decarboxylating malate. The resulting CO_2 is fixed by ribulose diphosphate carboxylase through the C_3 cycle. The essence of the two photosynthetic carbon cycles operating in the same plant is that together they might constitute a mechanism for the most efficient utilization of the environmental CO_2.

Recent findings indicate that C_4 plants may be broadly divided into three groups on the basis of their respective photosynthetic carbon fixation

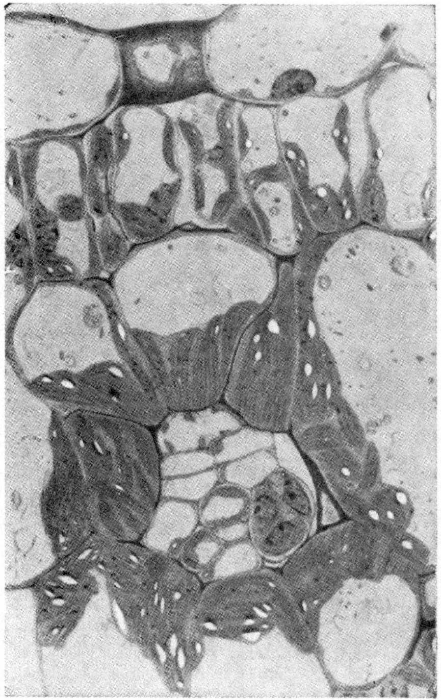

Fig. 7-4 Electron micrograph of a portion of a pigweed leaf in cross-section. (Top row of cells which is the epidermis is followed by mesophyll cells and then by very large bundle-sheath cell tightly surrounding the vascular tissue. White spots are starch grains.) (Courtesy : Professor C. C. Black, ×2200.)

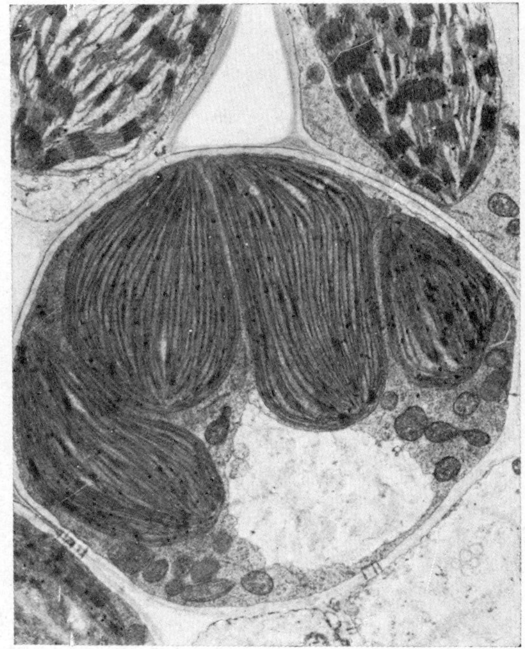

Fig. 7-5 Electron micrograph of a portion of a crab-grass leaf in cross-section showing a bundle-sheath cell with its chloroplasts having deficient grana development. (Two mesophyll cells abut the bundle-sheath cell and contain chloroplasts with well-developed grana.) (Courtesy : Professor C. C. Black, ×3900.)

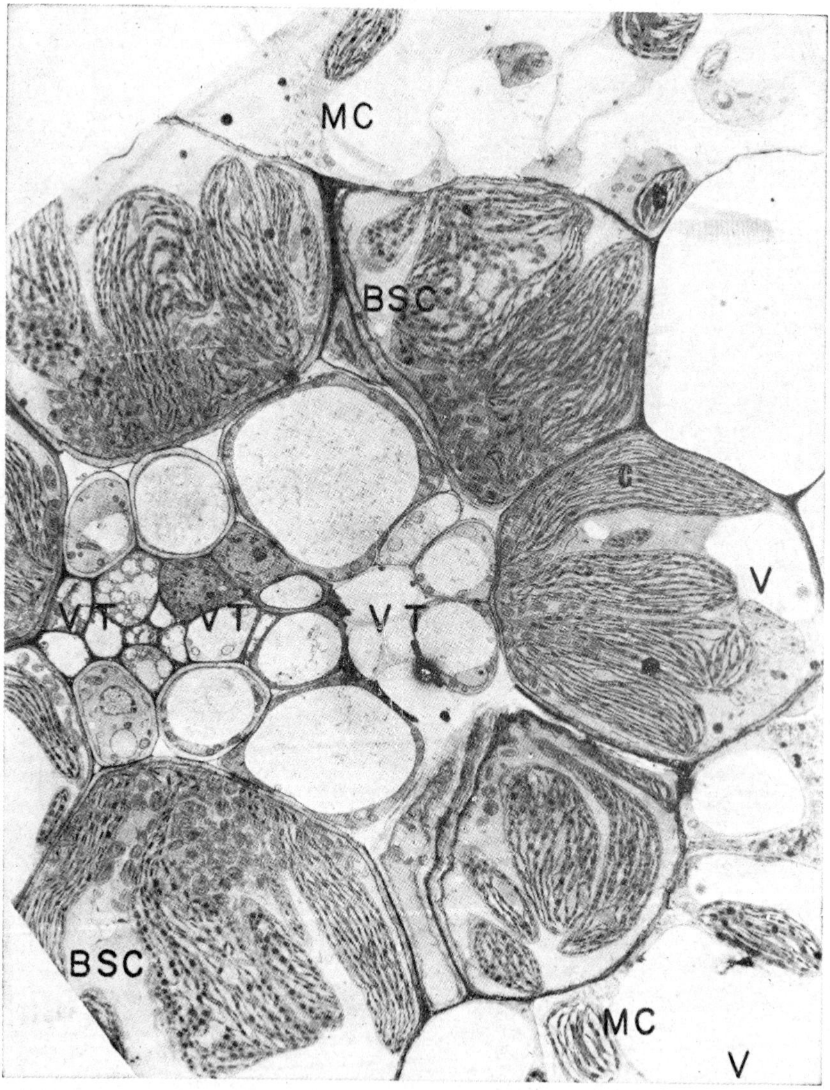

Fig. 7-6 Electron micrograph of a Bermuda grass leaf in cross-section showing vascular tissue (VT) tightly surrounded by bundle-sheath cells (BSC) with section of a few mesophyll cells (MC) visible but no epidermal tissue visible. (Chloroplasts (C) in both cell types clearly have highly-developed grana. Note the near-absence of a vacuole in these bundle-sheath cells and their prolific mitochondrial population.) (Courtesy : Professor C. C. Black, ×2400.)

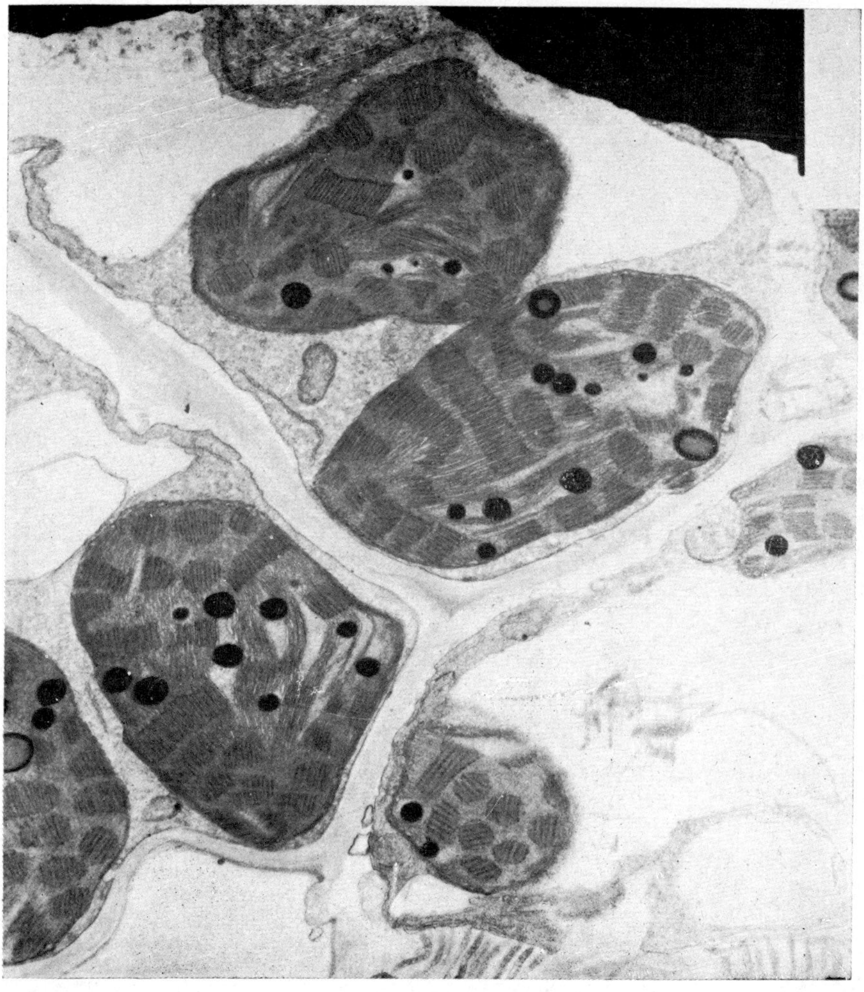

Fig. 7-7 Electron micrograph of a portion of a sugarcane leaf in cross-section passing through mesophyll cells. (Chloroplasts show good grana development and little starch.) (Courtesy : Professor W. M. Laetsch, ×14,700.)

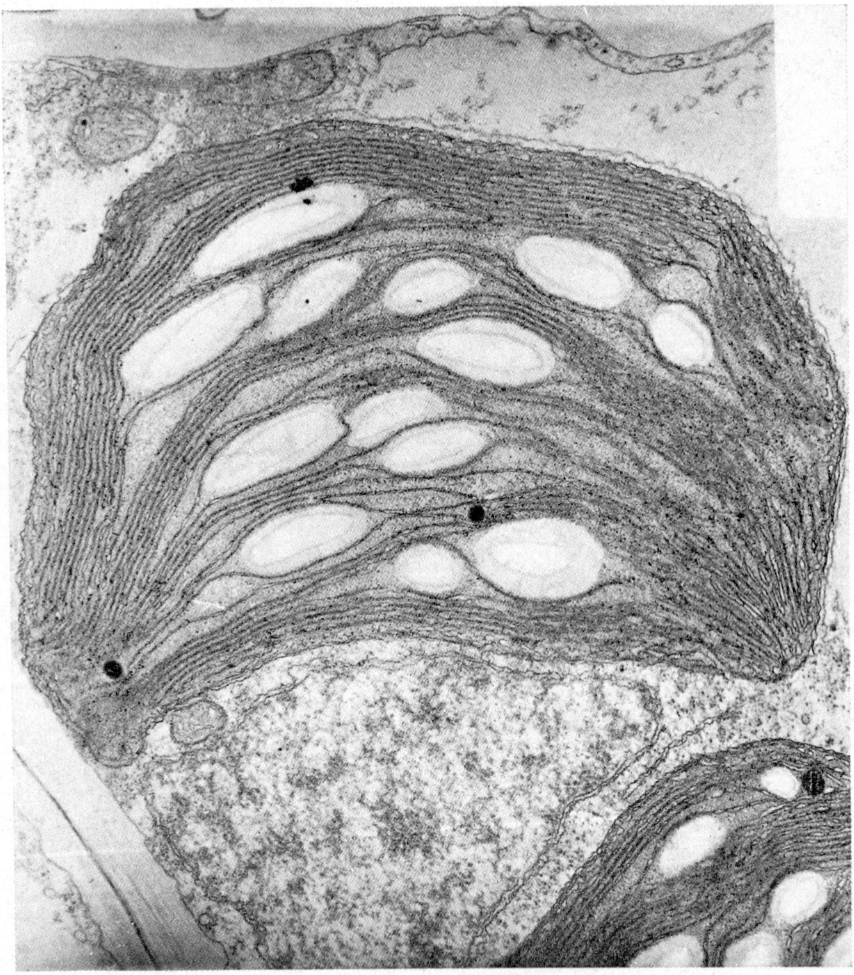

Fig. 7-8 Electron micrograph of a portion of a mature sugarcane leaf in cross-section passing through a bundle-sheath cell. (Chloroplast has prominent starch grains (white spots) and poor grana formation.) (Courtesy : Professor W. M. Laetsch, ×14,700.)

pathways (Chollet and Ogren, 1975). The first group includes plants such as maize and sugarcane. In this group, CO_2 is initially fixed in mesophyll cytoplasm by PEP carboxylase and oxaloacetate is formed; this oxaloacetate becomes reduced to malate by a $NADP^+$-specific malate dehydrogenase in mesophyll cell plastids. The malate is then transported to bundle-sheath chloroplasts and becomes converted into pyruvate and CO_2 by $NADP^+$-malic enzyme. The pyruvate is transported to mesophyll cell plastids and regenerates PEP through the action of pyruvate orthophosphate dikinase and the CO_2 is refixed by ribulose diphosphate carboxylase into PGA. The latter is cycled through the C_3 pathway to form carbohydrate in bundle-sheath chloroplasts.

C_4 plants of the second group are exemplified by *Panicum maximum* and *Chloris gayana*. In these, aspartate, rather than malate, is transported to bundle-sheath cells. There it is transaminated to oxaloacetate which becomes converted into pyruvate and CO_2 by PEP carboxykinase. The resulting pyruvate is transported back to mesophyll cells and CO_2 is fixed there through the enzymes of the Calvin cycle to generate carbohydrate. Plants of this group seem to have very low activity of $NADP^+$-specific malic enzyme.

In the third group, e.g., *Atriplex spongiosa*, the aspartate produced in mesophyll cells is transported to bundle-sheath cells where it is transaminated and reduced to malate which is decarboxylated to form pyruvate and CO_2.

It should be emphasized, however, that such schemes for the details of CO_2 fixation in C_4 plants are still rather tentative and research teams are engaged in establishing the veracity of the foregoing views.

Black and his coworkers (see Chen *et al.*, 1973; Rouhani *et al.*, 1973) successfully separated mesophyll cells and bundle-sheath cells from the leaves of certain C_4 plants, e.g., *Digitaria sanguinalis* and *Cyperus rotundus*. Comparative studies on photosynthesis in the two types of cells revealed the presence of a complete photosynthetic electron transport chain from oxygen liberation to $NADP^+$ reduction in both kinds of cells. However, the ratio of PSI : PSII activity was about three times higher in bundle-sheath cells than in mesophyll cells. In noncyclic electron transport, mesophyll cells were about three times more active than bundle-sheath cells whereas in cyclic flow, bundle-sheath cells were about equally more efficient than mesophyll cells. A clear compartmentation of the distinctive enzymes was found also between the two cell fractions, with most of the PEP carboxylase occurring in mesophyll cells and most of the ribulose diphosphate carboxylase and malic enzyme being located in bundle-sheath cells. Chen *et al.* (1973) concluded that most of the CO_2 uptake in mesophyll cells takes place by a β-carboxylation reaction of PEP carboxylase and that both a decarboxylation and a carboxylation reaction occur in bundle-sheath cells. The malate formed in mesophyll cells is thought to diffuse into bundle-sheath cells (see Fig. 7-9) where it is decarboxylated by malic enzyme to form pyruvate, CO_2, and NADPH. The CO_2 so produced is then reduced to

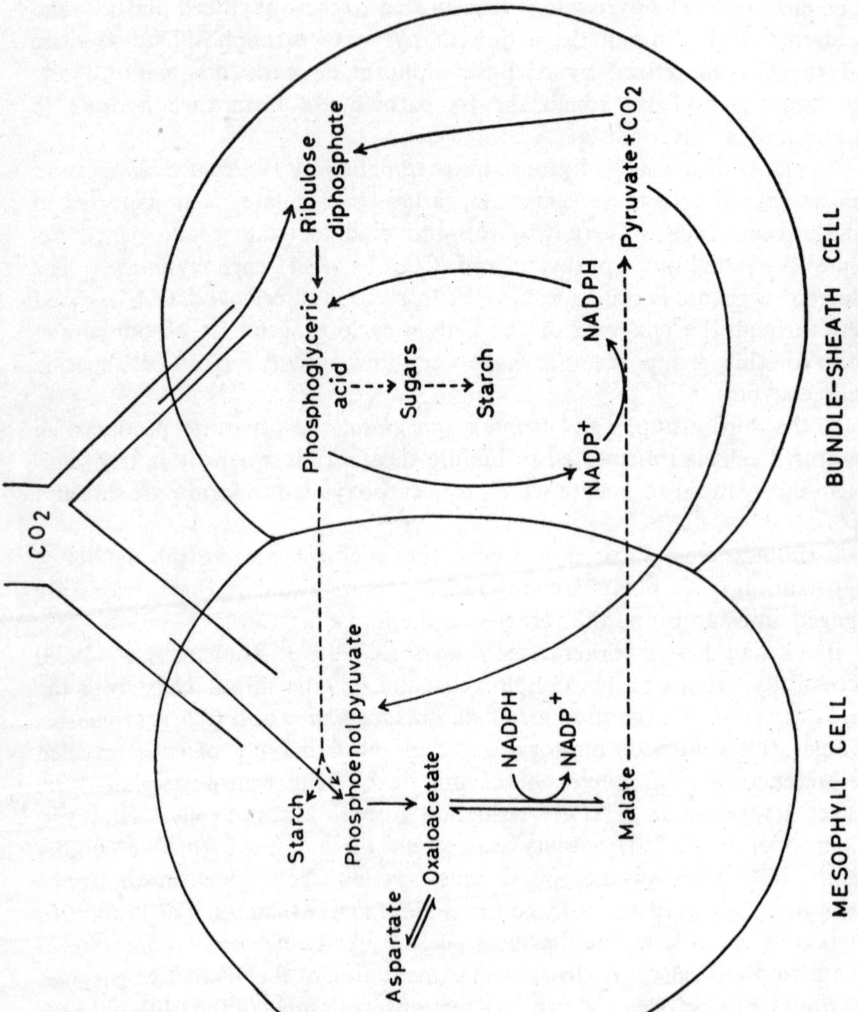

Fig. 7-9 **CO_2 fixation reactions in mesophyll and bundle-sheath cells.** (After Chen *et al.*, 1973.)

carbohydrate via the Benson-Calvin cycle.

The enzyme PEP carboxylase which plays a key role in the carbon metabolism of C_4 plants has recently been reported to occur at high levels in the guard cells in the leaf epidermal tissues of *Tulipa gesnariana* and *Commelina communis* (Willmer *et al.*, 1973). Thus, guard cells have an enzyme which can rapidly utilize CO_2, thereby effecting a steep fall in the intracellular CO_2 level which in turn would lead to stomatal opening. Another important significance of the work of Willmer and his team is that previously the incidence of high levels of PEP carboxylase was believed to be characteristic of only those plants that have either a C_4 dicarboxylic acid cycle or a crassulacean acid metabolism (CAM); the enzyme has now been found in the guard cells of plants that fall within neither of these two categories.

COMPARISON BETWEEN SPACE AND TIME FUNCTIONS IN C_4 PLANTS AND CAM PLANTS

Many plants belonging to the families Cactaceae, Aizoaceae and some others are xerophytic succulents and occur in dry habitats. In these, the stomata generally remain closed during the day and transpiration is minimized. Gas exchange occurs mostly during the night when stomata open, and CO_2 is fixed into carboxylic acids in much the same way as in C_4 plants during the day (see Laetsch, 1971). The titratable acidity of the cell sap rises during the night and falls during the day with deacidification. Plants that show dark acidification and light deacidification are styled CAM plants.

Recent studies have demonstrated a remarkable similarity between the basic biochemical pathways (especially carbon fixation and photosynthesis) of CAM plants and those of C_4 plants, mostly those belonging to the families Chenopodiaceae, Amaranthaceae, and Portulacaceae, and also many tropical grasses. In both cases, photosynthesis involves the fixation of CO_2 into C_4 dicarboxylic acids. These acids are subsequently decarboxylated, and the CO_2 so evolved is refixed, yielding PGA. The only difference is that in CAM plants the carboxylation reactions are separated from the decarboxylation and refixation reactions in time (the former occurring in the dark or the night, and the latter in the light or the day) whereas in C_4 plants the same reactions are separated spatially, with the carboxylation reaction mostly taking place in mesophyll cells and the carbohydrate synthesis occurring in bundle-sheath cells (both during the day). Thus, there is a close similarity between the reactions in the mesophyll cells of C_4 plants and those that take place in the mesophyll cells of CAM plants at night; likewise, the changes that occur in bundle-sheath cells are similar to those in the cells of CAM plants during the day. In CAM plants, the acid formation and acid degradation in each photosynthetic leaf cell are temporally separated whereas in C_4 plants the acid formation (in mesophyll cells) and acid degradation (in bundle-sheath cells) are spatially separated.

A new scheme elucidating the light and dark portions of net CO_2 fixation in metabolically-active mesophyll cells isolated from the leaves of the

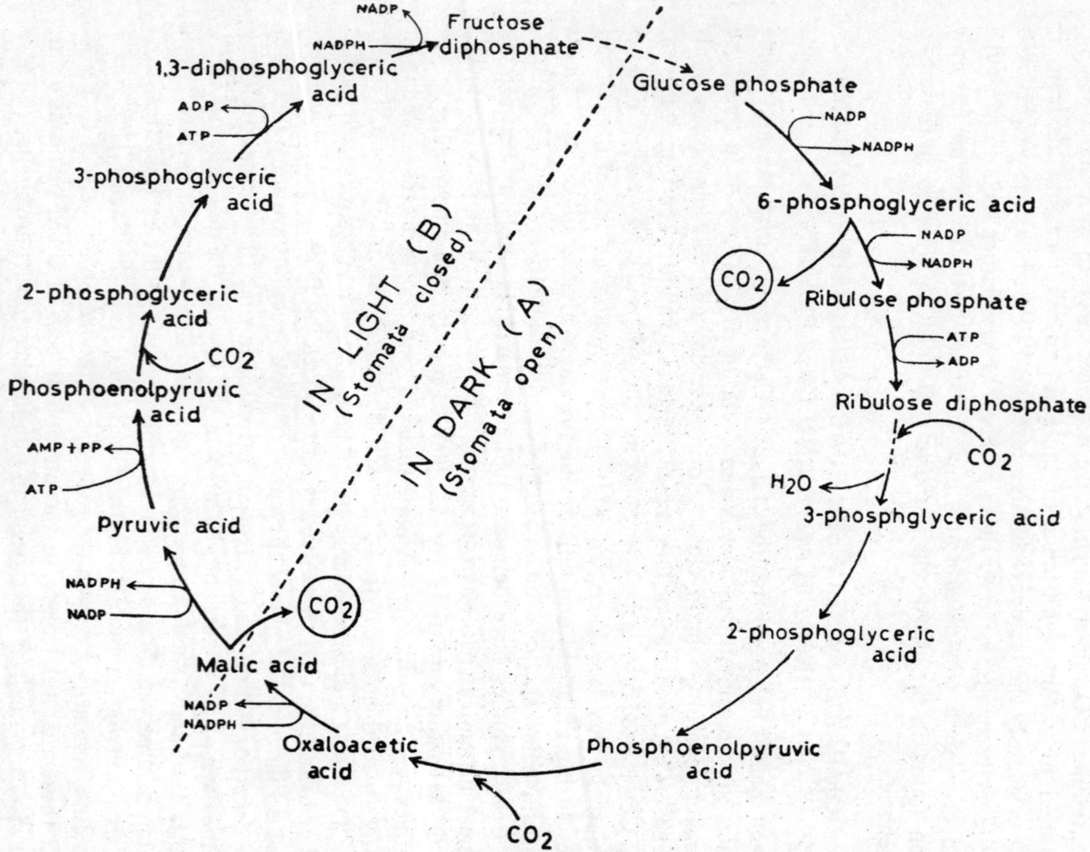

Fig. 7-10 **Scheme of CO_2 fixation reactions thought to occur in leaf cells of the CAM plant, *Sedum telephium*, in light and dark.** (After Rouhani *et al.*, 1973.)

CAM plant *Sedum telephium* has been proposed by Rouhani *et al.* (1973). The reactions shown in part A of Fig. 7-10 take place both in the light and the dark; the reactions shown in part B occur primarily in the light.

GLYCOLLATE PATHWAY

The Benson-Calvin cycle had initially been worked out in *Chlorella* at abnormally high CO_2 concentrations and light intensities which normally do not permit growth of the cells. At very low $^{14}CO_2$ concentrations, the short period of photosynthesis yields two labelled 2-carbon compounds: first, glycolic acid and then glycine. Subsequent analyses have shown that the production of glycolic acid is a general feature of most plants and algae and that this path works under conditions of low CO_2 and high oxygen concentrations. The mode of its origin is still largely unknown, but there are indications that it is derived from carbon-1 and carbon-2 of the Calvin cycle sugar phosphate intermediates and that the resulting 2-carbon fragment is believed to be an enzyme-bound glycolaldehyde unit. Bowes and Ogren (1972) assert that oxygen reacts with ribulose diphosphate carboxylase to make it function as oxygenase which reacts with ribulose diphosphate to form phosphoglycollate and PGA.

Glycollate formation occurs in illuminated chloroplasts and is inhibited by DCMU. These observations leave little doubt regarding its photosynthetic origin. The requirement of low CO_2 concentration for the formation of glycollate is explained thus: low CO_2 concentration leaves most of the ribulose diphosphate molecules free; since ribulose diphosphate may possibly be the source of the 2-carbon fragment that leads to glycollate formation, conditions favouring an increase in ribulose diphosphate content should naturally favour glycollate formation. The fate of the photosynthetic glycollate has been examined and followed enzymatically and isotopically, and one proposal of the sequence of reactions in this pathway is given in Fig. 7-11. (About four other schemes have been postulated for the synthesis of glycolic acid.)

Most of the enzymes of the glycollate pathway are located in particulate structures, called peroxisomes, which occur in the cell cytoplasm of photosynthetic tissues. This means that once glycollate is formed it must move out of the chloroplast to the exterior where the enzymes of the peroxisome will determine its fate. The first enzyme of the pathway is glycolic oxidase or dehydrogenase. The oxidase enzyme requires oxygen for oxidizing glycollate to glyoxylate and hydrogen peroxide whereas the dehydrogenase does not need oxygen but functions simply with the help of the coenzyme NAD^+ and oxidizes glycollate to glyoxylate and $NADH + H^+$. The peroxide produced is acted upon by the catalase enzyme which converts the toxic compound into water and oxygen:

$$2H_2O_2 \xrightarrow{\text{Catalase}} 2H_2O + O_2$$

Algal cells seem to have very little oxidase activity and a great deal of dehydrogenase activity whereas the reverse is true of higher plants.

The glyoxylate becomes aminated by an amino group from glutamate and this reaction is catalyzed by glutamate-glyoxylate transaminase. Two

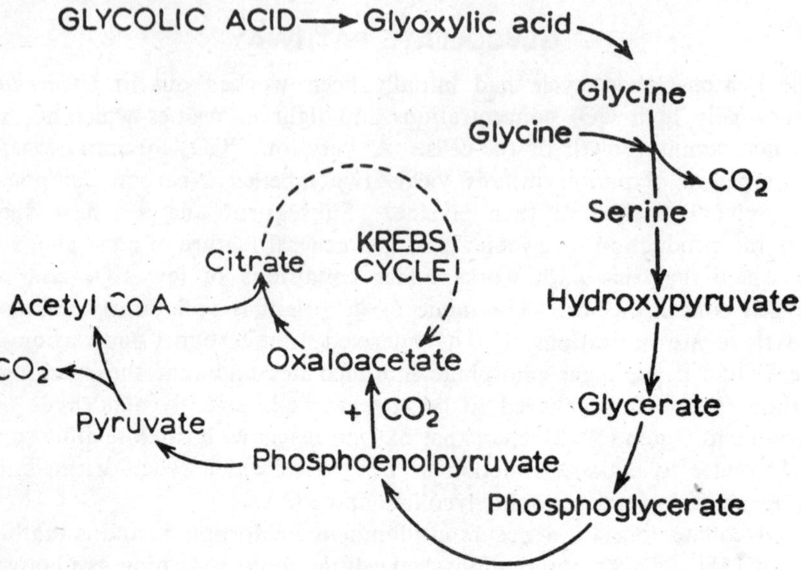

Fig. 7-11 Proposed sequence of reactions in the glycollate pathway.

molecules of the resulting glycine react with another enzyme, serine hydroxymethylase, to produce a molecule of CO_2 and serine. The amino group of the latter is removed by serine pyruvate transaminase and the resulting hydroxypyruvate gets reduced to glycerate by hydroxypyruvate reductase or glycerate dehydrogenase at the expense of a molecule of NADH + H^+. The glycerate may be phosphorylated by the kinase enzyme using a molecule of ATP and the phosphorylated intermediate may finally be converted into carbohydrate through the Benson-Calvin cycle.

The requirement of high oxygen for the glycollate pathway seems to be due to the activity of ribulose diphosphate carboxylase acting as oxygenase which is widespread in leaves.

In algae, most of the glycollate produced finds its way to the exterior; the reason for this phenomenon seems to be the high sensitivity of glycolic dehydrogenase to CO_2 concentration as the fate of glycolic acid in algae depends upon this key enzyme. In higher plants, glycollate is the substrate of photorespiration (the production of CO_2 dependent on photosynthetic light) and it is believed that photorespiratory CO_2 is produced during the conversion of two molecules of glycine into one molecule of serine. The significance of photorespiration is not very clear and the general opinion is that although the glycollate pathway seems necessary for the production of two key amino acids, on the whole it does not seem to be a very useful or efficient pathway (from our viewpoint) since a significant amount of the photosynthetically-fixed CO_2 is thrown back in this process; in case it can generate carbohydrate, it apparently does so at the cost of energy and by reducing power much greater than that used in the Benson-Calvin cycle.

CARBAMYL PHOSPHATE PATHWAY

Some blue-green algae, e.g., *Nostoc muscorum*, incorporate $^{14}CO_2$ in the light into carbamyl phosphate and subsequently into citrulline. The mechanism suggested is as follows:

$$CO_2 + NH_3 + ATP \xrightarrow{\text{Carbamyl phosphate synthetase}} H_2N\text{—}COOP + ADP$$

$$H_2N\text{—}COOP + \text{Ornithine} \xrightarrow{\text{Ornithine transcarbamylase}} \text{Citrulline} + P_i$$

This pathway operates possibly in all cellular organisms irrespective of whether or not they are photosynthetic. The enzyme carbamyl phosphate synthetase is widely distributed in living cells and all that is needed for the cycle to operate is ATP, CO_2, and NH_3.

GENERALITY OF PHOTOSYNTHETIC CARBON CYCLES

The C_3 and C_4 cycles have been shown to occur simultaneously, albeit (sometimes) in structurally and perhaps biochemically different chloroplasts, in some tropical grasses and dicotyledonous plants. The glycollate path-

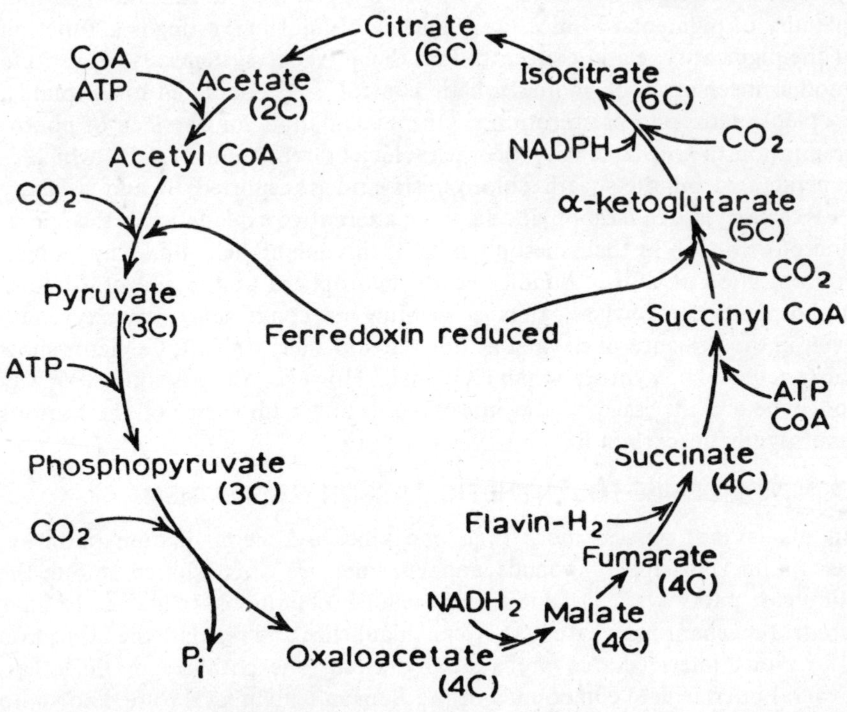

Fig. 7-12 Probable sequence of reactions in the carbon reduction cycle of some photosynthetic bacteria. (After Evans *et al.*, 1966.)

way seems to be a branch of the C_3 cycle. The C_4 pathway is unknown in bacteria, but some photosynthetic bacteria seem to have the glycollate pathway. *Chlorobium thiosulfatophilum* and some other photosynthetic bacteria have been shown to carry out a carbon reduction cycle in addition to the Benson-Calvin cycle. The new cycle is a reversal of the Krebs cycle and utilizes four molecules of CO_2 in one turn of the complete cycle or two molecules of CO_2 in a half turn of the complete cycle (see Fig. 7-12). One acceptor of the CO_2 is acetyl CoA, the second is phosphopyruvate, the third is succinyl CoA, and the fourth is α-ketoglutarate. NADPH + H^+ is required to produce isocitrate from CO_2 and α-ketoglutarate. ATP energizes pyruvate to accept CO_2 and thus to get converted into oxaloacetate. Its metabolic relation and significance to the Benson-Calvin cycle have not yet been determined. The net result of the operation of the cycle is to synthesize one molecule of oxaloacetate per four molecules of CO_2 fixed or one molecule of acetate for every two molecules of CO_2 utilized (see Evans *et al.*, 1966).

REGULATION OF PHOTOSYNTHESIS

All the factors participating in photosynthesis may be assumed to be regulatory mechanisms. Thus, a concentration of CO_2 or of oxygen can act as a switch for the type of carbon pathway operating in photosynthesis. The quality and intensity of the light may affect the type and quantity of pigments. Since the efficiency of light harvesting is a function of the pigment type and concentration, the pigment systems also constitute another mechanism of photosynthetic control. As compared to C_3 plants, C_4 plants lack photorespiration. One explanation for the lack of photorespiration in C_4 plants may be their efficient refixation of CO_2 which is generated in bundle-sheath chloroplasts and transported to and used by PEP carboxylase in mesophyll cells. An alternative explanation is the operation of PEP-CO_2 in their mesophyll cells: this might help build up a high concentration of CO_2 in bundle-sheath chloroplasts at the site of ribulose diphosphate carboxylase, thus preventing its conversion into oxygenase even in the presence of oxygen at the reaction site. In fact, C_4 plants show more active photosynthesis than C_3 plants. However, there is still a long way to go before the issue of the interrelation and significance of the various photosynthetic carbon fixation cycles is settled.

CONTROL OF PHOTOSYNTHETIC CARBON METABOLISM

In plants that possess more than one kind of carbon fixation pathway, certain intermediate compounds and enzymes are often shared among the different pathways. Depending on cellular conditions, cells seem to have control mechanisms whereby they can channelize and regulate the utilization of various intermediates preferentially along one pathway or the other. Several intermediate compounds of the Benson-Calvin cycle are known to be involved in diverse kinds of biosynthetic reactions in the cell and, depending on cellular conditions, such common intermediates may be shunted away from the cycle at different rates. This kind of shunting off along biosynthetic routes is remarkably regulated through the operation of precise and fine controls exerted on the activity of various enzymes and/or cell membranes. A good example of such regulation is described by Bassham (1973). He noticed that reversible changes in photosynthetic membranes induced by fatty acids in the light result in sudden increases in the steady state pools of dihydroxyacetone phosphate and fructose-1,6-diphosphate and a corresponding decline in the pools of fructose-6-phosphate and glucose-6-phosphate; such changes in pool sizes are obviously caused by a decrease in the rate of conversion of fructose diphosphate into fructose-6-phosphate (Bassham, 1973).

The transference of plant chloroplasts from the light to the dark or vice versa can switch over also the reductive pentose phosphate cycle to the oxidative pentose phosphate cycle, or vice versa. A major site of regulation in the Benson-Calvin cycle seems to be the carboxylase reaction catalyzed by ribulose diphosphate carboxylase.

Coenzyme II ($NADP^+$) is another substance that may play an important regulatory role. Considerable amounts of reduced coenzyme ($NADPH + H^+$) are generated in illuminated chloroplasts through photophosphorylation and some of it, in the form of triosephosphate, may diffuse from the chloroplast into the cytoplasm where it may become involved in diverse anabolic or catabolic reactions.

According to Bassham (1973), the regulation of the Benson-Calvin cycle in the light involves a control of the levels of intermediate compounds and this is achieved by a delicate balance between the activities of certain key enzymes, such as ribulose diphosphate carboxylase.

PHOTOSYNTHESIS *IN VITRO*

Complete photosynthesis by whole chloroplasts isolated from spinach leaves was demonstrated by Arnon and his associates in the mid-1950s. Recently, Bassham *et al.* (1974) have reported this process in a system consisting of broken spinach chloroplasts and the enzymes or other soluble components from previously isolated whole chloroplasts. They have further established that CO_2 fixation by these broken chloroplasts occurs only in the presence of light and that the resulting products are mainly intermediate components of the reductive pentose phosphate cycle. Such observations indicate that the delimiting action of the membranes of intact chloroplasts may not be an absolute requirement for the operation of the Benson-Calvin cycle at high rates. The reconstituted system containing broken chloroplasts plus enzymes may prove useful in the study of the regulatory mechanisms proposed for photosynthetic carbon metabolism (Bassham *et al.*, 1974).

WARBURG EFFECT

The Warburg effect is the result of competitive interaction of CO_2 and oxygen in photosynthesis, with CO_2 promoting photosynthesis and inhibiting glycollate formation, and oxygen inhibiting photosynthesis and promoting glycollate formation. This effect was observed by Warburg and Krippahl in 1960 in the green alga *Chlorella*; they noted that more than 75% of the total CO_2 fixed by the alga during photosynthesis accumulates in glycollate under certain conditions. Recent papers suggest that the Warburg effect reflects the balance between photorespiration and photosynthesis in photorespiring plants and is controlled at the level of ribulose diphosphate carboxylase which shows a dual activity, viz., carboxylase action when CO_2 is present in high amounts, and oxygenase action when oxygen is in excess (see Chollet and Ogren, 1975).

REFERENCES

Bassham, J. A., Control of photosynthetic carbon metabolism. *S. E. B. Sympos.*, No. 27, 461–83, Cambridge Univ. Press, London (1973).

Bassham, J. A., Levine, G., Forger III, J., Photosynthesis *in vitro*. I. Achievement of high rates. *Plant Sci. Letters*, **2**, 15–21 (1974).

Benson, A. A., Calvin, M., The path of carbon in photosynthesis. VII. Respiration and photosynthesis. *J. Exp. Bot.*, **1**, 63–68 (1950).

Bowes, G., Ogren, W. L., Oxygen inhibition and other properties of soybean carboxylase. *J. Biol. Chem.*, **247**, 2171–76 (1972).

Chen, T. M., Campbell, W. H., Dittrich, P., Black, C. C., Distribution of carboxylation and decarboxylation enzymes in isolated mesophyll cells and bundle sheath strands of C_4 plants. *Biochem. Biophys. Res. Commun.*, **51**, 461–67 (1973).

Chollet, R., Ogren, W. L., Regulation of photorespiration in C_3 and C_4 species. *Bot. Rev.*, **41**, 137–79 (1975).

Evans, M. C. W., Buchanan, B. B., Arnon, D. I., A new ferredoxin-dependent carbon reduction cycle in a photosynthetic bacterium. *Proc. Natl. Acad. Sci.*, USA, **55**, 928–34 (1966).

Hatch, M. D., Slack, C. R., Photosynthesis by sugarcane leaves. A new carboxylation reaction and the pathway of sugar formation. *Biochem. J.*, **101**, 103–11 (1966).

Kortschak, H. P., Hartt, C. E., Burr, G. O., Carbon dioxide fixation in sugarcane leaves. *Plant Physiol.*, **40**, 209–13 (1965).

Laetsch, W. M., Relationship between chloroplast structure and photosynthetic carbon fixation pathways. *Sci. Prog.*, Oxford, **57**, 323–51 (1969).

Laetsch, W. M., Chloroplast structural relationships in leaves of C_4 plants. In Hatch, M. D., Osmond, C. B., Slatyer, R. O. (eds.), *Photosynthesis and Photorespiration*, 323–49, Wiley Interscience, New York (1971).

Laetsch, W. M., The C_4 syndrome: a structural analysis. *Ann. Rev. Plant Physiol.*, **25**, 27–52 (1974).

Rouhani, I., Vines, H. M., Black, C. C., Isolation of mesophyll cells from *Sedum telephium* leaves. *Plant Physiol.*, **51**, 97–103 (1973).

Warburg, O., Krippahl, G., Glykolsauerbildung in *Chlorella*. *Z. Naturforsch.*, **15B**, 197–99 (1960).

Willmer, C., Kanai, R., Pallas, J. E., Black, C. C., Detection of high levels of phosphoenolpyruvate carboxylase in leaf epidermal tissue and its significance in stomatal movements. *Life Sciences*, **12**, Part II, 151–55 (1973).

8: Respiration

INTRODUCTION

Enzyme-mediated burning or the oxidation of organic substances is termed respiration. The substrates of respiration may include carbohydrates, fatty acids, amino acids, or other organic acids, but its products are always CO_2, water, and ATP. Since every activity of the cell, whether synthetic, osmotic, or mechanical, needs usable energy (ATP), the cell must have some means for the efficient production of ATP in order to maintain its life activities and to enable it to grow and multiply. The synthesis of proteins, fats, and nucleic acids also consumes large amounts of energy. In the dark, cells meet this demand through the activity of mitochondria whose principal role is to synthesize ATP aerobically. But most of the ATP produced in green plant cells during periods of light comes from photosynthetic phosphorylation and not from mitochondrial activity. The starting material for the oxidative activity of mitochondria is an active form of acetate, called acetyl CoA, which may arise through glycolysis or from β-oxidation of fatty acids (Beevers, 1961; Stiles and Leach, 1960; James, 1957).

The term respiration used to be applied earlier to the characteristic process whereby living organisms absorbed oxygen and released CO_2 (e.g., breathing in animals). It is now generally restricted to processes in which the respiratory substrate is broken down through oxidation (using molecular oxygen), with an accompanying release of energy and its storage or conservation in the form of ATP. Such processes occur in all living cells and they do so independent of light. In addition to this kind of 'normal' or 'dark' respiration, a new kind of respiratory process which occurs only in some green plants in the presence of light (photorespiration) has been noticed in recent years. In this chapter, only dark respiration is described; photorespiration is discussed in detail in Chapter 9.

RESPIRATORY QUOTIENT

Respiratory quotient (R.Q.) is defined as the ratio of the moles of CO_2 evolved to those of oxygen absorbed:

$$\text{R.Q.} = \frac{\text{moles of } CO_2 \text{ evolved}}{\text{moles of } O_2 \text{ absorbed}}$$

The respiratory equation is:

$$\underset{\text{Glucose}}{C_6H_{12}O_6} + 6O_2 \rightarrow 6CO_2 + 6H_2O + 678 \text{ kcal}$$

When fat is a substrate, then it is:

$$\underset{\text{Glycerol tripalmitate}}{2C_{51}H_{98}O_6} + 145O_2 \rightarrow 102CO_2 + 98H_2O + \text{Energy}$$

The R.Q. in the case of glucose is 1.0 whereas in glycerol tripalmitate it is nearly 0.7. If an organic acid, such as oxalic acid, is respired, the R.Q. observed is nearly 2.0. The R.Q. of proteins varies between 0.5 and 1.0, depending upon the oxygen content. In succulent plants, the respiratory CO_2 may be assimilated in the dark and, hence, the R.Q. observed is always less than 1.0.

The significance of determining R.Q. is that it indicates the chemical nature of the substrate being respired. In general, substances with low oxygen content show an R.Q. of less than 1.0 whereas those with high oxygen content have an R.Q. of more than 1.0. Carbohydrates generally have an R.Q. of 1.0 because they contain carbon and oxygen in equal proportions.

MITOCHONDRIA

The mitochondrion is the organ of respiration in eucaryotic cells; it is therefore worthwhile to examine its structure in relation to its function. In procaryotic organisms, mitochondria are absent and the respiratory enzymes are located along the cell membrane. As procaryotic cells are believed to be the precursors of eucaryotic cells, the structural segregation of important cellular functions seems to have been the springboard of evolution beyond the procaryotic level of organization.

Although extremely variable in shape, size, and number, mitochondria from different sources display a remarkable uniformity in their ultrastructure. Each mitochondrion is bound by a double membrane. The inner membrane, by repeated infoldings, forms membranous finger-like or plate-like projections in the mitochondrial lumen which is filled with a homogeneous matrix of liquid nature (Fig. 8-1). Each projection is called a crista (plural cristae).

The chemical study of the mitochondrial membrane suggests that it consists of lipids and proteins in the ratio of one protein molecule to two lipid molecules. All the oxidative activity of mitochondria is membrane-bound. The inner membrane contains the complete electron transport chain, viz., the enzymes and factors, for coupling the oxidation to the phosphorylation, and a contractile protein. These enzymes were taken to be located within certain knob-like structures along the inner membrane, but recent findings suggest that the knobs may be mere artefacts. Thus, nothing precise is known about the disposition of various enzymes and the electron transport chain components along the mitochondrial membrane. The growing belief, however, is that the inner membrane participates in the electron transport and associated phosphorylation whereas the outer membrane contains certain repetitive structures, each including all the enzymes involved in the oxidations of fatty acids, of the citric acid cycle, and of amino acids.

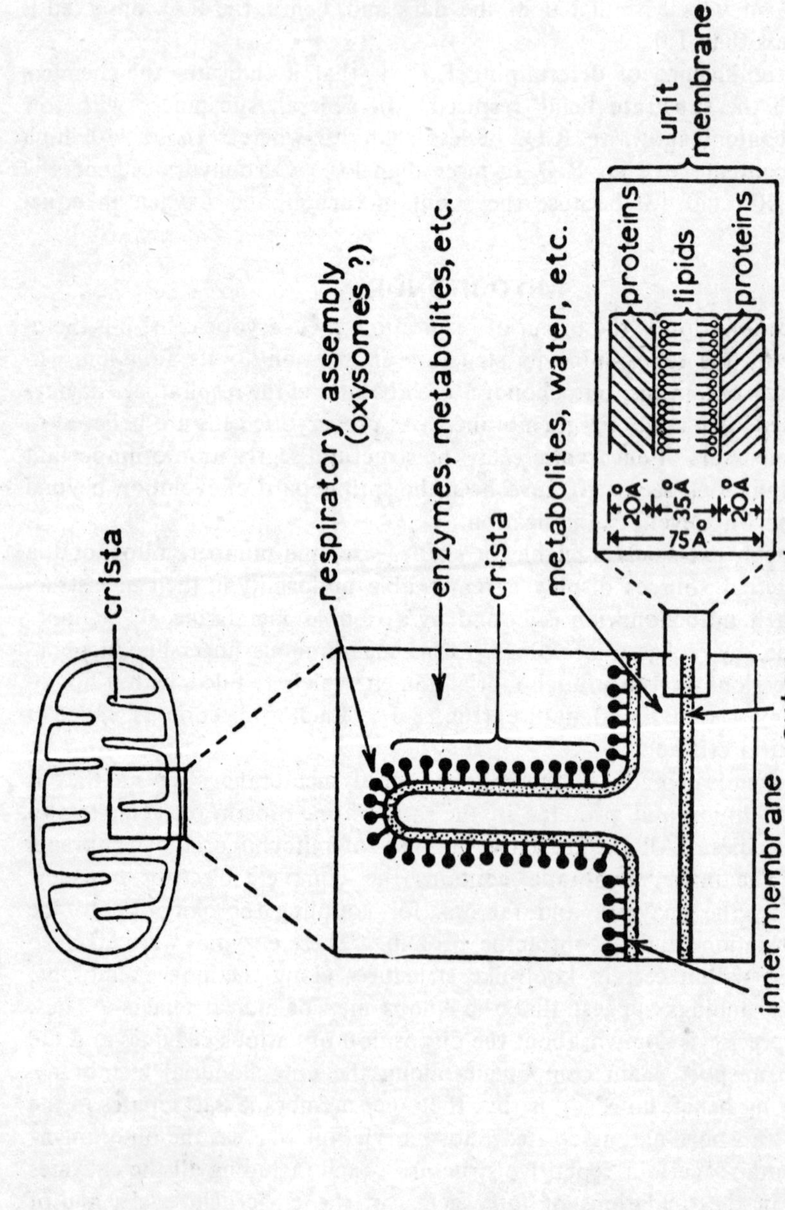

Fig. 8-1 Fine structure of a mitochondrion. (Structure of inner membrane is similar to that of outer.)

It is believed that the change in the permeability of the mitochondrial membrane controls its size and morphology as follows: The membrane expands with increase in permeability because of the presence of a contractile protein. As a consequence of such membrane expansion, the mitochondrion swells and this swelling is assumed to be controlled by the ATP level: High ATP content in other parts of the cell makes the mitochondrion lose water, thus leading to shrinkage of the membrane. This in turn results in the blockage of the passage of its energy and other materials to the cell; when the ATP decreases, the permeability of the membrane to mitochondrial ATP increases and the ATP released then refuels the cell cytoplasm.

BIOLOGY

Like chloroplasts, mitochondria are semiautonomous and divide to give rise to new mitochondria which have their own DNA (histone-free), RNA, and ribosomes (70S type). Mitochondria have mutations, called *petite*, and these mutant mitochondria are incapable of carrying out substrate oxidation. Petite mutants have structurally deformed mitochondria; this means that the structural integrity of a mitochondrion is a prerequisite for its proper functioning. Some functional components of the mitochondrion are coded for by the nuclear genes. One such well-known case is the synthesis of cytochrome-c. The gene determining this enzyme behaves as a Mendelian gene. This enzyme has been shown to be synthesized in the cell cytoplasm before it migrates to the mitochondrion where it forms one of the components of the electron transport chain.

The organization of mitochondrial DNA and mitochondrial protein-synthesizing machinery is very similar to that in bacteria, even though bacteria do not have organized mitochondria as such. This fact and the ability of mitochondria to form new mitochondria have led to the suggestion that they are intracellular symbionts which had existed as free-living procaryotic heterotrophs in the remote past.

FUNCTION

Mitochondria oxidize dicarboxylic and tricarboxylic acids, especially succinate, fumarate, malate, and citrate, and the resulting oxidation products include CO_2, H_2O, and ATP. If the photosynthetic reactions are compared with the oxidative reactions of mitochondria, it becomes apparent that whereas chloroplasts route the environmental CO_2 and H_2O through various cellular substances at the cost of light energy, the mitochondria release them back into the environment through oxidations in which the liberated energy is conserved in newly-formed ATP; this ATP is used for powering the biosynthetic reactions of the cell. It seems appropriate to regard the chloroplast as a source of substrates for the activity of mitochondrial enzymes.

The basic plan of the mechanism by which mitochondria achieve the oxidation of various cellular substances was originally worked out by Krebs and his research team (see James, 1957). Subsequently, workers found the cycle operating in all aerobic organisms and therefore named it Krebs cycle or tricarboxylic acid (TCA) cycle.

Figure 8-2 shows that, in addition to generating ATP, the mitochondrion produces substances that are precursors of fatty acids, amino acids, and sugars. Therefore, the earlier notion that the TCA cycle is a catabolic pathway is not strictly accurate. In recent times, such pathways as glycolysis, pentose shunt, TCA cycle, and glyoxylate cycle, which serve to regenerate ATP and to supply the intermediates for biosynthetic reactions, have been designated as amphibolic. In Krebs cycle, a molecule of acetyl

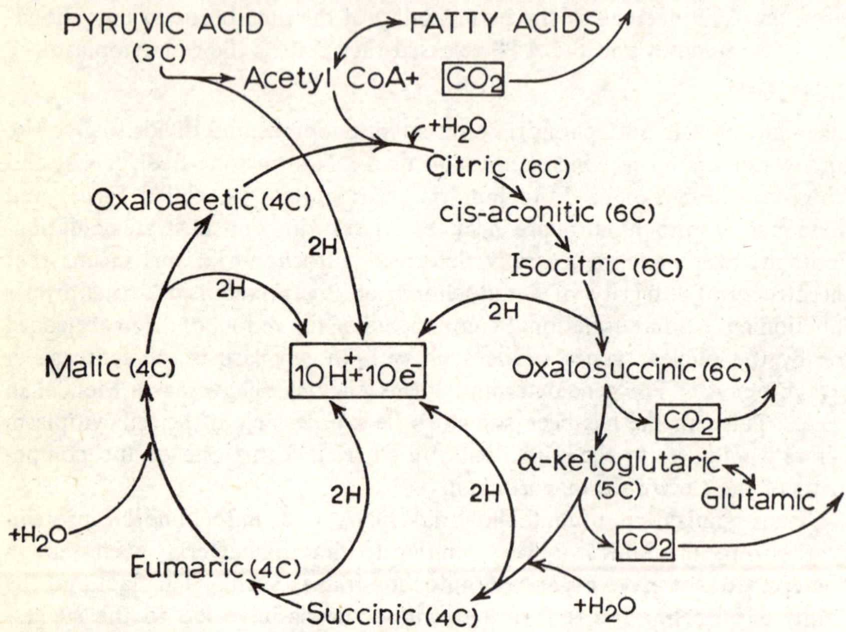

Fig. 8-2 Some chemical transformations and metabolic reactions known to occur in mitochondria.

CoA and a molecule of oxaloacetate react to produce citrate from which oxaloacetate is regenerated through four dehydrogenation steps and two decarboxylations. Thus, the most crucial components for the operation of the cycle are acetyl CoA and oxaloacetate. If oxaloacetate is not consumed in any other biosynthetic reaction, then the only substance required would be acetyl CoA, which comes from β-oxidation of fatty acids or from pyruvate which in turn can result from glycolysis or from the intermediates of the Benson-Calvin cycle or the C_4 cycle or from deamination of alanine. Another important source is the pentose shunt pathway under aerobic atmosphere.

ORIGIN OF PYRUVATE

GLYCOLYSIS

Glycolysis literally means the breaking down of sugar. It generates pyruvate and ATP through a series of enzymatic steps (Fig. 8-3). The fate of pyruvate depends upon the presence of oxygen, and the organism or tissue system involved. Under anaerobic conditions, the pyruvate may be reduced to lactic acid, e.g., during muscular contraction in animals, or it may form ethyl alcohol, as in yeast. Under aerobic conditions, it gets completely oxidized to water and CO_2 through the TCA cycle. Glycolysis, in the absence of oxygen, is known also as anaerobic respiration or fermentation.

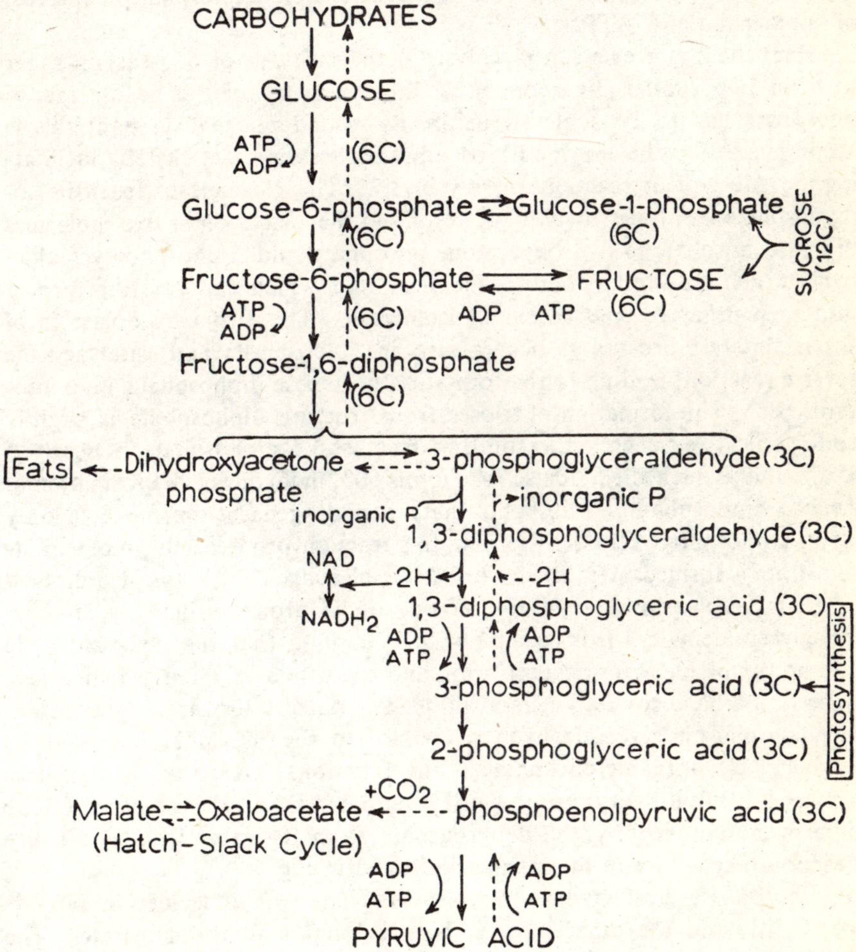

Fig. 8-3 Glycolysis. (Pyruvate formation from carbohydrates.)

Glucose is the substrate of glycolysis and occurs in cell cytoplasm where all its enzymes are known to be present. It can arise from cellular starch or glycogen by hydrolytic reactions or it can be supplied to the cell culture from outside. The cell can absorb glucose only after phosphorylating it with a molecule of ATP. The phosphorylated glucose may enter the glycolytic reaction chain directly. Glucose exists in solution as a six-membered ring configuration, and it is in this form that its phosphorylation with ATP occurs. Since energy conservation through ATP synthesis is regarded as one of the net results of glucose assimilation, a situation arises where a certain amount of energy of the cell must first be expended in order that more of it be produced. Glucose-6-phosphate is then converted into its isomer fructose-6-phosphate, a five-membered ring, by the enzyme hexose phosphate isomerase. Next, fructose-6-phosphate is phosphorylated, under the influence of phosphofructokinase, to fructose-1,6-diphosphate at the cost of another mole of ATP.

Since the major event in glycolysis is the cleavage of a 6-carbon sugar to form two similar 3-carbon units, the symmetry of the hexose monophosphate has to be achieved before its second phosphorylation; this is accomplished by the enzyme hexose phosphate isomerase, thereby facilitating the subsequent reactions of glycolysis. The cleavage of fructose-1,6-diphosphate catalyzed by aldolase results in the formation of two molecules of triosephosphate (dihydroxyacetone phosphate and 3-phosphoglyceraldehyde) which are similar, but not identical, and which can get transformed into each other by the action of isomerase. The enzyme aldolase is of special interest because it occurs also in chloroplasts and catalyzes the reverse reaction, leading to the formation of hexose diphosphate in photosynthesis. The formation of trioses from fructose diphosphate is slightly endergonic (free energy for splitting fructose-1,6-diphosphate is nearly 5 kcal), and as such the aldolase reaction is 90% in favour of the formation of hexose diphosphate. Triosephosphate dehydrogenase enzyme acts only upon 3-phosphoglyceraldehyde and this reaction proceeds the moment its substrate is formed. In other words, the cleavage of fructose-1,6-diphosphate is coupled to the oxidation of one of its products, namely, 3-phosphoglyceraldehyde. It is because of this coupling that the glyceraldehyde has no opportunity for accumulation and the whole reaction which otherwise would have come to a standstill is thus pushed forward. The oxidation of phosphoglyceraldehyde is coupled to the phosphorylation of its carbon-1 by inorganic phosphate. The functional group of the dehydrogenase is a sulph-hydryl group (—SH), and NAD^+ is the acceptor of the hydrogen removed by the dehyrogenase from its substrate. The entire reaction takes place in the manner indicated in Fig. 8-3.

Phosphoglyceraldehyde dehydrogenase binds two molecules of NAD^+, one tightly and the other loosely, both belonging to the substrate. The pair of hydrogen removed during oxidation is passed on first to the tightly bound NAD^+ and then to the loosely bound form which gets dissociated from the enzyme molecule after its reduction. The enzyme diphosphoglycerate kinase makes diphosphoglyceric acid donate its 1-phosphoryl group to ADP; thus, one molecule of ATP is produced per molecule of 3-phosphoglyceraldehyde oxidized. Phosphoglyceromutase catalyzes the subsequent reaction in which 3-phosphoglycerate changes to 2-phosphoglycerate and the ordinary phosphate bond of the latter is transformed into an energy-rich form by the activity of another enzyme, called enolase, which removes a molecule of water from 2-phosphoglycerate. In this way, the most energetically active compound, phosphoenolpyruvate, is produced.

Phosphopyruvate kinase transfers the phosphate group of phosphoenolpyruvate to ADP and, ultimately, one molecule of ATP is produced from each molecule of phosphoenolpyruvate dephosphorylated.

Thus, in atmosphere containing oxygen, the reactions of glycolysis are completed with the production of two molecules of pyruvate for each hexose molecule degraded. It is possible to estimate the energetic efficiency of the process upto the level of pyruvate. During complete combustion of a glucose molecule, nearly 673 kcal of energy is liberated, and during that of two molecules of pyruvate, almost 546 kcal of energy is set free. This means 127 kcal of energy escapes in the glycolytic pathway. As a molecule of diphosphoglycerate gives rise to two molecules of ATP (one at the step between di- and monoglycerophosphate, and the other during the formation of pyruvate from phosphoenolpyruvate) and since two molecules of diphosphoglycerate result from a molecule of fructose diphosphate, the total ATP formation amounts to four molecules. As two molecules of ATP are consumed during the conversion of glucose into fructose diphosphate, the net gain in the process is these two ATP molecules. Besides, the formation of a molecule of NADH + H^+ from NAD^+ during the oxidation of phosphoglyceraldehyde involves an expenditure of 52 kcal of energy. Since two molecules of NADH + H^+ are produced with the oxidation of two molecules of phosphoglyceraldehyde, the net storage energy in this reaction totals 104 kcal. The formation of eight molecules of ATP from ADP expends nearly 7 kcal/mole of energy. The additional energy storage in terms of ATP is nearly 14 kcal. Thus, nearly 118 kcal of the total 127 kcal liberated during glycolysis is readily conserved in two compounds, namely, ATP and NADH + H^+. Perhaps the remaining energy is lost in the form of heat.

In the absence of oxygen, pyruvic acid is reduced to lactic acid with the reductant NADH + H^+ which is produced during the oxidation of phosphoglyceraldehyde. The enzyme lactic dehydrogenase mediates the transfer of hydrogen from NADH + H^+ to pyruvate. The formation of ethyl alcohol under similar conditions is supported by two enzymes, pyruvic decarboxylase, which converts pyruvate into acetaldehyde, and alcohol dehydrogenase, which, with the hydrogen from NADH + H^+, reduces aldehyde to alcohol. The formation of lactic acid and alcohol occurs in different organisms. Energetically, such formation seems wasteful, firstly, because neither lactic acid nor alcohol is further used biologically and, secondly, because the energy stored in NADH + H^+ is also lost as it is used up in their formation. Consequently, the energy yield from one molecule of glucose under anaerobic conditions is only two moles of ATP or 14 kcal.

Glycolysis (in aerobic organisms) under anaerobic conditions is therefore definitely a very inefficient process. The energy made available through this process is just sufficient to keep such aerobic organisms alive or at the most slow growing. (However, it should be understood that those organisms that are normally anaerobic often grow quite rapidly.) Petite mutants of yeast, which lack mitochondrial or aerobic respiration, grow very slowly, possibly because they have only glycolysis to serve the need of ATP. Intermediates of glycolysis, such as phosphoenolpyruvate, are used as the CO_2-acceptor molecule in C_4 plants. Also, pyruvate can give rise to alanine by transamination reactions. Thus, glycolysis generates both ATP and the intermediates required in subsequent biosynthetic reactions.

PENTOSE SHUNT PATHWAY

Plants also possess another pathway for degrading glucose to pyruvate. This is called the pentose shunt pathway as it involves a number of pentose sugars and generates pyruvate by short-circuiting the glycolysis. It operates strictly under aerobic conditions (Fig. 8-4). The enzymes in this pathway are, like the glycolytic enzymes, located in the cytoplasm. As in glycolysis, the first step here too involves the phosphorylation of glucose in the carbon-6 position by the enzyme hexokinase. Soon, glucose-6-phosphate dehydrogenase oxidizes its substrate to 6-phosphogluconate and reduces $NADP^+$ to $NADPH + H^+$. The phosphogluconate dehydrogenase causes oxidative decarboxylation of the phosphogluconate, and a pentose phosphate is thus produced from glucose-6-phosphate with the liberation of a CO_2 molecule and the formation of two molecules of $NADPH + H^+$. The reaction sequence is shown in Fig. 8-4. The product of phosphogluconate dehydrogenase action is ribulose-5-phosphate which changes to ribose-5-phosphate, thus forming a molecule of sedoheptulose-7-phosphate and another of 3-phosphoglyceraldehyde. The enzyme transaldolase converts a molecule of sedoheptulose phosphate, in reaction with a molecule of xylulose-5-phosphate, into two molecules of fructose-6-phosphate. The hexose phosphate isomerase converts fructose-6-phosphate into glucose-6-phosphate.

In the pentose shunt pathway, two of the three molecules of glucose phosphate are regenerated; the third is used up in the formation of six molecules of $NADPH + H^+$, three molecules of CO_2, and one molecule of a 3-carbon compound, 3-phosphoglyceraldehyde. The formation of pyruvate from glyceraldehyde follows the glycolytic pathway. In between the reactions from phosphoglyceraldehyde to pyruvate, two molecules of ATP and one molecule of $NADH + H^+$ are formed. The net energy conservation in the form of ATP in the pentose shunt pathway is 7 kcal or one molecule of ATP. However, energy is also stored during the formation of the reductants $NADH + H^+$ and $NADPH + H^+$ (52 kcal). Seven molecules of the reductant (6 $NADPH + H^+$ and 1 $NADH + H^+$) are produced during the transformation of three glucose phosphate molecules to two molecules of glucose phosphate, three molecules of CO_2, and one pyruvate in this pathway, the energy stored in this form amounting to ($52 \times 7 =$) 364 kcal. Thus, the total energy stored in the form of ATP and reductant is nearly ($364 + 7 =$) 371 kcal. Three complete turns of the cycle make a net expenditure of one glucose molecule to produce three molecules of CO_2 and one pyruvate. The free energy of glucose is nearly 673 kcal. The total amount of energy generated during the shunt pathway would then be 410 kcal; of this total, 371 kcal gets stored, thereby suggesting a 90% efficiency for the process. In

$$\text{Glucose-6-phosphate} + \text{NADP}^+ \underset{}{\overset{\text{Glucose phosphate dehydrogenase}}{\rightleftharpoons}} \text{6-phosphoglucono-}\delta\text{-lactone} + \text{NADPH} + \text{H}^+$$

$$\text{6-phosphoglucono-}\delta\text{-lactone} + \text{H}_2\text{O} \xrightarrow[\text{Mg}^{++}]{\text{Lactonase}} \text{6-phosphogluconate}$$

$$\text{6-phosphogluconate} + \text{NADP}^+ \longrightarrow \text{3-keto-6-phosphohexonic acid} + \text{NADPH} + \text{H}^+$$

$$\text{3-keto-6-phosphohexonic acid} \xrightarrow{\text{Dehydrogenase}} \text{ribulose-5-phosphate} + \text{CO}_2$$

$$\text{Ribulose-5-phosphate} \rightleftharpoons \text{Ribose-5-phosphate}$$

$$\text{Ribulose-5-phosphate} \rightleftharpoons \text{Xylulose-5-phosphate}$$

$$\text{Ribulose-5-phosphate} \underset{\text{ATP} \to \text{ADP}}{\rightleftharpoons} \text{Ribulose-1,5-diphosphate}$$

$$\text{Ribose-5-phosphate} + \text{Xylulose-5-phosphate} \overset{\text{Transketolase}}{\rightleftharpoons} \text{Sedoheptulose-7-phosphate} + \text{Glyceraldehyde-3-phosphate}$$

$$\text{Sedoheptulose-7-phosphate} + \text{Glyceraldehyde-3-phosphate} \overset{\text{Transaldolase}}{\rightleftharpoons} \text{Fructose-6-phosphate} + \text{Erythrose-4-phosphate}$$

$$\text{Fructose-6-phosphate} \rightleftharpoons \text{Glucose-6-phosphate}$$

$$\text{Fructose-6-phosphate} \xrightarrow[\text{ATP} \quad \text{ADP}]{} \text{Fructose-1,6-diphosphate}$$

Fig. 8-4 The oxidative pentose shunt pathway. (Hexose monophosphate shunt.)

comparison, glycolysis is slightly more efficient as it conserves nearly 93% of glucose energy.

C_3 CYCLE

Phosphoglyceric acid is the first stable product of photosynthetic CO_2 fixation through the C_3 cycle which occurs in chloroplasts and which involves pentose phosphates, hexose phosphates, triosephosphates, and sedoheptulose phosphates as intermediates. Except for dehydrogenases, the remaining enzymes of the pentose shunt pathway operating in the cytoplasm have been noticed in chloroplasts as well. The photosynthetic PGA can form pyruvate in three steps involving phosphoglycerate mutase, enolase, and pyruvate kinase. However, for this process, PGA has to move to the cell cytoplasm where the enzymes are located.

C_4 CYCLE

The enzyme phosphoenolpyruvate carboxylase carboxylates its substrate to oxaloacetate which gets reduced to malate by malic dehydrogenase at the cost of photosynthetic $NADPH + H^+$. C_4 cells are found to have malic enzyme which converts malate into pyruvate by successive dehydrogenation and decarboxylation. The pyruvate so produced may be recycled to regenerate phosphoenolpyruvate or may escape to the mitochondria through the cell cytoplasm (Zelitch, 1971). It is now generally agreed that the initial carboxylation yielding oxaloacetate occurs in the cytoplasm rather than the chloroplasts.

DEAMINATION OF ALANINE

Another minor pathway is the oxidative deamination of alanine which forms ammonia and pyruvate (or the transamination reaction between alanine and alpha-ketoglutaric acid, catalyzed by the enzyme transaminase). Two hydrogen atoms are probably removed to produce an imino acid intermediate which subsequently undergoes hydrolysis, producing NH_3 and the corresponding alpha-keto acid. The oxidative deamination proceeds under aerobic conditions and the hydrogen removed in the process goes first to NAD^+ and finally to oxygen via the respiratory cytochromes.

ACETYL CoA FORMATION

By oxidative decarboxylation in the presence of CoA pyruvate forms an active acetate compound called acetyl CoA. The entire reaction is mediated by the enzyme pyruvate dehydrogenase and the cofactors NAD^+, diphosphothiamine, TPP, and lipoic acid. The substrate is dehydrogenated and the free hydrogen is passed on to NAD^+. TPP acts as the decarboxylating agent, and lipoic acid as a carrier of the acetyl group from TPP to CoA. The functional group of lipoic acid is two sulph-hydryl groups and that of CoA is one sulph-hydryl group. It accepts the decarboxylated pyruvate from TPP in its oxidized state and, during the transfer of the accepted molecule to CoA, itself gets reduced. It is reoxidized by NAD^+

in the presence of the dehydrogenase enzyme and thus becomes ready to repeat the cycle. The reactions are indicated in Fig. 8-5.

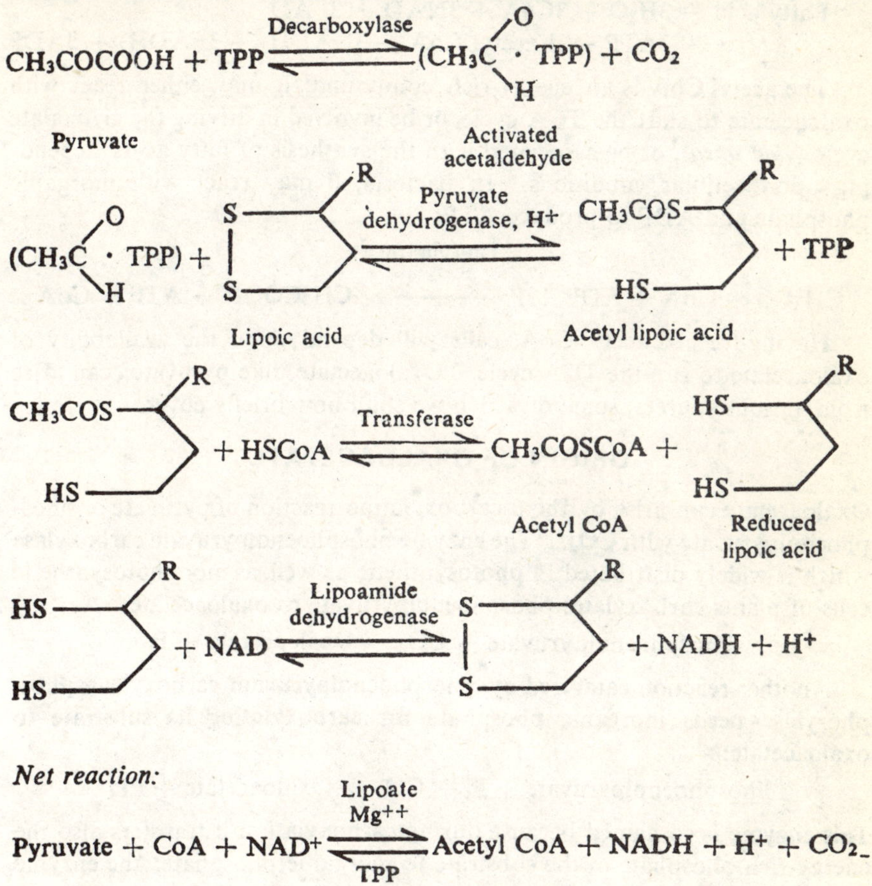

Fig. 8-5 Acetyl CoA formation.

Acetyl CoA can arise also from fatty acids (Fig. 8-6) as follows:

Fatty acid (chain length X) + CoA + NAD + H_2O
+ FAD + ATP → Acetyl CoA + Fatty acid (chain length X − 2)
+ $NADH_2$ + $FADH_2$ + ADP

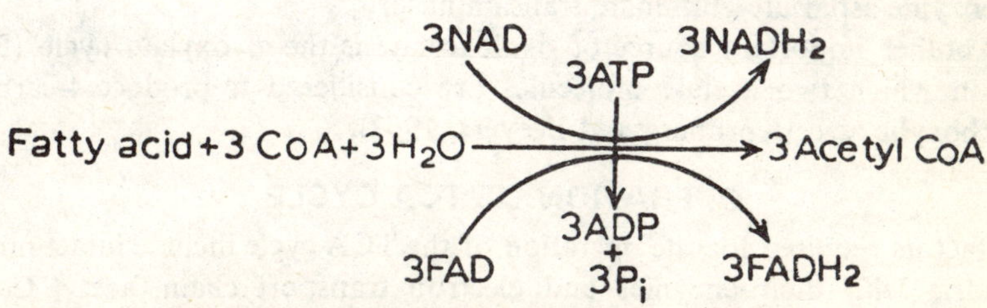

Fig. 8-6 Acetyl CoA formation from a fatty acid.

Specifically, suppose that the fatty acid $CH_3\cdot(CH_2)_5\cdot COOH$ is used in the foregoing kind of reaction. Then two carbon atoms are simultaneously cleaved off from the fatty acid and attached to CoA, and the conversion of this fatty acid into acetyl CoA may be summarized as follows:

$$\text{Fatty acid} + 3H_2O + 3CoA + 3NAD + 3FAD + 3ATP \rightarrow 3\ \text{acetyl CoA} + 3NADH_2 + 3FADH_2 + 3ADP$$

The acetyl CoA is an energy-rich compound; it may either react with oxaloacetate to start the TCA cycle, or be involved in driving the glyoxalate cycle (*vide infra*), or be associated with the synthesis of fatty acids depending upon cellular conditions. In bacteria, it may react with inorganic phosphate and ADP to produce ATP:

$$CH_3CO \sim CoA + ADP + P_i \xrightarrow{\text{Phosphotrans-acetylase}} CH_3COOH + ATP + CoA$$

Having formed acetyl CoA, cells will depend upon the availability of oxaloacetate to run the TCA cycle. Oxaloacetate, like pyruvate, can arise from various sources, some of which we shall now briefly cover.

ORIGIN OF OXALOACETATE

Oxaloacetate can arise by the β-carboxylation reaction of pyruvate or phosphoenolpyruvate with CO_2. The enzyme phosphoenolpyruvate carboxylase which is widely distributed in photosynthetic as well as nonphotosynthetic cells of plants carboxylates phosphoenolpyruvate to oxaloacetate:

$$\text{Phosphoenolpyruvate} + CO_2 \rightarrow \text{Oxaloacetate} + P_i$$

Another reaction catalyzed by phosphoenolpyruvate carboxytransphosphorylase needs inorganic phosphate for carboxylating its substrate to oxaloacetate:

$$\text{Phosphoenolpyruvate} + P_i + CO_2 \rightarrow \text{Oxaloacetate} + PP_i$$

This enzyme is so named because during carboxylation it transfers also the energy-rich phosphate of the substrate to inorganic phosphate; the enzyme pyruvate carboxylase which is also very common in cells makes pyruvate and CO_2 react in the presence of ATP to produce oxaloacetate:

$$\text{Pyruvate} + ATP + CO_2 \xrightarrow{\text{Carboxylase}} \text{Oxaloacetate} + ADP$$

Oxaloacetate can also arise from aspartate by transamination, involving the enzyme aspartate glutamate transaminase.

Another important source of oxaloacetate is the glyoxylate cycle (Fig. 8-7) in which two acetate molecules are considered to produce 4-carbon dicarboxylic acid (Kornberg and Beevers, 1957).

OPERATION OF TCA CYCLE

The factors required for the operation of the TCA cycle include intact mitochondria with their enzymes and electron transport chain, acetyl CoA, oxaloacetate, and oxygen. Given these, the cycle proceeds with the condensation of oxaloacetate and acetyl CoA to form citrate, followed by the loss of a molecule of water to produce aconitate (Varner, 1971). The latter uses a molecule of water to form isocitrate. The other reactions involved, along with the enzymes concerned, are shown in Fig. 8-2. Ultimately, oxaloacetate is regenerated and, provided acetyl CoA is available, initiates the cycle again.

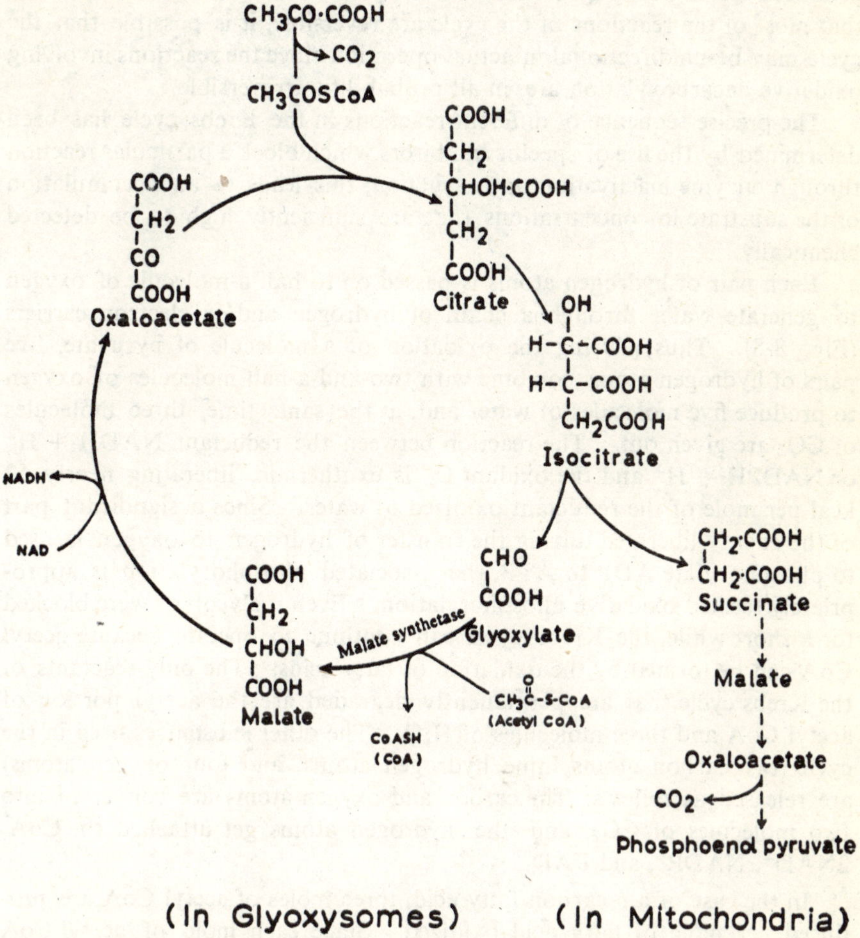

Fig. 8-7 Glyoxylate cycle.

Four pairs of hydrogen atoms are transferred to the appropriate acceptor molecules (Fig. 8-2) in reactions catalyzed by isocitric dehydrogenase, α-ketoglutaric dehydrogenase, succinic dehydrogenase, and malic dehydrogenase. The hydrogen acceptor of an isocitric dehydrogenase reaction is $NADP^+$, that of a succinic dehydrogenase reaction is a flavoprotein, and that of ketoglutaric and malic dehydrogenase reactions is NAD^+.

Although the formation of acetyl CoA from pyruvate is not one of the reactions of the Krebs cycle, it occurs in mitochondria since, in this reaction too, a pair of hydrogen atoms is removed to NAD^+ by the enzyme pyruvic dehydrogenase. The total number of pairs of hydrogen atoms produced during a complete turn of the cycle is five (see Fig. 8-2). The cycle also evolves two molecules of CO_2 during complete oxidation of acetyl CoA at steps between isocitrate and ketoglutarate and those between ketoglutarate and succinate. With pyruvate, an extra molecule of CO_2 is produced during its conversion into acetyl CoA. Although it is certain that most of the reactions of the cycle are reversible, it is possiblc that the cycle may be unidirectional in actual operation since the reactions involving oxidative decarboxylation are, in all probability, irreversible.

The precise sequence of different reactions in the Krebs cycle has been determined by the use of specific inhibitors which block a particular reaction through enzyme inactivation or destruction; this leads to an accumulation of the substrate in concentrations that are sufficiently high to be detected chemically.

Each pair of hydrogen atoms is passed on to half a molecule of oxygen to generate water through a chain of hydrogen and/or electron carriers (Fig. 8-8). Thus, during the oxidation of a molecule of pyruvate, five pairs of hydrogen atoms combine with two-and-a-half molecules of oxygen to produce five molecules of water and, at the same time, three molecules of CO_2 are given out. The reaction between the reductant $NADH + H^+$ or $NADPH + H^+$ and the oxidant O_2 is exothermic, liberating nearly 52 kcal per mole of the reductant oxidized to water. Since a significant part of the energy liberated during the transfer of hydrogen to oxygen is used to phosphorylate ADP to ATP, the associated phosphorylation is appropriately termed oxidative phosphorylation. Even if glycolysis were blocked for a short while, the Krebs cycle can continue to operate because acetyl CoA can be formed by the oxidation of fatty acids. The only reactants of the Krebs cycle that are permanently degraded are the acetyl portion of acetyl CoA and three molecules of H_2O. The other substances used in the cycle (two carbon atoms, nine hydrogen atoms, and four oxygen atoms) are released as follows: The carbon and oxygen atoms are converted into two molecules of CO_2, and the hydrogen atoms get attached to CoA, $2NAD^+$, $NADP^+$, and FAD.

In the case of a 6-carbon fatty acid, three moles of acetyl CoA are produced per mole of fatty acid oxidized. Since each mole of acetyl CoA degraded via the Krebs cycle yields two moles of $NADH + H^+$ plus one mole of $NADPH + H^+$ plus one mole of $FADH_2$, a total of six moles of $NADH + H^+$, i.e., three of $NADPH + H^+$ and three of $FADH_2$, are generated for one mole of fatty acid used. In addition, three moles of $NADH + H^+$ and three moles of $FADH_2$ are produced when the fatty acid is converted into acetyl CoA. The total number of reduced hydrogen acceptors is nine moles of $NADH + H^+$ plus three moles of $NADPH + H^+$ plus six moles of $FADH_2$. Since each mole of $NADPH + H^+$ and of $NADH + H^+$ regenerates three moles of ATP, and each mole of FAD regenerates two moles of ATP, the total amount of ATP produced during the oxidation of the fatty acid via the respiratory electron transport chain is 48 moles. This may be compared with the catabolization of a

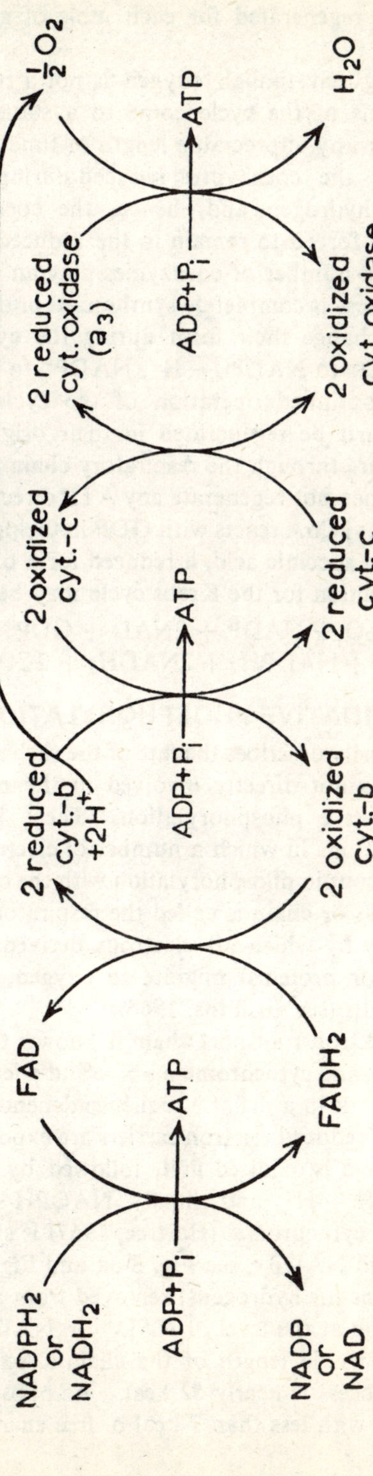

Fig. 8-8 Respiratory electron transport chain.

mole of glucose: 38 moles of nucleoside triphosphate, instead of 48 moles for the fatty acid, are regenerated for each mole of glucose broken down to $6CO_2 + 6H_2O$.

It is also interesting that though oxygen is not a reactant in the Krebs cycle yet the reactions of the cycle come to a standstill if the cells are deprived of oxygen for any appreciable length of time. This is because in anaerobic conditions, the coenzymes reduced during the Krebs cycle are unable to lose their hydrogen and, hence, the coenzyme pool (NAD^+, $NADP^+$, and FAD) is forced to remain in the reduced form.

Further, although a number of coenzymes play an important role in the Krebs cycle, none of these is completely synthesized or degraded. However, some coenzymes do change their form during the cycle, e.g., acetyl CoA changes to CoA, NAD^+ to $NADH + H^+$, $NADP^+$ to $NADPH + H^+$, and FAD to $FADH_2$. Continued operation of the cycle requires that such coenzymes must in turn be regenerated in their original (oxidized) form. This regeneration occurs through the respiratory chain and ATP formation.

The Krebs cycle does not regenerate any ATP directly, but it regenerates some GTP when succinyl CoA reacts with GDP in the presence of phosphoric acid and water to yield succinic acid, a reduced form of CoA and GTP.

The net overall reaction for the Krebs cycle may be written as follows:

$$\text{Acetyl CoA} + 2H_2O + \text{NADP} + 2\text{NAD} + \text{GDP} + P_i + \text{FAD} \rightarrow \text{CoA} + \text{NADPH}_2 + 2\text{NADH}_2 + 2CO_2 + \text{GTP} + \text{FADH}_2$$

OXIDATIVE PHOSPHORYLATION

The Krebs cycle primarily describes the fate of the carbon skeleton of pyruvic or acetic acid, but it is not directly involved in the mechanism of energy conservation or oxidative phosphorylation. There is another series of enzyme-mediated reactions in which a number of electron carriers and other factors participate to couple phosphorylation with the oxidation of hydrogen by oxygen. This series or chain is called the respiratory chain and constitutes the final pathway by which all electrons derived from different fuels (carbohydrates, fats, or proteins) migrate to oxygen, the final acceptor of electrons in aerobic cells (see Griffiths, 1965).

The respiratory electron transport chain is known to contain NAD^+ or $NADP^+$, flavoproteins, and cytochrome-a, -b, -c and a few other cytochromes. These are known to work in a definite sequence depending upon their redox potential. If the fully reduced electron carriers are exposed to an atmosphere of oxygen, cytochrome-a is oxidized first, followed by cytochrome-c, cytochrome-b, FP, $NADH + H^+$, and, finally, $NADPH + H^+$. The general chemical structure of cytochromes (Hartree, 1957) is shown in Fig. 8-9 (for structure of NAD^+ and $NADP^+$, see Fig. 3-3a and Fig. 3-3b).

When two electrons (or hydrogens) removed from a substrate enter the electron transport chain at the level of NAD^+ or $NADP^+$ and, eventually, are passed along the whole length of the chain to oxygen, the free energy released during the process is nearly 52 kcal. As phosphorylation of ADP to ATP cannot occur with less than 7 kcal of free energy, the sites for ATP synthesis would be those that liberate more than 7 kcal of energy during electron transfer. One such step is between NAD^+ and flavoprotein, another

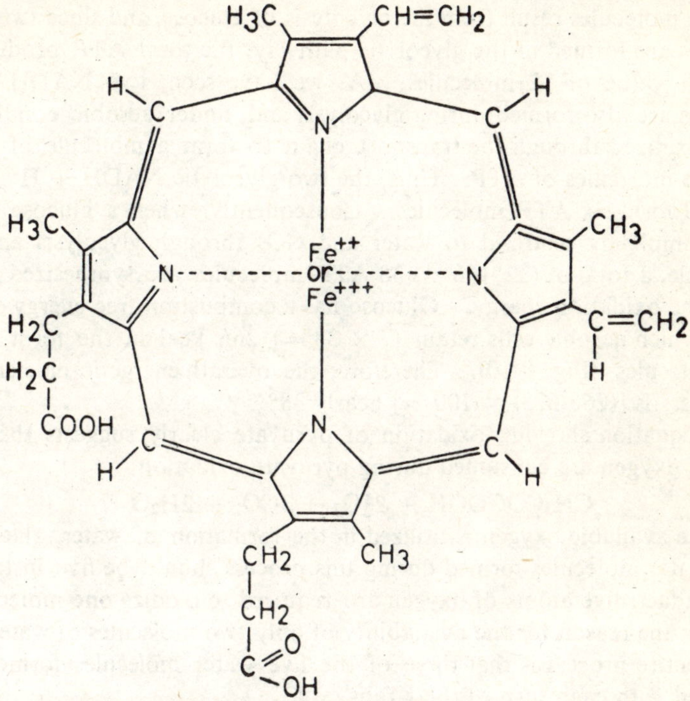

Fig. 8-9 Nonproteinaceous heme portion common to all cytochromes.

between cytochrome-b and cytochrome-c, and a third between cytochrome-a and oxygen (see Fig. 8-8).

Thus, it may be estimated that 3 moles of ATP will be formed per pair of electrons transferred to oxygen. This has been experimentally shown to be the case. In all dehydrogenation reactions, electrons enter the chain at NAD^+ or $NADP^+$ level, except in the case of succinate dehydrogenase which employs coenzyme-Q as the initial acceptor of its electrons. As coenzyme-Q joins the electron transport chain at the level of cytochrome-b and, as the potential gap between the coenzyme and the cytochrome is not sufficient to form a molecule of ATP, the only phosphorylation sites remaining in this process are those between cytochrome-b and cytochrome-c, and between cytochrome and oxygen. In fact, only two molecules of ATP are synthesized during the oxidation of succinate with oxygen. However, since the formation of succinate from succinyl CoA yields a molecule of ATP, it may be said that the formation of fumarate from α-ketoglutarate generates three molecules of ATP as does the formation of acetyl CoA from pyruvate or of α-ketoglutarate from isocitrate. Thus, in total energy economy, every oxidation reaction of the TCA cycle has the same potential.

The actual estimation of the production of the total number of ATP molecules during complete oxidation of a pyruvate molecule under aerobic conditions is placed at 15; this agrees well with the concept of three ATP molecules produced per pair of electrons transported to oxygen. Since two pyruvate molecules result from the glycolysis of glucose, and since two ATP molecules are formed in the glycolytic pathway, the total ATP production reaches a value of 32 molecules. As we have seen, two $NADH + H^+$ molecules are also formed during glycolysis and, under aerobic conditions, each is oxidized through the transport chain to form a molecule of water and three molecules of ATP. Thus, the two glycolytic $NADH + H^+$ molecules will form six ATP molecules. Consequently, when a glucose molecule is completely oxidized to water and CO_2 through glycolysis and the TCA cycle, a total of (32 + 6 =) 38 ATP molecules are synthesized at the cost of the oxidation energy. Glucose has a combustion-free energy of 673 kcal of which aerobic cells retain (7 × 38 =) 266 kcal in the form of 38 ATP molecules (Fig. 8-10). Therefore, the overall energetic efficiency of the process is [(266/673) × 100 =] nearly 38%.

The equation showing oxidation of pyruvate clearly suggests that five atoms of oxygen are consumed during pyruvate oxidation:

$$CH_3COCOOH + 2\tfrac{1}{2}O_2 \rightarrow 3CO_2 + 2H_2O$$

As all the available oxygen is utilized in the formation of water, the number of water molecules formed during this process should be five instead of two. In fact, five atoms of oxygen are required to oxidize one molecule of pyruvate; the reason for the availability of only two molecules of water during the entire process is that three of the five water molecules formed are consumed with each turn of the Krebs cycle.

TCA CYCLE AS A METABOLIC POOL

A critical evaluation of the TCA cycle and glycolysis in aerobic atmosphere would reveal that glycolysis is merely a source of carbon skeletons or fuel to mitochondria as it yields only two molecules of ATP as compared with 36 molecules of ATP produced by mitochondria. Since the source of oxidative ATP is reduced NADP or NAD, any reaction that utilizes the reduced coenzymes for biosynthesis will limit the amount of oxidative ATP. The major biosynthetic reactions consuming $NADH + H^+$ or $NADPH + H^+$ involve the formation of alpha-amino acids, such as glutamic acid, aspartic acid, and alanine by reductive amination, and the synthesis of fatty acids through the reversal of β-oxidation. The cells seem to achieve the requisite partitioning of the reduced coenzyme by molecular controls whose precise nature is unknown.

In addition, the cycle is a potential source of raw materials for the formation of polysaccharides, fatty acids, and amino acids (Fig. 8-11). For instance, if a large amount of radioactive succinate is added to an actively metabolizing organism, only the catalytic amount remains in the cycle; the rest finds its way mostly into glycogen and fatty acids. The explanation is that most of the added succinate is converted into oxaloacetate which yields phosphoenolpyruvate (or pyruvate + ATP) through enzyme action. Phos-

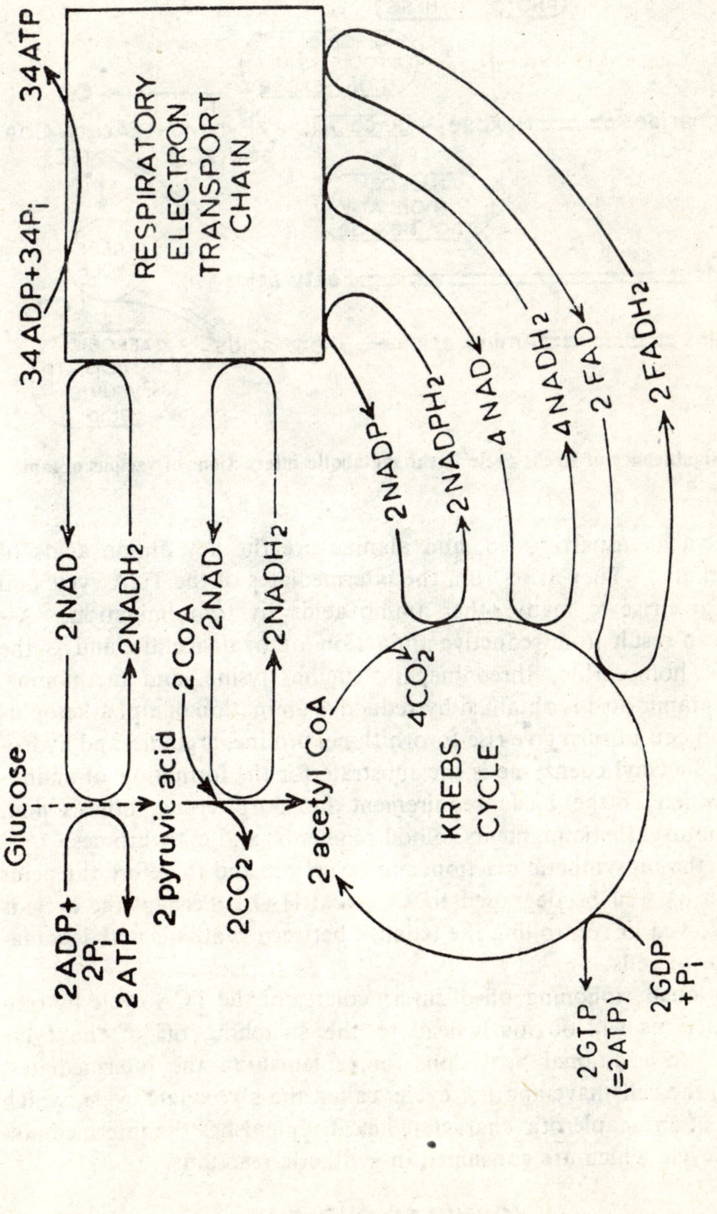

(Total ATP produced = 34+2+2 = 38)

Fig. 8-10 ATP production during respiration.

phoenolpyruvate is routed through glycolysis to form monosaccharides and polysaccharides. Pyruvate generates fatty acids through its oxidative decarboxylation product acetyl CoA.

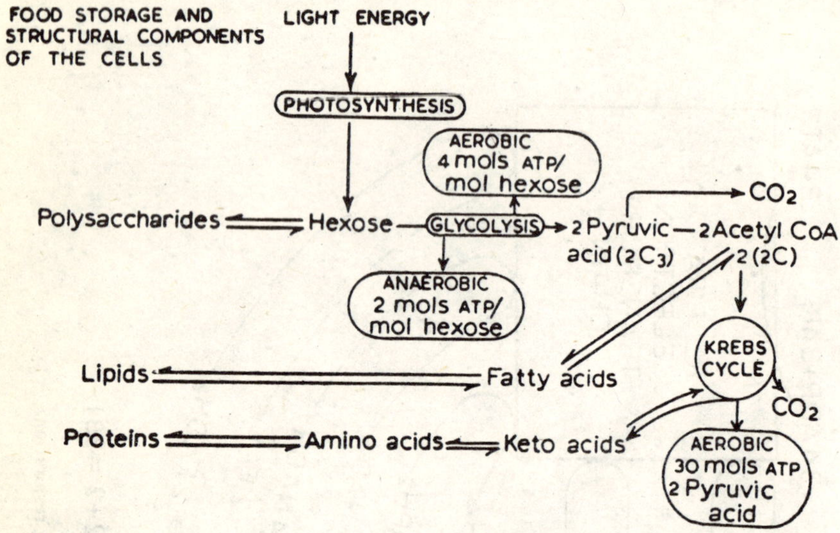

Fig. 8-11 Significance of Krebs cycle in the metabolic interactions of various organic substrates.

Glutamic acid, aspartic acid, and alanine are the key amino acids of protein synthesis. They arise from the intermediates of the TCA cycle and can in turn give rise to many other amino acids by transamination. Aspartic acid can result from reductive amination of oxaloacetate, and is the precursor of homoserine, threonine, isoleucine, lysine, and methionine. Similarly, glutamic acid is obtained by reductive amination of alpha-ketoglutaric acid and can in turn give rise to ornithine, proline, arginine, and hydroxyproline. Succinyl coenzyme is the substrate for the formation of aminolevulinate which is the basic requirement for porphyrin synthesis and, hence, of photosynthetic pigments, blood pigments, and cytochromes.

Most of the biosynthetic reactions are reversible and therefore the acids involved can as well be degraded to CO_2 and H_2O. Perhaps the cycle is basically involved in controlling the balance between synthetic and degradative reactions of cells.

The continued siphoning off of intermediates of the TCA cycle by biosynthetic reactions will obviously lead to the switching off of the cycle unless there are additional provisions for replenishing the intermediates. Fortunately, the cells have another cycle, called the glyoxylate cycle, which is primarily of an anaplerotic character, i.e., it replenishes the intermediates of the TCA cycle which are consumed in synthetic reactions.

GLYOXYLATE CYCLE

The observation that bacteria, when grown on labelled acetate, mainly synthesize labelled citrate and malate, or that fatty seeds, when germinating, show a significant decrease in lipid content with a concomitant rise in sucrose content led to the discovery of the glyoxylate cycle (see Fig. 8-7) by Kornberg and Beevers in 1957. The two key enzymes of this pathway are malate synthetase which condenses acetate and glyoxylate to malate, and isocitratelyase which fragments isocitrate into glyoxylate and succinate. The isocitrate arises from citrate (Krebs cycle).

No CO_2 is produced in the glyoxylate cycle. On the other hand, the cycle has the provision for the synthesis of ATP through the formation of NADPH + H^+ or of NADH + H^+. During each turn of the cycle, two acetate molecules disappear and one mole of succinate reappears; this mole can go to the TCA cycle for use in polysaccharide synthesis. The malate and oxaloacetate produced through this cycle can then be utilized in the Krebs cycle. Glyoxylate is also the precursor of two important amino acids, glycine and serine; the synthesis of these two amino acids in heterotrophic organisms depends perhaps upon the glyoxylate cycle. In fact, the glyoxylate cycle also seems to link carbohydrate, lipid, and protein metabolisms.

DICARBOXYLIC ACID CYCLE

Yet another cycle, whose main function is to provide energy at the expense of glyoxylate oxidations, is found to operate in microorganisms. This cycle, also discovered by Kornberg, is named the dicarboxylic acid cycle as it involves two dicarboxylic acids of the tricarboxylic acid cycle, viz., oxaloacetic acid and malic acid. The cycle takes up a molecule of glyoxylate and degrades it completely to CO_2 and H_2O. Therefore, the potential use of this cycle is mainly for generating oxidative ATP.

REGULATION OF CARBON PATH IN RESPIRATION

Heterotrophs derive their nucleic acids, protein, fat, and carbohydrate carbons from the assimilation of CO_2 or organic compounds, such as glucose, pyruvate, and acetate. Even the amino acids can serve as a source of carbon and nitrogen in several organisms. Phototrophs synthesize all their cellular carbon from photosynthetic CO_2 assimilation. The pentose shunt pathway, glycolysis, TCA cycle, and glyoxylate cycle are the routes for the partitioning of the external carbon source into various cellular substances. In order to achieve proper partitioning, cells obviously possess subtle mechanisms, some of which have been revealed by researches carried out during the past 10 years.

Respiratory pathways are the routes by which interconversion of the carbon of fats, proteins, and carbohydrates takes place. These metabolic routes serve two functions, viz., the formation of ATP, and the generation of intermediates which can be channelled to proteins, fats, or carbohydrates, or to CO_2 and H_2O. For example, phosphoenolpyruvate may follow the reverse path of glycolysis to generate glucose, may be acted upon by its carboxylase to form oxaloacetate, or pyruvate kinase and pyruvate dehydrogenase, and may convert it into acetyl CoA, thus finally degrading it to

CO_2 and H_2O. The course actually followed at any particular time is determined primarily by the energy budget of the cells; with low ATP (or high ADP or AMP) it is degraded through the TCA cycle to generate more ATP, but if the ATP content is relatively high, then it may follow one or the other of the synthetic routes or both the routes.

The means by which cells achieve a proper partitioning of key intermediates into synthetic and degradative pathways are problems that are just being clarified in some detail. We shall now briefly summarize the various control mechanisms known to adjust the entire intermediary metabolism of cells in a most economical manner.

As already explained, carbohydrates, proteins, and fats, all can be degraded to the components of glycolysis, the TCA cycle, and the glyoxylate cycle. Also, the components of these cycles can go to form any one of these substances, or two, or all three of them. This means that the metabolic routes leading to their synthesis or degradation are reversible although a few of the key reactions, especially those of oxidative decarboxylations, are considered irreversible. But cells have evolved mechanisms to overcome this irreversibility by maintaining carboxylases which recapture the CO_2 produced and replace it in the channel of intermediary metabolism.

CONTROLS OF GLYCOLYSIS, TCA CYCLE, AND GLYOXYLATE CYCLE

ATP seems to be involved in the controls of intermediary metabolism. A century ago, Louis Pasteur had observed the inhibition of glycolysis by oxygen. This so-called Pasteur effect remained a mystery for a long time. The cells degrading glucose for energy were observed to consume more of glucose under anaerobic than aerobic conditions, implying that glycolysis is inhibited by oxygen.

Another explanation of the Pasteur effect may be with regard to the availability of inorganic phosphate for triosephosphate dehydrogenase reaction. Aerobically, glucose is oxidized to CO_2 and water by the TCA cycle, and ATP is formed from ADP and inorganic phosphate. This will lead to a lowering of the concentration of inorganic phosphate, ADP, or AMP in the cells. Since the triosephosphate dehydrogenase reaction does not occur in the absence of inorganic phosphate and since both the glyceryl-kinase- and pyruvate-kinase-mediated reactions depend primarily on the availability of ADP, the assimilation of aerobic glucose will deplete the concentration of these substances; thus, glycolysis will be inhibited or forced to slacken its rate which will be determined by P_i and ADP concentrations. AMP is involved in the regulation of the activity of phosphofructokinase, and both ATP and AMP regulate the activity of another important enzyme, viz., fructose diphosphate phosphatase. Phosphofructokinase is inhibited by a low level of AMP and activated by a high level of AMP whereas fructose diphosphate phosphatase is activated by ATP and inhibited by AMP. As aerobic respiration lowers the AMP level and increases the ATP level, the net effect would be the inhibition of glycolysis due to the inactivation of its key enzyme phosphofructokinase and the activation of fructose diphosphate phosphatase. The latter enzyme prevents glycolysis by hydrolyzing fructose diphosphate to monophosphate. In addition, ATP inhibits the phosphofructokinase reaction, and its concentration thus provides another mechanism of glycolysis inhibition. In the language of cellular activity, a high AMP level will trigger the cellular machinery towards more synthesis of ATP whereas a high level of ATP will oppose this. Further, a low level of AMP depresses the activity of the TCA cycle by inhibiting the action of isocitric dehydrogenase, the consequence of which will be a decrease in ATP production and an increase in citrate concentration.

Oxaloacetate is one of the key components of both the TCA cycle and the glyoxylate cycle, but when its concentration exceeds the actual requirement, it inhibits the TCA cycle by inhibiting the activity of succinic dehydrogenase. The excess of phosphoenolpyruvate acts as a potent inhibitor of the glyoxylate cycle because it inhibits the activity of isocitritelyase. Therefore, only the constant siphoning off of oxaloacetate and phosphoenolpyruvate in biosynthetic reactions can provide the remedy for meaningful operation of the amphibolic pathways.

β-oxidation of fatty acids leads to the accumulation of acetyl CoA and ATP. Under these conditions, the glyoxylate cycle is a major route for the conversion of fatty acids or dicarboxylic acids of the TCA cycle into sugar. This cycle runs more effectively in the presence of excess acetyl CoA (β-oxidation product of fatty acids). If the concentration of oxaloacetate goes up, the TCA cycle may stop operating and the reactions of sugar synthesis will be promoted. As the main job of the TCA cycle is to generate ATP, and since enough ATP becomes available during the β-oxidation of fatty acids and the operation of the glyoxylate cycle, the inhibition of the TCA cycle in no way affects the energy-requiring synthetic reactions. The conversion of acetyl CoA into sugar does not seem to occur through the TCA cycle as this cycle degrades acetyl CoA to CO_2 and H_2O. The level of acetyl CoA is also known to affect the activity of phosphoenolpyruvate carboxylase. The enzyme activity increases with acetyl CoA concentration, leading to the formation of oxaloacetate. This provides another control for the routing of glucose carbon into energy-producing and synthetic reactions.

It is expected that the discovery of many more control mechanisms in the future will make possible a full appreciation of their biological and evolutionary significance.

REFERENCES

Beevers, H., Respiratory Metabolism in Plants. Harper and Row, New York (1961).

Griffiths, D. E., Oxidative phosphorylation. *Essays in Biochemistry*, **1**, 91–120 (1965).

Hartree, E. F., Cytochrome in higher plants. *Adv. Enzymol.*, **18**, 1–64 (1957).

James, W. O., Reaction paths in the respiration of higher plants. *Adv. Enzymol.*, **18**, 281–318 (1957).
Kornberg, H. L., Beevers, H., The glyoxylate cycle as a stage in the conversion of fat to carbohydrate in castor beans. *Biochim. Biophys. Acta*, **26**, 531–37 (1957).
Stiles, W., Leach, W., Respiration in Plants. Methuen, London (1960).
Varner, J. E., Plant metabolism. McGraw-Hill Encyclopaedia of Science and Technology, Vol. 10, 402–5, McGraw-Hill, New York (1971).
Zelitch, I., Photosynthesis, Photorespiration and Plant Productivity. Academic Press, New York (1971).

9: Photorespiration and Glycollate Metabolism

INTRODUCTION

Until about a decade ago, it was generally believed that only one kind of respiration, namely, 'dark' respiration, occurs in plants, animals, and bacteria. This belief was responsible for the widespread assumption that respiration during photosynthesis (in green plants) was approximately equal to respiration of the green tissues in darkness. It was only in the late 1950s and the early 1960s that a new and interesting aspect of respiration in some green plants began to emerge as a result of Decker's (1955, 1959) pioneering researches on tobacco, *Mimulus*, and certain other plants. These researches indicated that respiration of plants such as tobacco could be much more under photosynthetic conditions and in the light than respiration in darkness. When such plants were illuminated, both CO_2 evolution and oxygen uptake were stimulated, and this took place simultaneous with photosynthesis (in which oxygen is evolved and CO_2 consumed). Respiration that occurs in photosynthetic tissues in the presence of light and results in CO_2 evolution has been termed photorespiration; it is effected by reactions quite different from those that take place during normal, dark respiration.

Photorespiration seems to have a direct bearing on the photosynthetic production capacities of various plants. In general, the rate of net photosynthesis in many tropical plants is as much as about four times higher than in many temperate plants. Such differences in net photosynthesis seem to depend (in some but not all cases) upon whether photorespiration occurs.

COMPARISON BETWEEN PHOTORESPIRATION AND DARK RESPIRATION

In animals and bacteria, only one kind of dark respiration (actually unaffected by the presence or absence of light) occurs. In most green plants, however, there are two distinct kinds of respiration: photorespiration and dark respiration. Table 9-1 gives the comparison between these two.

Table 9-1 Comparison between photorespiration and dark respiration

Photorespiration	Dark respiration
Takes place only in photosynthetic cells in the presence of light	Encountered in all living cells in the mitochondria
Involves light-dependent uptake of oxygen and evolution of CO_2	Oxygen uptake and CO_2 evolution are light-independent
Depends upon the photochemical activity of chloroplasts and biochemical activities of peroxisomes and of mitochondria	Is a function of mitochondria alone
Increases with rise in oxygen concentrations from 0% to 100%	Is saturated at very low oxygen concentrations (about 2%)
Glycollate is the substrate	Acetyl CoA, or any substance giving rise to acetyl CoA, is the substrate
Not associated with ATP formation	Associated with ATP formation
Biological significance not clearly known; many workers believe it may be a wasteful process (since it is not linked with phosphorylation and since it leads to the removal of photosynthetically fixed CO_2 from green cells)	Biological significance is well established

EVIDENCES

(1) When Decker (1955) transferred tobacco leaves from the light to darkness, he observed a burst of CO_2 evolution before the rate of CO_2 output resumed a lower, steady pace. This CO_2 release was thought to represent an overshoot of the greater CO_2 evolution that occurred in the light and, hence, is evidence of photorespiration. No such burst occurs in species that do not exhibit photorespiration.

(2) Forrester *et al.* (1966) observed that CO_2 production by excised soybean leaf discs was much greater when these were illuminated than when in the dark. The rate of CO_2 output in illuminated discs rose with the increase in oxygen concentrations whereas the oxygen content had very little effect on the rate of CO_2 output in leaf discs kept in the dark.

As early as 1920, Warburg had demonstrated that oxygen inhibits photosynthesis in *Chlorella*. This Warburg effect has since been described in several algae and higher plants. The effect seems to depend upon not only oxygen but also the CO_2 concentration.

Plants lacking photorespiration are not inhibited by high oxygen concentrations.

(3) Photorespiration occurs only in photosynthetic cells in the light. If, by mutation, the capacity for chlorophyll synthesis is lost in a photorespiratory plant, then photorespiration does not take place. Thus, chlorophyll-free mutant leaves of the sunflower do not exhibit photorespiration and they have the same rate of CO_2 evolution both in the light and the dark. On the other hand, normal green leaves from the same plant show photorespiration at rates as much as five times higher than the rate of dark respiration.

(4) A normally nonphotorespiring plant, such as maize, can be made to evolve CO_2 in the light after treatment with the photosynthetic inhibitor DCMU which does not affect mitochondrial respiration. It is assumed that the CO_2 so released represents respiratory CO_2 produced during normal dark respiration. There is no release of such CO_2 in the absence of the inhibitor because all respiratory CO_2 is recycled in photosynthesis. When photosynthesis is blocked by DCMU, recycling cannot take place and, hence, CO_2 is released.

SITE

Early research stated that photorespiration occurs in chloroplasts. Subsequent work, leading to the isolation and characterization of a new cellular organelle from green plants, has demonstrated that this organelle, now termed peroxisome, may be the chief site of photorespiration (though different parts of the cycle occur in different organelles).

First described by Mollenhauer *et al.* (1966), peroxisomes were later observed in several plants and isolated free from chloroplasts. Peroxisomes are small bodies and commonly appear closely adpressed to the chloroplast. They have been shown to contain the enzymes of glycollate metabolism. The observation that peroxisomes are often in intimate contact with the chloroplast suggests that materials, enzymes, substrates, or products of certain reactions may well be exchanged between chloroplasts and peroxisomes.

Interestingly, peroxisome-like bodies have also been reported in various animal cells.

MEASUREMENT

Since both photosynthesis and photorespiration take place in the light, the measurement of photorespiration can often prove difficult and confusing.

The uptake and release of CO_2 by detached leaves of photorespiratory plants is generally measured in a closed system. Hitachi Horiba EJA-1 IR CO_2 Gas Analyzer is a suitable instrument for this purpose. By means of this instrument, photosynthesis is estimated as the initial rate of CO_2 uptake at a CO_2 concentration of 300 ppm. Photorespiration is measured as the initial rate of CO_2 liberation into air that is free of CO_2 after 20 minutes of photosynthesis (see Hew *et al.*, 1969).

BIOCHEMICAL MECHANISM

It is generally acknowledged that glycolic acid serves as the primary substrate for photorespiration and is the main source of CO_2 within leaves of photorespiratory plants.

Zelitch (1964) first suggested that the CO_2 produced in photorespiration could result from glycollate as follows:

$$CH_2\cdot{}^*COOH + O_2 \xrightarrow[\text{oxidase}]{\text{Glycollate}} CHO\cdot{}^*COOH$$

$$+ H_2O_2 \xrightarrow{\text{Nonenzymatic}} H\cdot COOH + {}^*CO_2 + H_2O$$

The formation of glycollate during photosynthesis (see Fig. 9-1) is stimulated by low levels of CO_2, high levels of oxygen, and high light intensities. Glycollate is derived from some intermediate of the photosynthetic carbon reduction cycle and its formation seems linked with the photosynthetic electron transport chain in view of the fact that it is produced from CO_2 only in the light. There are interesting differences between glycollate

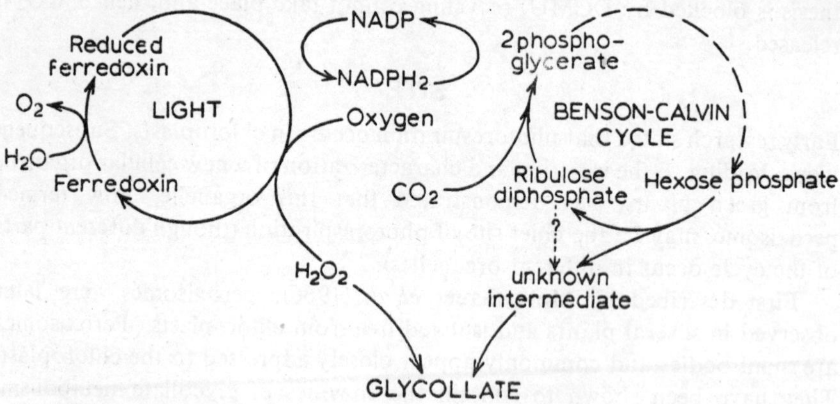

Fig. 9-1 Probable scheme showing glycollate formation in photosynthesis.

metabolism in algae and in higher plants. The oxidation of glycollate to glyoxylate, mediated by glycollate oxidase, is obligatory in higher plants but this may or may not be so in algae. Many algae are known to excrete the unoxidized glycollate into the ambient medium, and the extracellular glycollate thus excreted may often be of considerable ecological and nutritional importance in aquatic habitats. Unlike higher plants, very few algae seem to contain the enzyme glycollate oxidase, but further research is necessary to verify this distinction.

Kisaki and Tolbert (1969) have indicated that oxidation of glycollate to glyoxylate takes place in peroxisomes whereas oxidation of glyoxylate to CO_2 probably occurs within the chloroplasts. Oxidation of glycollate to glyoxylate is generally accepted, but it is doubtful whether glyoxylate oxidation yields formic acid. An alternative possibility which has gained considerable credibility is the conversion of glyoxylate into the amino acid glycine. This is believed to occur within peroxisomes and is catalyzed by the transaminase enzyme glutamate-glyoxylate aminotransferase. Two molecules of glycine then interact (in mitochondria) to give rise to a molecule of serine and a molecule of CO_2 (Fig. 9-2). The latter is then released in photorespiration. In the cell, some glyoxylate may diffuse from the peroxisome to the chloroplast where either it may be oxidized to CO_2 or it may be reduced to glycollate by a glyoxylate reductase (Devlin and Barker, 1971).

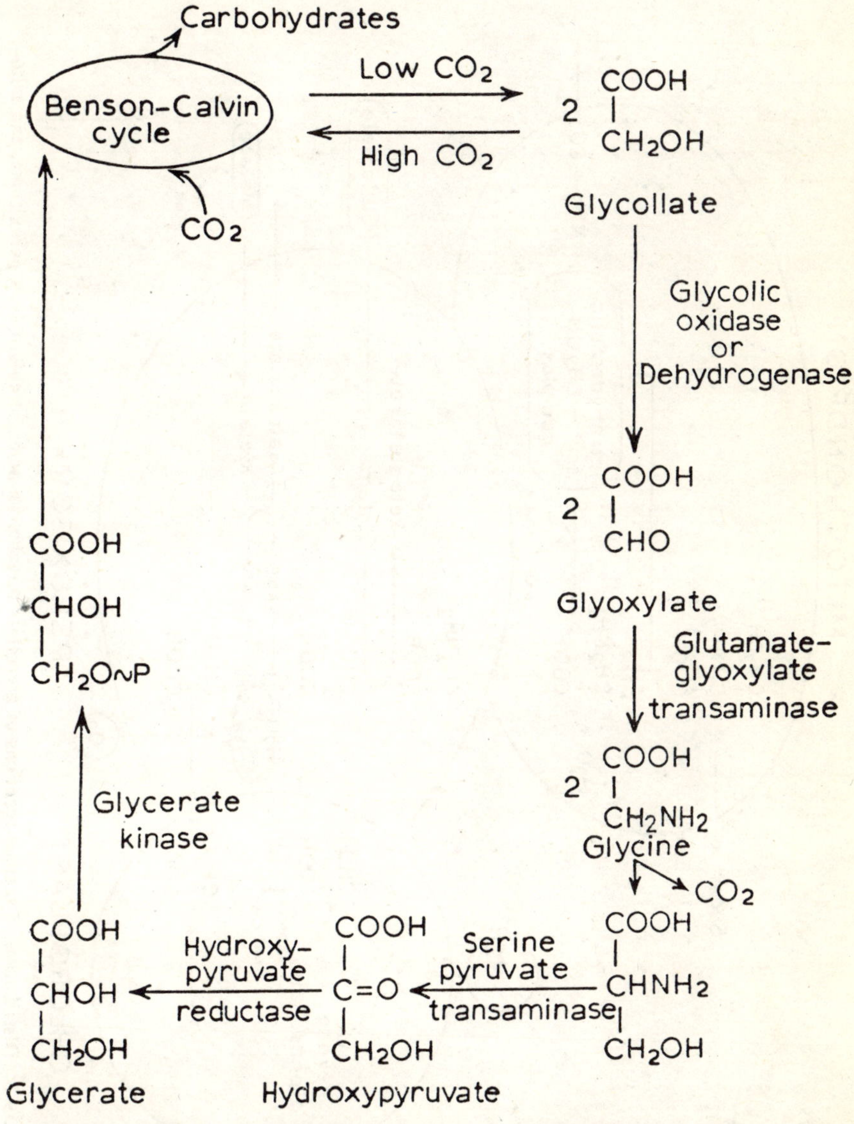

Fig. 9-2 Chemical reactions involved in the formation and metabolism of glycollate.

Kisaki *et al.* (1971) have studied the intracellular localization of photorespiratory enzymes in tobacco and spinach leaves. They have shown that oxygen uptake and CO_2 evolution occur in different particles and that glycine is decarboxylated to serine in mitochondria (Fig. 9-3). Photorespiratory gas exchange may be explained by assuming that oxygen uptake takes place

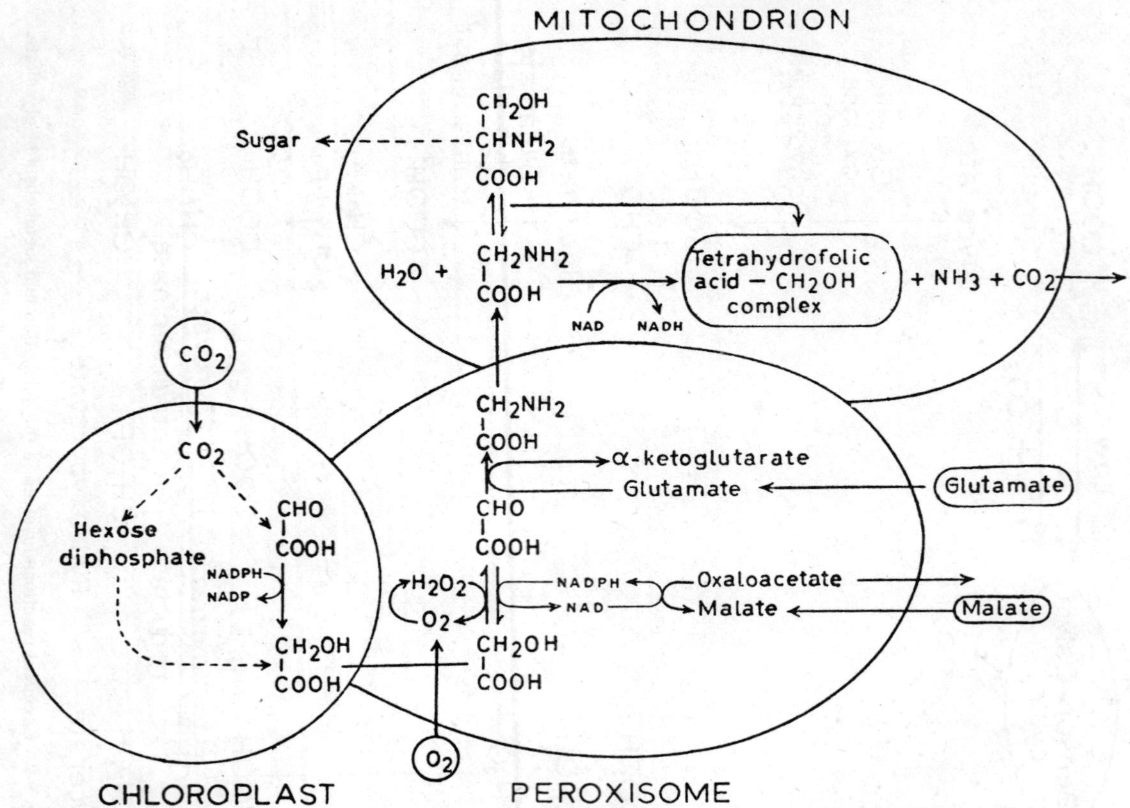

Fig. 9-3 Distribution of various reactions of glycollate metabolism amongst chloroplast, peroxisome, and mitochondria. (After Kisaki *et al.*, 1971.)

in peroxisomes whereas CO_2 output occurs in mitochondria (Fig. 9-3). It is assumed that the H_2O_2 produced during glycollate oxidation is degraded by peroxisomal catalase and that the serine formed from glycine in mitochondria returns to peroxisomes. The enzyme serine hydroxymethyltransferase has been found in both mitochondria and chloroplasts of green leaves. On the other hand, glycine decarboxylase is confined exclusively to mitochondria. It is further believed that leaf glycine decarboxylase is associated with mitochondrial membranes whereas serine hydroxymethyltransferase occurs in a soluble form in mitochondria. The serine hydroxymethyltransferase seems readily accessible to glycine decarboxylase in mitochondria.

An elegant mechanism for controlling the reactions of the glycollate pathway seems to operate by confining the glyoxylate reductase to peroxisomes; the reductase retards or controls the transformation of glyoxylate into glycine. This enzyme is dependent on NADH. Similarly, the glycine decarboxylase is confined to mitochondria and generates the NADH required for the activation of peroxisomal glyoxylate reductase. The control is based on the stimulation of glycine decarboxylase by NAD^+ and on its inhibition by NADH.

Kisaki *et al.* (1973) have studied the effect of the age of leaves on the comparative activities of enzymes related to photosynthesis and photorespiration in tobacco. The maximum photorespiratory activity was observed in young rather than old and mature leaves. On the other hand, the enzyme ribulose diphosphate carboxylase was more or less equally active in all leaves, except in the youngest leaves where its activity was considerably low.

Though tobacco is a photorespiratory plant, Kisaki *et al.* (1973) detected the operation of the C_4 dicarboxylic acid cycle in young leaves and observed that the maximum relative enzyme activity for the C_4 cycle was detectable in the uppermost (i.e., the youngest) leaf. These studies have indicated that the activities of enzymes of the C_4 dicarboxylic acid cycle as well as of the glycollate pathway may largely determine the efficiency of photosynthesis in the case of photorespiratory plants.

ANATOMICAL DIFFERENCES BETWEEN PHOTORESPIRING AND NONPHOTORESPIRING PLANTS

Significant differences in respect of the anatomy of leaves and bundle-sheath cells and the ultrastructure of subcellular organelles have been reported between plants that photorespire and plants that do not. The fact that the vascular bundles in the leaves of plants such as *Digitaria sanguinalis*, *Cyperus* sp., and *Saccharum* sp. are surrounded by prominent bundle-sheath cells (Fig. 9-4) containing chloroplasts was described as early as 1914 by Haberlandt, though the physiological significance of such bundle-sheath cells was not known at the time. Such conspicuous bundle-sheath cells were found to be lacking in other plants such as *Triticum vulgare*. It is now recognized that the presence or absence of bundle-sheath cells is intimately correlated with the absence or presence of photorespiration and also with high or low photosynthetic capacities.

Another remarkable feature of bundle-sheath cells is that they contain an unusually high concentration of chloroplasts, peroxisomes, and mitochondria. This has been lucidly demonstrated in many genera, e.g., *Amaranthus, Atriplex, Saccharum, Cynodon, Digitaria, Eragrostis*, and *Cyperus* (see Black, 1971). A marked dimorphism between the chloroplasts of bundle-sheath cells and those of the mesophyll cells is often noticeable. The chloroplasts of bundle-sheath cells are generally large and contain conspicuous starch grains (Fig. 9-5) whereas those of mesophyll cells are smaller, have grana, and lack starch.

Plants that have remarkable photosynthetic efficiency, e.g., *Amaranthus edulis*, have been found to contain a well-developed peripheral reticulum of anastomosing microtubules contiguous with the inner chloroplast membrane in the chloroplasts of their bundle-sheath as well as their mesophyll cells (Laetsch, 1968). Incidentally, peripheral reticulums of apparently the same type are seen also in electron micrographs of certain blue-green algal cells.

Chloroplast dimorphism, in at least some C_4 plants, is well established now, although no great mitochondrial dimorphism or peroxisomal dimorphism has been demonstrated. Nevertheless, it has been confirmed that, in photosynthetically efficient plants, the mitochondria of bundle-sheath cells are commonly (not always) much larger than those in the mesophyll. There are some indications that the enzymes catalyzing the generation of phosphoenolpyruvate, the primary CO_2 acceptor, and the primary carboxylation reaction are located in mesophyll cells whereas those concerned with the later steps of triose conversion into hexose occur in bundle-sheath cells (Fig. 9-6).

Some important differences between photorespiratory and nonphotorespiratory plants are enumerated in Table 9-2.

Table 9-2 Distinction between photorespiring and nonphotorespiring plants

Character	Photorespiring plants	Nonphotorespiring plants
Geographical distribution	Generally temperate	Generally tropical
Photosynthetic efficiency	Low	High
CO_2 evolution in light	High	Nil
Leaf bundle-sheath	Absent	Present
Leaf chloroplasts	Mostly similar	Mostly dimorphic
Carbon pathway	Benson-Calvin cycle	C_4 pathway
Reassimilation of respiratory CO_2	Mostly none	Probably occurs

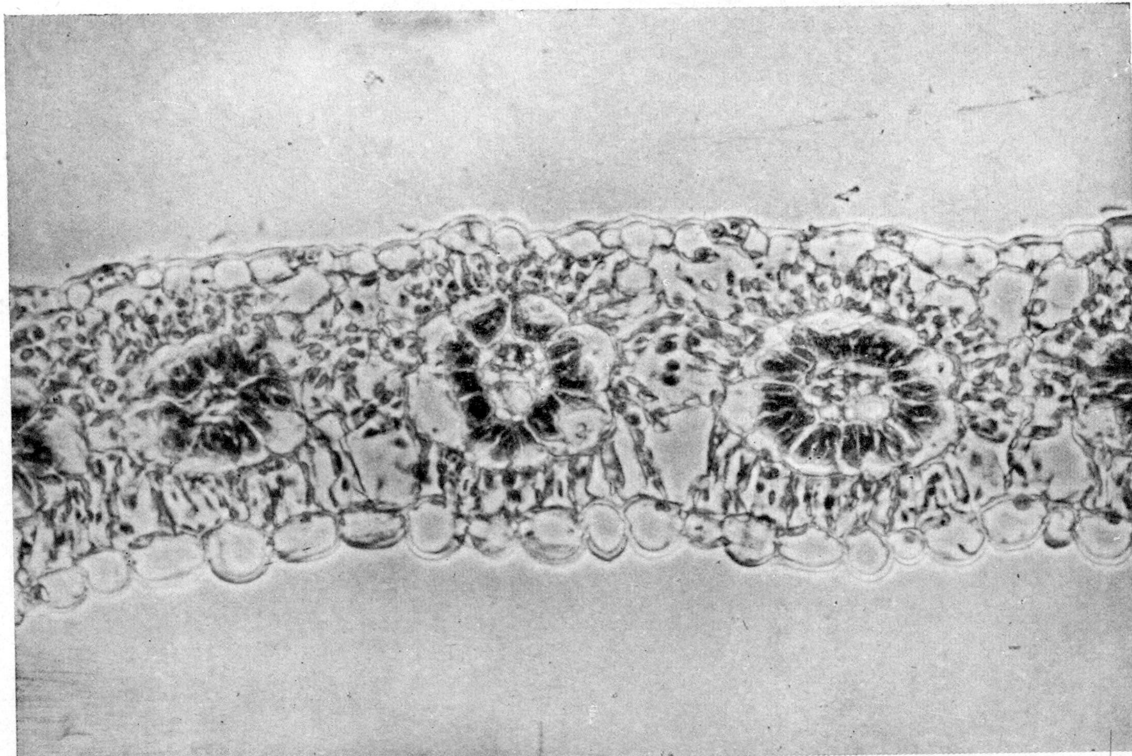

Fig. 9-4 Photomicrograph of transverse section passing through a leaf of *Amaranthus* showing conspicuous sheath cells around vascular bundles. (Courtesy : Professor W. M. Laetsch.)

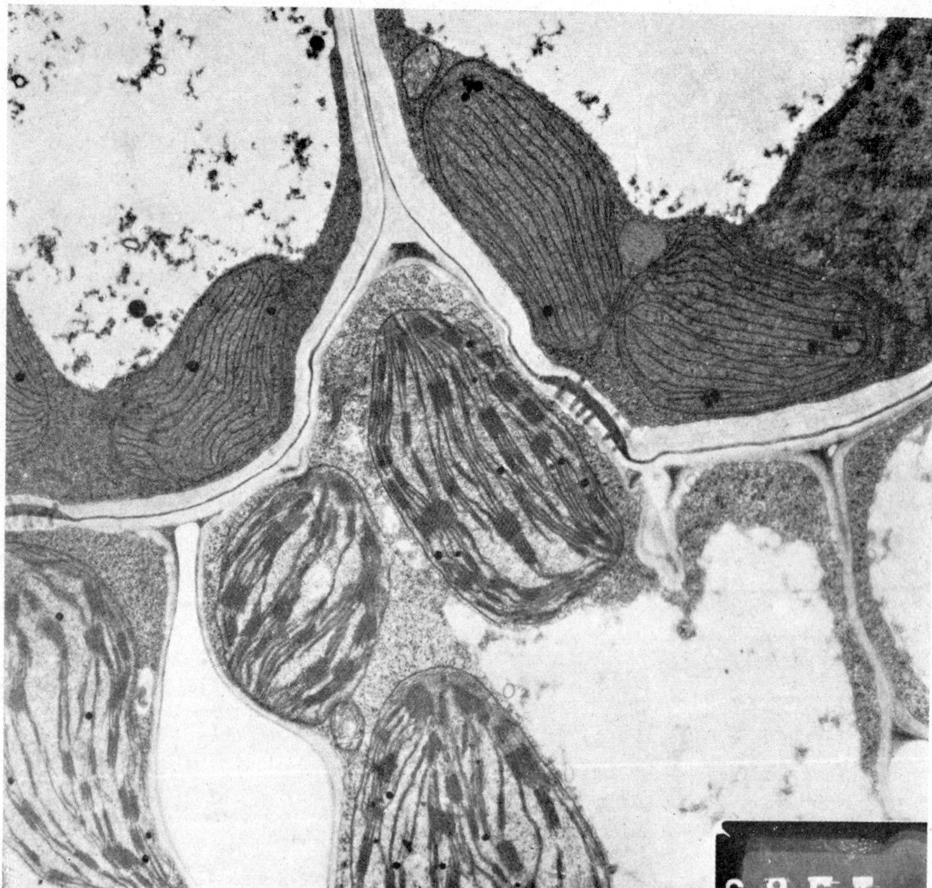

Fig. 9-5 Fine structure of chloroplasts from bundle-sheath cells of sugarcane stem. (Courtesy : Professor W. M. Laetsch, ×10,900.)

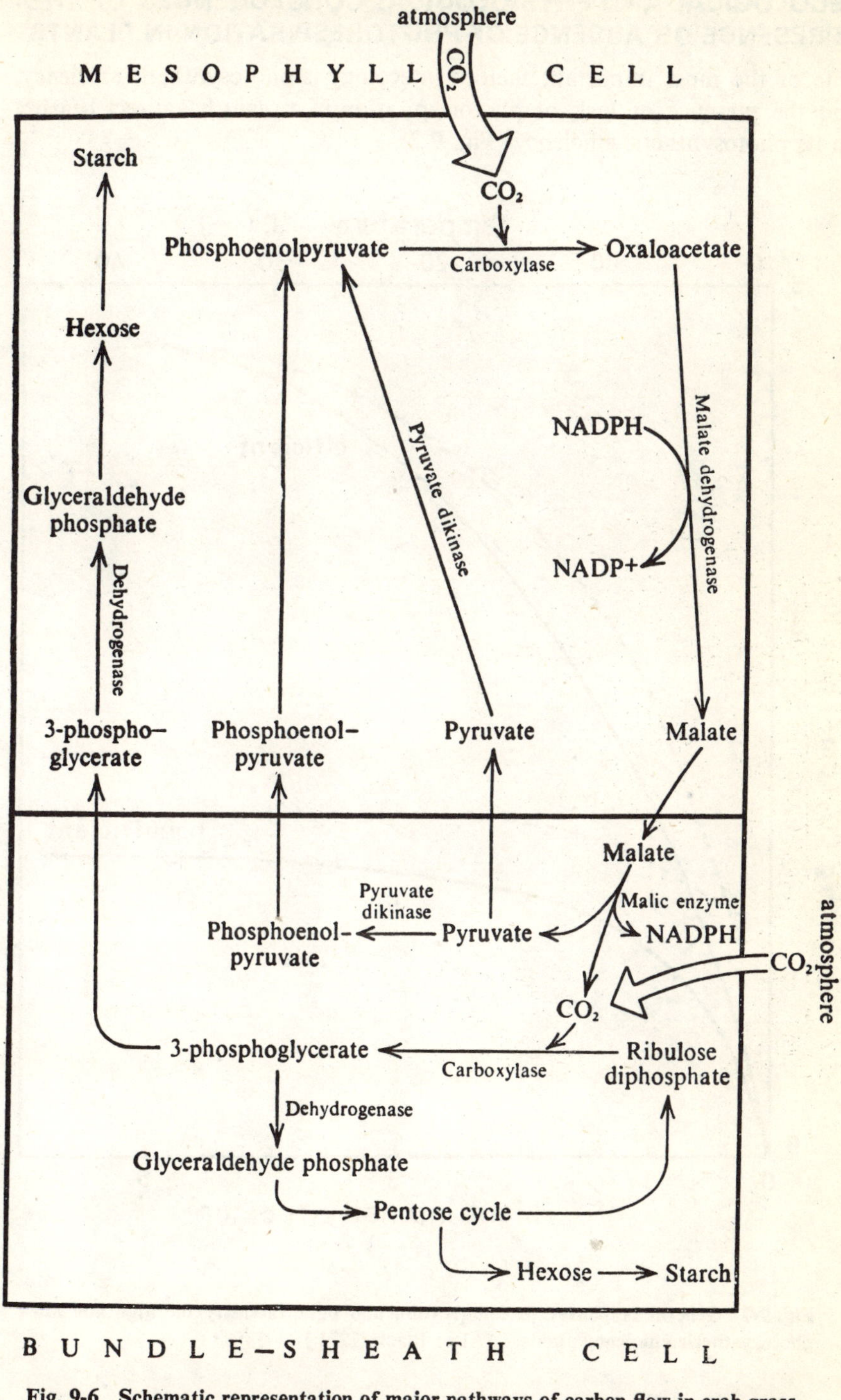

Fig. 9-6 Schematic representation of major pathways of carbon flow in crab grass. (After Black, C. C., Photosynthetic carbon fixation in relation to net CO_2 uptake. *Ann. Rev. Plant Physiol.*, **24**, 253–86, 1973.)

ECOLOGICAL AND PHYSIOLOGICAL CONSEQUENCES OF THE PRESENCE OR ABSENCE OF PHOTORESPIRATION IN PLANTS

One of the most important factors in ecology is photosynthetic efficiency, and the presence or lack of photorespiration in a plant has direct bearing on its photosynthetic efficiency (Fig. 9-7).

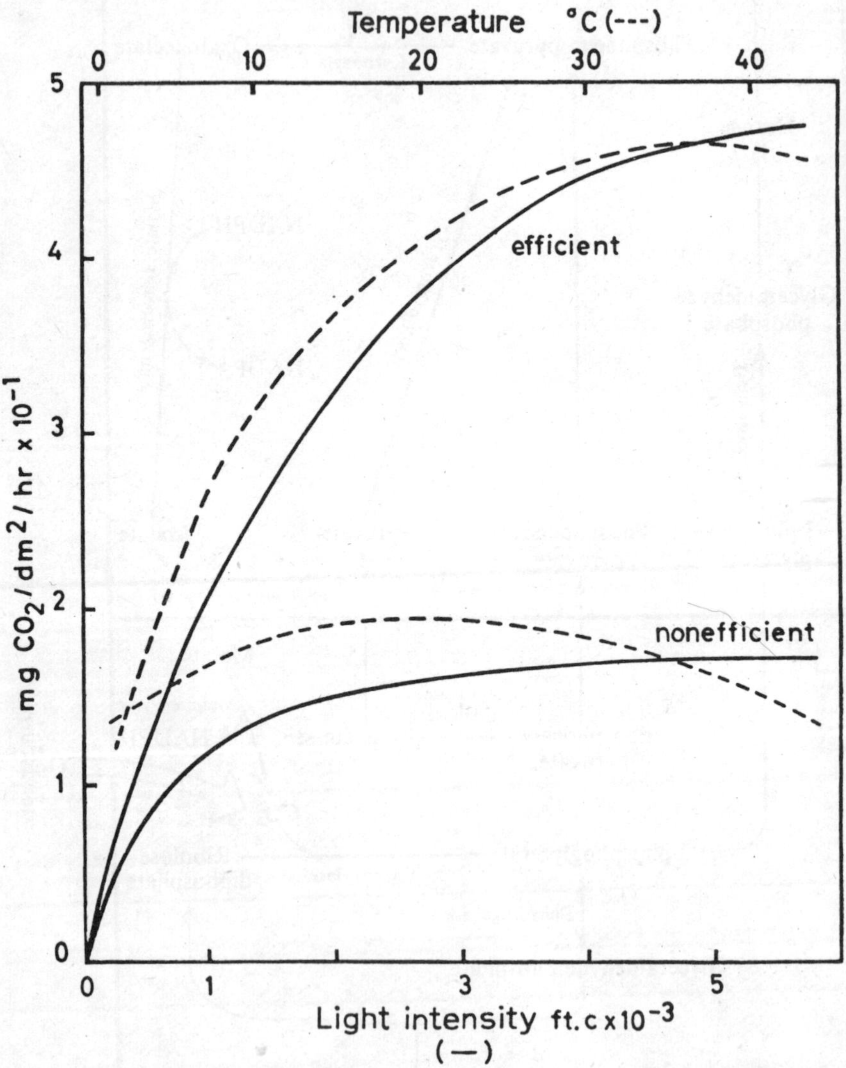

Fig. 9-7 General responses to temperature and light intensity for high and low photosynthetic efficiency plants. (After Black, 1971.)

Some examples of plants characterized by photosynthetic efficiency (mostly lacking photorespiration) are listed here.

HIGH PHOTOSYNTHETIC EFFICIENCY PLANTS (GENERALLY TROPICAL PLANTS)

Monocots *Cynodon dactylon, Digitaria sanguinalis, Cyperus rotundus, Panicum virgatum, Saccharum officinarum, Sorghum vulgare*, and *Zea mays.*

Dicots *Amaranthus edulis, Atriplex rosea, Kochia scoparia, Portulaca grandiflora*, and *Salsola kali.*

LOW PHOTOSYNTHETIC EFFICIENCY PLANTS (GENERALLY TEMPERATE PLANTS)

Monocots *Avena sativa, Eichhornia crassipes, Panicum lindheimevi, Hordeum vulgare, Cyperus papyrus, Oryza sativa*, and *Triticum aestivum.*

Dicots *Arachis hypogaea, Atriplex hastata, Beta vulgaris, Daucus carota, Helianthus annuus*, and *Spinacea oleracea.*

From these lists it may be seen that different species of the same genus may sometimes vary in respect of photosynthetic efficiency.

Plants lacking photorespiration exhibit increasing growth and net photosynthesis with increase in light intensity, temperature, and the content of oxygen and CO_2. On the other hand, in photorespiratory plants, photosynthesis becomes saturated at a much lower light intensity, temperature, and content of oxygen and CO_2, with photorespiration being stimulated at higher levels of these factors. Since sunlight is the major source of energy in the biosphere and since nonphotorespiratory plants can capture and transfer, into an ecosystem, energy upto about four times faster than photorespiratory plants, the presence or lack of photorespiration is a factor of vital importance in understanding the energy and nutrient budgets and also the productivity of a given ecosystem.

Photorespiration has recently been demonstrated in a blue-green alga by Lex *et al.* (1972). This hitherto unreported process in photosynthetic bacteria, in contrast to its presence in cyanophytes and higher plants, is suggestive of some close link between the origin of the process and oxygen evolution in photosynthesis.

REFERENCES

Black, C. C., Ecological implications of dividing plants into groups with distinct photosynthetic production capacities. *Adv. Ecol. Res.*, **7**, 87–114 (1971).

Decker, J. P., A rapid postillumination deceleration of respiration in green leaves. *Plant Physiol.*, **30**, 82–84 (1955).

Decker, J. P., Some effects of temperature and carbon dioxide concentration on photosynthesis of *Mimulus*. *Plant Physiol.*, **34**, 103–6 (1959).

Devlin, R. M., Barker, A. V., Photosynthesis. Van Nostrand Reinhold, New York (1971).

Forrester, M. L., Krotkov, G., Nelson, C. D., Effect of oxygen on photosynthesis, photorespiration and respiration in detached leaves. I. Soybean. *Plant Physiol.*, **41**, 422–27 (1966).

Haberlandt, G., Physiological Plant Anatomy. Macmillan, London (1914).

Hew, C. S., Krotkov, G., Canvin, O. T., Determination of the rate of CO_2 evolution by green leaves in light. *Plant Physiol.*, **44**, 662–70 (1969).

Kisaki, T., Tolbert, N. E., Glycolate and glyoxylate metabolism by isolated peroxisomes or chloroplasts. *Plant Physiol.*, **44**, 242–50 (1969).

Kisaki, T., Imai, A., Tolbert, N. E., Intracellular localization of enzymes related to photorespiration in green leaves. *Plant Cell Physiol.*, **12**, 267–73 (1971).

Kisaki, T., Hirabayashi, S., Yano, N., Effect of the age of tobacco leaves on photosynthesis and photorespiration. *Plant Cell Physiol.*, **14**, 505–14 (1973).

Laetsch, W. M., Chloroplast specialization in dicotyledons possessing the C_4 dicarboxylic acid pathway of photosynthetic CO_2 fixation. *Amer. J. Bot.*, **55**, 875–83 (1968).

Lex, M., Silvester, W. B., Stewart, W. D. P., Photorespiration and nitrogenase activity in the blue-green alga, *Anabaena cylindrica*. *Proc. R. Soc. Lond. B.*, **180**, 87–102 (1972).

Mollenhauer, H. H., Morre, D. J., Kelly, A. G., The widespread occurrence of plant cytosomes resembling animal microbodies. *Protoplasma*, **62**, 44–52 (1966).

Warburg, O., Über die Geschwindigkeit der photochemischen Kohlensaurezersetzung in lebender Zeilen. *Biochem. Z.*, **103**, 188–217 (1920).

Zelitch, I., Organic acids and respiration in photosynthetic tissues. *Ann. Rev. Plant Physiol.*, **15**, 121–42 (1964).

10: Nitrogen and Sulphur Metabolism

INTRODUCTION

Nitrogen and sulphur are, like carbon, hydrogen, oxygen, and phosphorus, essential constituents of living matter, i.e., no organism can survive in the absence of these elements. Some organisms, e.g., photosynthetic plants, can derive all their cellular carbon, nitrogen, and sulphur from the respective inorganic forms of these elements; i.e., they can obtain organic carbon from CO_2, organic nitrogen from nitrate, nitrite, or ammonium, and organic sulphur from sulphate, sulphite, or thiosulphate. Other organisms, e.g., fungi, have carbon auxotrophy, and nitrogen and sulphur autotrophy, i.e., they cannot synthesize organic carbon from CO_2, but they synthesize organic nitrogen and organic sulphur from inorganic nitrogenous and sulphur compounds, respectively. Nitrogen enters the organic form as ammonia, and sulphur enters as sulphide. Since the most prevalent inorganic compounds of nitrogen and sulphur in nature are nitrate and sulphate—and the plants depend on these for their nitrogen and sulphur requirements—, these substances must first undergo reduction before they can be converted and incorporated into the structural and functional components of plant cells. We know that plants in general reduce nitrate or nitrite to ammonia, and sulphate or sulphite to sulphide or thiol, before these reduced substances are consumed in the formation of amino acids (the formation of only cysteine, methionine, glutathione, and cystine involves the use of reduced sulphur).

It is instructive to compare the reactions of nitrate and sulphate with those of phosphate. Although phosphate is as important a nutrient as nitrate or sulphate, it does not suffer valency changes as do the inorganic compounds; rather it enters organic matter by forming ester compounds with a number of cell substances. But sulphate, like phosphate, also enters esterification reactions with organic substances, thus forming a wide variety of sulphate esters. Also, it is the only nutrient for which all organisms, including animals, are autotrophic. Cells avoid the reduction of phosphate probably because it may lead to phosphine production which is a highly lethal substance.

Our biosphere is an inexhaustible reservoir of nitrogen. However, most organisms (except some blue-green algae) that are both free living and associative symbiotic, and certain free-living and associative or obligatory symbiotic bacteria, cannot make use of this free nitrogen. This handicap is attributed to their lack of the nitrogenase enzyme which is essential for biological nitrogen fixation and has the ability to reduce molecular nitrogen to ammonia. Some blue-green algae and bacteria can fix nitrogen because they possess the nitrogenase enzyme. Nitrogen-fixing organisms are of practical interest mainly for two reasons: (1) they recycle the atmospheric nitrogen in organic form to the soil, thus replenishing the soil's nitrogen lost through cultivation, bacterial activity, and other causes; and (2) a knowledge of the formation and activity of the nitrogenase enzyme may lead to its successful introduction into crop plants, thus making them self-sufficient in nitrogen. The realization of this goal, though extremely unlikely at the moment, will be an outstanding achievement and will undoubtedly solve the nitrogen starvation problem forever (Dixon and Postgate, 1971; Stewart, 1973).

NITROGEN CYCLE

The main nitrogenous compounds of the environment are ammonia, organic amines, dinitrogen (N_2), nitrous oxide (N_2O), nitric oxide (NO), nitrite (NO_2), nitrogen dioxide (NO_2), and nitrate (NO_3). Other types of nitrogen compounds are insignificant in relation to the nitrogen cycle.

The nitrogen cycle of the biosphere includes two major components, the chemical transformation of nitrogen by biological or abiological agents, and the physical transfers. The conversion of molecular nitrogen or dinitrogen into ammonia (nitrogen fixation), the assimilation of ammonia into organo-nitrogen compounds, the ammonification of organo-nitrogen to ammonia, the nitrification of ammonia to nitrate, and the denitrification of nitrate to mainly dinitrogen or molecular nitrogen are instances of biological-chemical transformations of nitrogen (Fig. 10-1).

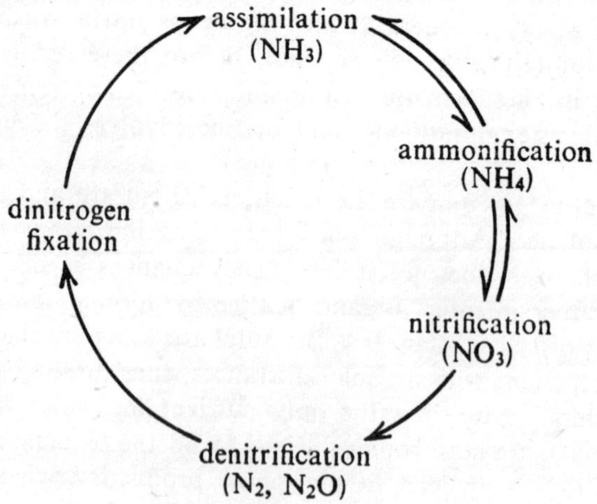

Fig. 10-1 Biological nitrogen cycle.

Abiological chemical transformation of nitrogen in the atmosphere is composed of the following processes: (1) lightning, in which (i) O^+ and N_2 react forming $NO^+ + N$, and (ii) NO^+ and H_2O react forming $2H^+ + NO_2^-$; the net result is the formation of nitrite from molecular and atomic oxygen; (2) photolysis, in which N_2O is either decomposed to molecular nitrogen and atomic oxygen ($N_2O \rightarrow N_2 + O$) or to NO and N; (3) ozonization, in which (i) molecular nitrogen is oxidized to nitrous oxide ($N_2 + O_3 \rightarrow N_2O + O_2$), and (ii) nitric oxide is oxidized to nitrogen dioxide ($NO + O_3 \rightarrow NO_2 + O_2$); (4) oxidation of nitric oxide to nitrogen dioxide ($2NO + O_2 \rightarrow 2NO_2 + O_2$) and of ammonia to nitrate ($NH_3 + O \rightarrow NO_3^-$); and (5) hydration, in which nitrogen dioxide is hydrated to nitric acid ($3NO_2 + H_2O \rightarrow 2HNO_3 + NO$).

Abiological denitrification also occurs in soil. It is evident that abiological interconversions of N_2, N_2O, NO, and NO_2 in the upper layers of the atmosphere proceed with the consumption of ozone. The intensity of this interconversion may lead to the depletion of ozone in the higher strata of the atmosphere, the biological consequence of which would be the exposure of the earth's surface to lethal and mutagenic ultraviolet radiation. The Haber-Bosch process, another major means of abiological nitrogen fixation, is an industrial process through which molecular nitrogen is converted into ammonia at extremely high pressures and temperatures; the ammonia is finally used in the production of nitrogenous fertilizers and other useful compounds.

The physical transfers of nitrogen are achieved by (1) volatilization, in which ammonium nitrogen escapes to the atmosphere; (2) precipitation (the reverse of volatilization), in which ammonium, nitrate, and nitrite nitrogen are returned to the land and sea through rain and mist; (3) sedimentation and solubilization, in which a major part of the sea's nitrogen is transferred to marine sediments from which some amounts are brought back to the sea through solubilization; (4) leaching and run-off, in which the land-cum-atmospheric inorganic nitrogen finds its way into the sea through streams, rivers, and rain; and (5) volcanic eruptions, in which the nitrogen confined in primary rocks is released to the atmosphere in the form of molecular nitrogen. Physical transfers, such as leaching and run-off, are generally considered to be the main cause of increasing eutrophication of natural water bodies.

Nitrogen fixation is, according to currently accepted views, any reaction of dinitrogen or molecular nitrogen that forms a compound in which nitrogen is covalently bonded to another atom, normally oxygen or hydrogen. On this basis, the abiological means of nitrogen fixation are just as important as the biological means. In contrast to most abiological fixation processes (except the Haber-Bosch process), biological nitrogen fixation is always a reductive process. The resulting ammonia is most readily assimilated into organic form.

The next component of the nitrogen cycle is ammonia and nitrate assimilation in which both the nitrogen sources get converted into amino acids (nitrate is first reduced to ammonia through a series of reductases).

The ability to assimilate ammonia and nitrate nitrogen is common to all nonanimal organisms.

The assimilatory stage is followed by ammonification in which the organic nitrogen locked up in the body of dead organisms is mineralized to ammonia through the microbial agency of heterotrophic bacteria and fungi.

Nitrification involves the oxidation of ammonia to nitrate by chemosynthetic bacteria: (1) *Nitrosomonas*, which while converting ammonia into nitrite consumes energy to synthesize ATP from ADP and inorganic phosphate:

$$NH_4^+ + \tfrac{3}{2}O_2 \rightarrow NO_2^- + 2H^+ + H_2O; \qquad \Delta F = -65 \text{ kcal}$$

and (2) *Nitrobacter*, which converts nitrite into nitrate and also uses the liberated energy for the synthesis of its own ATP:

$$NO_2^- + \tfrac{1}{2}O_2 \rightarrow NO_3^-; \qquad \Delta F = -18 \text{ kcal}$$

The occurrence of nitrification reactions in aquatic habitats at any appreciable rate is likely to create major biological oxygen demand because of the resulting anoxia; such a possibility requires a critical examination of water bodies for nitrifying bacteria in terms of their numbers and the conditions conducive to the nitrification process.

Denitrification involves the conversion of nitrate or nitrite into molecular nitrogen. The biological process of denitrification is brought about through the activity of anaerobic organisms in which nitrate substitutes for molecular oxygen as the terminal acceptor for respiratory electrons. *Pseudomonas denitrificans*, *Thiobacillus denitrificans*, and *Micrococcus denitrificans* are examples of denitrifying bacteria in which, under anaerobic conditions, denitrification provides a mechanism for the coupling of ATP synthesis with the migration of respiratory electrons from organic substrates to nitrate or nitrite. It has been observed that denitrification proceeds in four steps, NO_3^- to NO_2^- to NO to N_2O to N_2 (Payne, 1971). This process of nitrate or nitrite reduction is termed dissimilatory in order to distinguish it from the assimilatory reduction of nitrate to ammonia. Both assimilatory and dissimilatory nitrate reductases have been isolated and characterized, and the two reductases differ in respect of oxygen sensitivity which is greater in dissimilatory reductases than in assimilatory ones. Denitrification is restricted to anaerobic environments whereas nitrification occurs in aerobic conditions.

According to Burns' and Hardy's (1973) estimation of the global rate of nitrogen fixation in sea and land, the total amount of nitrogen fixed per annum is nearly 175×10^6 tons—almost 30×10^6 tons to 31×10^6 tons through the Haber-Bosch process and about 35×10^6 tons by other abiological processes of fixation. Nearly 15×10^6 tons of nitrogen are transported annually from the land to the sea through run-off, and approximately the same quantity is transferred to deep sea sediments; of this amount, nearly 5×10^6 tons are returned to the sea through solubilization. The loss of nitrogen in ammonium form is placed at approximately 165 tons per year, and 140×10^6 tons of ammonia and 60×10^6 tons of nitrate/nitrite nitrogen are brought back to the sea and land through precipitation.

Quantitative estimates of nitrogen lost through denitrification are not available.

BIOLOGICAL NITROGEN FIXATION

On the basis of their occurrence in natural habitats, nitrogen-fixing organisms are divided into two groups: symbiotic and nonsymbiotic. The symbiotic forms are subdivided into two types: associative (e.g., lichens) and obligatory (e.g., root nodules of legumes in which the interaction of host and bacterial genomes is necessary for fixation to occur). In this context, the distinction between nonsymbiotic and associative symbiotic forms is purely ecological as the associative symbiotic forms of blue-green algae or bacteria can fix nitrogen also when separated from their counterparts which may be an angiosperm, a gymnosperm, a fern, or a fungus. Obligatory symbiotic nitrogen fixers include leguminous plants with the root-nodule forming species of *Rhizobium*; and symbiotic azotrophs include nonleguminous plants, such as *Hippophae* and *Casuarina*, with nodule-forming actinomycetes, and lichens consisting of a fungal partner and a nitrogen-fixing blue-green alga. The free-living nonsymbiotic nitrogen-fixing agents include (1) heterocystous blue-green algae, such as *Nostoc*, *Anabaena*, *Aulosira*, and *Cylindrospermum*; (2) photosynthetic bacteria, for instance, *Rhodospirillum*, *Chromatium*, and *Chlorobium*; (3) aerobic nonphotosynthetic bacteria, e.g., *Azotobacter* and *Beijerinckia*; and (4) anaerobic nonphotosynthetic bacteria, such as *Clostridium* and *Desulphovibrio*.

The mechanism of nitrogen fixation appears to be the same in all organisms. Legume nodules contain a pink pigment very much like the haemoglobin of animals; this pigment is called leghaemoglobin. It is apparently essential for nitrogen fixation in nodules since those nodules that lack this pigment have been found inactive in nitrogen fixation. The actual function of leghaemoglobin is still open to conjecture: it is suggested that it is concerned with the transport of intracellular oxygen away from the nitrogen-fixing site; the enzyme nitrogenase is highly sensitive to oxygen and its activity is lost in the presence of oxygen. Such a mechanism may be applicable to leguminous plants, but not to all nitrogen-fixing plants since many of these that may not have the leghaemoglobin may yet fix nitrogen. Recently, a strain of the cowpea *Rhizobium* has been shown to fix nitrogen in the absence of its host, i.e., in the free-living state (Pagan *et al.*, 1975; McComb *et al.*, 1975). The essential requirements for nitrogen fixation in the free-living state appear to be a pentose sugar, a fixed form of nitrogen such as ammonium and glutamine, and a dicarboxylic acid of the TCA cycle. Should this situation turn out to be applicable to other strains of *Rhizobium* as well, then the concept of obligatory symbiosis, in which the host is regarded as indispensable for nitrogen fixation (in Rhizobial strains), would no longer be valid; in this case, legume symbiosis would be similar to nonlegume symbiosis in respect of the ability of bacterial symbionts to fix nitrogen in the absence of the host. Phylloplane and rhizosphere associations involving free-living diazotrophs and higher plants have recently been studied and significant levels of nitrogenase activity have been reported for rhizosphere of sugarcane (a C_4 plant). Positive evidences are now available to show that there exists an association between nitrogen-fixing bacteria and some tropical C_4 grasses. Von Bülow and Dobereiner (1975) have reported that, in Brazil, maize roots fix nitrogen and contain a nitrogen-fixing *Spirillum*.

The general requirements for nitrogen fixation are (1) a protective mechanism for nitrogenase against oxygen; (2) a hydrogen source; (3) an energy source, ATP; (4) the enzyme nitrogenase in an active form; and (5) a carbon compound, generally a keto acid, for trapping the ammonia resulting enzymatically from the reduction of molecular nitrogen. The hydrogen is utilized in the conversion of molecular nitrogen into ammonia with the help of ATP. Both hydrogen and ATP may come from photochemical reactions of photosynthesis or from aerobic respiration. In fact, the basic requirements for nitrogen fixation are ATP and a suitable reductant (Fig. 10-2), e.g., reduced ferredoxin or flavodoxin, irrespective of whether these result from respiration or photosynthesis.

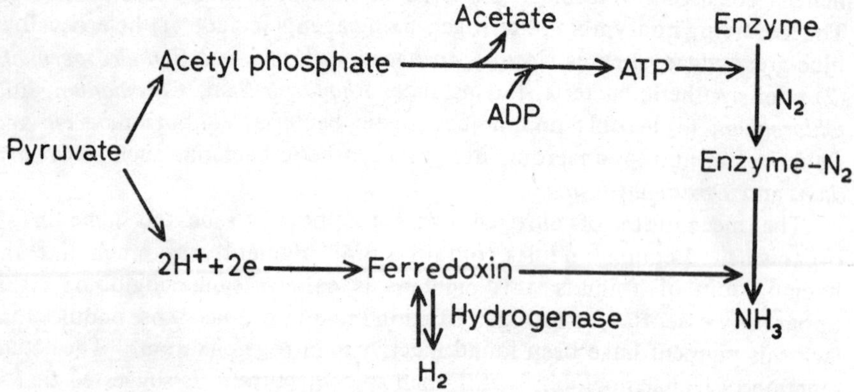

Fig. 10-2 Electron and energy transfer in nitrogen fixation.

That ammonia is the key intermediate of nitrogen fixation is no longer questioned. All the intermediates between nitrogen and ammonia have not yet been identified perhaps because molecular nitrogen is converted into ammonia on the surface of the enzyme itself thus avoiding the need for free intermediates.

Since, in tracer experiments, glutamate and glutamine derive the highest label from either $^{15}N_2$ or $^{15}NH_3$, these have generally been thought to result by reductive amination of alpha-ketoglutaric acid by glutamate dehydrogenase (GDH), leading to the formation of glutamate. The alternative possibility that the primary route of ammonia assimilation might be through amide nitrogen of glutamine has not received sufficient attention mainly because of the lack of a mechanism for the cyclical formation of glutamate from glutamine. Recently, another route of ammonia assimilation has been discovered in which glutamine synthetase (GS) and glutamate synthetase were shown to participate in the formation of glutamate through glutamine as the primary aminating reaction. GS seems to have a very high affinity for ammonia in comparison to GDH. Both GS and glutamate synthetase are produced in nitrogen-fixing organisms in a low-ammonium medium or in a nitrogen-free medium. In contrast, GDH formation occurs invariably in a high-ammonium medium. Thus, GS is the key enzyme of free-living nitrogen-fixing organisms for the organic incorporation of ammonia that results from nitrogen fixation. In *Klebsiella*, GS acts as a positive regulator of the *nif* operon (see Shanmugam *et al.*, 1975). The high ammonia pathway and the low ammonia pathway may be represented as follows:

High ammonia pathway:

$$NH_3 + \alpha\text{-ketoglutarate} + NADPH + H^+ \xrightarrow{GDH} \text{Glutamate} + NADP^+ + H_2O$$

Low ammonia pathway:

$$\text{Glutamate} + NH_3 + ATP \xrightarrow{GS} \text{Glutamine} + ADP + P_i$$

$$\text{Glutamine} + \alpha\text{-ketoglutarate} + NADPH + H^+ \xrightarrow[\text{synthetase}]{\text{Glutamate}} \text{Glutamate} + NADP^+$$

Figure 10-3 outlines the accepted mechanism of nitrogen fixation upto the formation of ammonia.

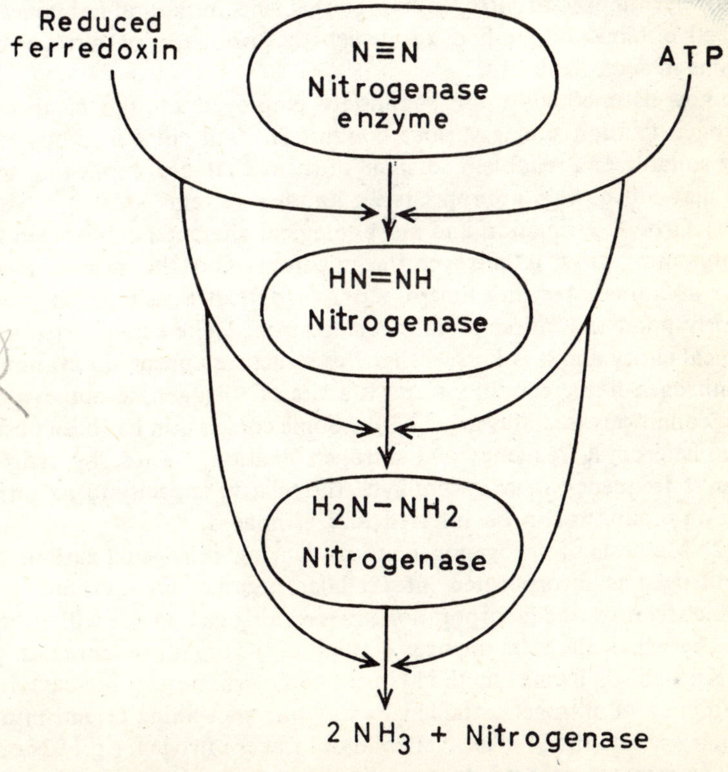

Fig. 10-3 Mechanism of nitrogen reduction upto the level of ammonia.

The enzyme nitrogenase is composed of two proteins, a Mo-Fe protein and an Fe-protein, each named on the basis of its metal component. The nitrogenases from different biological sources seem to be identical in their composition, functional requirements, and biological properties. They are extremely sensitive to oxygen, and the Fe-protein is also cold-labile. Because of the extreme sensitivity of nitrogenase to oxygen, different nitrogen-fixing organisms have evolved essentially varying biochemical and morphological structures for protection against oxygen toxicity. Examples are nodules in legumes, heterocysts in blue-green algae, extremely high respiratory rates in *Azotobacter*, and anaerobic photosynthesis and respiration in various bacteria. Nitrogenase has a very high requirement of energy which amounts to 15–20 ATP molecules per molecule of nitrogen fixed to the ammonia level. The reaction is composed of two steps: (1) electron activation, and (2) substrate reduction. Its maximum activity is around neutral pH (7) and its end product, ammonia, shows no inhibition of nitrogenase activity; however, it blocks nitrogenase synthesis by repression control.

Nitrogenase is a very versatile enzyme and attacks most compounds containing triple bonds, e.g., nitrogen ($N{\equiv}N$), acetylene ($HC{\equiv}CH$), and hydrogen cyanide ($HC{\equiv}N$). When the enzyme acts upon acetylene, the latter is reduced to ethylene ($H_2C{=}CH_2$); this has proved to be a very sensitive and inexpensive method for determining, quantitatively as well as qualitatively, the role of nitrogen-fixing organisms in natural habitats and as an effective bioassay method to detect the presence of nitrogen-fixing organisms in such habitats.

We now list methods more commonly employed for the measurement of nitrogen fixation under various conditions. All these methods are not equally suitable and reliable, and some of them that are applicable to one system may not be very appropriate for another system.

(1) Growth estimation and morphological characters: Nitrogen-fixing microorganisms grow in nitrogen-free medium, and the measurement of increase in biomass per unit time in terms of optical density or dry weight is a fairly good indication of nitrogen fixation. The heterocyst is a morphological entity and it is known that this structure appears in cyanophytes under nitrogen-fixing conditions and is a site of nitrogenase activity under aerobic conditions (see Stewart, 1973). Some correlation has been observed between heterocyst frequency and nitrogen fixation. Thus, by estimating heterocyst frequency in a cyanophyte, the relative capacity to fix nitrogen in such an organism can be inferred and estimated.

(2) Methods of nitrogen estimation: During nitrogen fixation, molecular nitrogen is incorporated into cellular organic nitrogen and the rate with which fixation and incorporation into cell nitrogen occurs will determine the rate by which the total nitrogen content of the organism increases. The micro-Kjeldahl or Dumas method is used to determine the increase in the total amount of nitrogen attained by any nitrogen-fixing organism over a particular span of time. The availability of tracer nitrogen, e.g., ^{15}N or ^{13}N, makes the measurement of nitrogen fixation more sensitive and accurate.

Normal nitrogen is ^{14}N, and organisms growing with ^{14}N will contain ^{14}N as the only cellular nitrogen. Such organisms, if allowed to utilize labelled nitrogen, will fix the isotopic nitrogen, with the consequence that their cellular nitrogen will get progressively enriched with ^{15}N or ^{13}N. When ^{15}N which is a stable isotope is used, its proportion in an ^{15}N-^{14}N mixture can be detected and estimated in a mass spectrometer.

Titrimetric or colorimetric methods can also be employed. All the cellular nitrogen of the organism is chemically digested into ammonia which is then determined titrimetrically or colorimetrically.

(3) Acetylene reduction technique: The nitrogenase enzyme found in the nitrogen-fixing organism can reduce acetylene to ethylene. This reaction can be followed quantitatively in a gas chromatograph (Hardy *et al.*, 1968).

Alternatively, the ethylene so produced can be converted into a coloured compound and the latter estimated quantitatively by a colorimetric method (LaRue *et al.*, 1972).

The nitrogenase enzyme can also reduce nitriles and isonitriles and the extent of such reduction can be measured by gas chromatography (Fuchsman and Hardy, 1971).

(4) When supplied with electrons and ATP, nitrogenase catalyzes the union of electrons with protons with the release of molecular hydrogen. This property can be utilized for estimating fixation manometrically, gas chromatographically, or mass spectrometrically (Fuchsman and Hardy, 1971).

NITRATE REDUCTION

Nitrate is reduced to ammonia and water before being converted into organic form. The reduction involves a gain of electrons (and hydrogen). Three oxygen atoms and one nitrogen atom of nitrate will respectively require six and three electrons (or hydrogens) to form three molecules of water and one molecule of ammonia, viz.,

$$HNO_3 + 8(H) \rightarrow NH_3 + 3H_2O$$

and the nitrate anion has one extra electron or negative charge; the complete reduction of the anion will therefore involve a net change of eight electrons.

Since each biological reduction step proceeds with the consumption of two electrons, the formation of NH_3 from nitrate will occur in four steps:

$$NO_3^{(+5)} \longrightarrow NO_2^{(+3)} \longrightarrow X_1^{(+1)} \longrightarrow X_2^{(-1)} \longrightarrow NH_3^{(-3)}$$

The valency of the nitrogen atom is zero, that of NH is -3, and that of NO is 5. Thus, in terms of valency too, the conversion of nitrate into ammonia is represented as a change of valency state from 5 to -3. Much controversy still exists regarding the nature of the intermediates at the oxidation levels 1 and -1, i.e., X_1 and X_2. It is well substantiated that nitrate is reduced first to nitrite and its final reduction product is ammonia. Some workers suggest that hydroxylamine is also one of the intermediates and, since its oxidation level is -1, it is believed to represent the hypothetical X_2 in the foregoing equation.

One way to ascertain whether a particular substance is the intermediate in a postulated multistep pathway is to feed the substance to the organism and study its growth behaviour and enzymatic make-up. For example, plants can use both nitrate and ammonium; when both are available, ammonium is preferred to nitrate. Such ammonium-grown organisms lack the entire enzyme systems that reduce nitrate to ammonia. Similarly, plants also use nitrite as a nitrogen source; when they do so, the enzyme that reduces nitrate to nitrite, nitrate reductase, is lacking. When nitrate is the only nitrogen source available, the plants use it to make all the enzymes required to form ammonium. This process of nitrate reduction has also been confirmed by the use of isotopic nitrate or nitrite, i.e., $^{15}NO_3$ or $^{15}NO_2$.

Hydroxylamine is identified as the intermediate in nitrate reduction, but this appears doubtful when it is noticed that hydroxylamine is toxic to plant growth. If hydroxylamine is actually an intermediate, then perhaps its rate of turnover is sufficiently high to prevent it from accumulating in physiologically toxic concentrations. It is believed that X_1 and X_2 intermediates always remain enzyme-bound and, hence, the difficulty in isolating and identifying them. Mutants of organisms lacking nitrate reductase do not grow with nitrate, but can grow like the wild-type parent when nitrite is supplied. This is proof that nitrite is the first step in the reduction of nitrate to ammonia.

MECHANISM OF REDUCTION

The enzyme that reduces nitrate to nitrite is reported to be formed in all organisms using nitrate. The enzyme-catalyzed reduction depends upon a suitable electron donor which may be $NADPH_2$ (produced in photosynthesis), reduced ferredoxin (formed during photosynthesis or respiration), or $NADH_2$ (generated during respiration). It is observed that nitrate reductases from various sources differ in respect of the electron donor and the cofactors. In this respect, nitrate reductases fall into various categories. Nevertheless, all these different reductases share the common function of reducing nitrate to nitrite. The enzyme is made up of a protein part (apoenzyme) which is specific for the substrate and the two prosthetic groups FAD and Mo. Its mode of action is as follows:

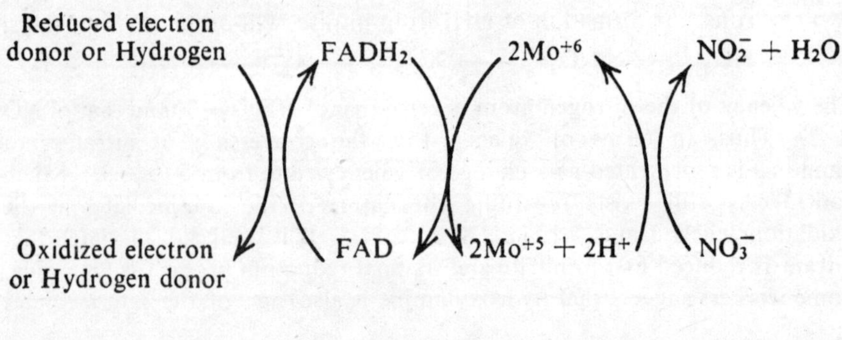

The electron donor may be either water (through NADPH formation), as in plants, or hydrogen, in case the plants contain the enzyme hydrogenase, or pyruvate which by oxidative decarboxylation generates $NADH_2$. Reduced ferredoxin can also cause the reduction of nitrate.

The nitrite reductase catalyzes the reduction of nitrite to an unknown intermediate which by further reduction in two steps forms ammonia. The details of the enzymatic steps between nitrite and ammonia are not yet clear. Perhaps a single complex enzyme reduces nitrite to ammonia on its own surface without giving the intermediate an opportunity to arise freely.

We have seen that both inorganic molecular nitrogen and inorganic combined nitrogen (e.g., nitrate and nitrite) get converted into ammonia by different enzymatic mechanisms. Cells seem to have the discretionary power to use the most economical of inorganic nitrogen sources when all these forms are available. For example, ammonia is the level at which inorganic nitrogen enters the organic form. When cells come in contact with inorganic nitrogen sources that include ammonia, they consume ammonia preferentially by repressing the production of nitrate- or nitrite-reducing enzyme systems. The availability of a variety of enzyme systems performing the same function, viz., ammonia formation, is ecologically advantageous to the cells in that the cells will be able to grow in normal environment with any kind of inorganic nitrogen that may be available.

SULPHUR METABOLISM

Like nitrogen and carbon, sulphur is known to undergo extensive metabolic transformations in the biosphere. Sulphate, the highly oxidized form of sulphur, is the most readily available source of sulphur in nature, and procaryotes as well as eucaryotes (excluding protozoa and most animals) derive all their sulphur from sulphate. Protozoa and animals, in general, lack the enzyme systems (required for the reduction of sulphate to the thiol level) found in cysteine and methionine, the key amino acids required for the formation of protein.

Sulphate enters two major kinds of biochemical reaction: (1) esterification, and (2) reduction. In (1), sulphate esters of polysaccharides, phenols, steroids, and other organic substrates result. It is significant to note that whereas sulphate esterification is widespread among procaryotic and eucaryotic microorganisms and higher animals, it is very limited in higher plants. The notable sulphated polysaccharides of animals include chondroitin sulphate and hyaluronic acid. Carrageenin and agar are the most common sulphated polysaccharides found in cell walls of diverse red algae. In (2), sulphate is reduced to the thiol level. This reduction is characteristic of all organisms except protozoa and most animals. The sulphate-reducing groups of organisms (algae, fungi, higher plants) and the sulphate-nonreducing groups (protozoa and animals) resemble each other in the oxidation of reduced sulphur compounds back to the level of sulphate.

During the oxidation of cellular fuels, certain anaerobic bacterial species belonging to the *Desulphovibrio* group use sulphate as the terminal electron acceptor in place of oxygen. The net result of such a mechanism is the coupling of ATP formation with the energy released during the migration of electrons from respiratory substrates to sulphate which ultimately gets reduced to sulphide. The reverse process which is also energy-yielding and in which the reduced sulphur is oxidized to sulphate is carried out by another group of autotrophic sulphur bacteria, called chemosynthetic sulphur bacteria. Figure 10-4 gives an outline of the sulphur cycle in the biosphere. In this, the sulphate reduction leading to the formation of cysteine and methionine is termed assimilatory in contrast to that leading to H_2S formation which is called dissimilatory. In assimilatory reduction, sulphur is reduced to the thiol form which is trapped by organic acids, thus leading to the formation of sulphur-containing amino acids. Besides, there are numerous other biological reactions involving an intermolecular transfer of thiol groups in order to sustain the needs of fundamental cellular processes. In dissimilatory reduction, sulphate reduction is accompanied by the release of energy which is consumed in the formation of ATP from ADP and inorganic phosphate. The oxidation of reduced sulphur compounds leads also to the liberation of biologically useful energy which chemosynthetic sulphur bacteria use as the sole means of ATP synthesis.

SULPHATE UPTAKE

Sulphate uptake is an active process and almost all studies on this aspect of sulphate support its carrier-mediated transport in which the carrier may be an enzyme (Schiff and Hodson, 1973). Sulphite, thiosulphate, methionine, and cysteine are known to inhibit the absorption of sulphate. In addition, the carrier protein seems to be specific for all the anions of Group VI of the Periodic Table, since selenium analogues of sulphur competitively antagonize the uptake of the corresponding sulphur compound.

SULPHATE REDUCTION

The reduction of sulphate to sulphite is an endergonic reaction which proceeds at the expense of nearly 14 kcal of energy. This may be contrasted with nitrate reduction in which nitrate is reduced to nitrite with the liberation of a large amount of energy, thereby becoming exothermic. Since the usual source of energy for endothermic reactions is ATP, sulphate must react with this compound before it undergoes reduction; i.e., the cells should have a mechanism to activate sulphate prior to its reduction to sulphite. Recently, an enzyme called ATP sulphurylase has been isolated from the cells of sulphate-metabolizing organisms. This enzyme catalyzes, both *in vivo* and *in vitro*, the formation of adenosine-5′-phosphate sulphate (APS) from sulphate and ATP:

$$ATP + SO_4 \rightleftharpoons \underset{\text{APS}}{AMP\text{-}SO_4} + \underset{\substack{\text{Inorganic}\\ \text{pyrophosphate}}}{PP_i}$$

However, this reaction overcomes nearly 8–9 kcal of the 14 kcal energy barrier of the sulphate reduction and, hence, still has a remarkable tendency

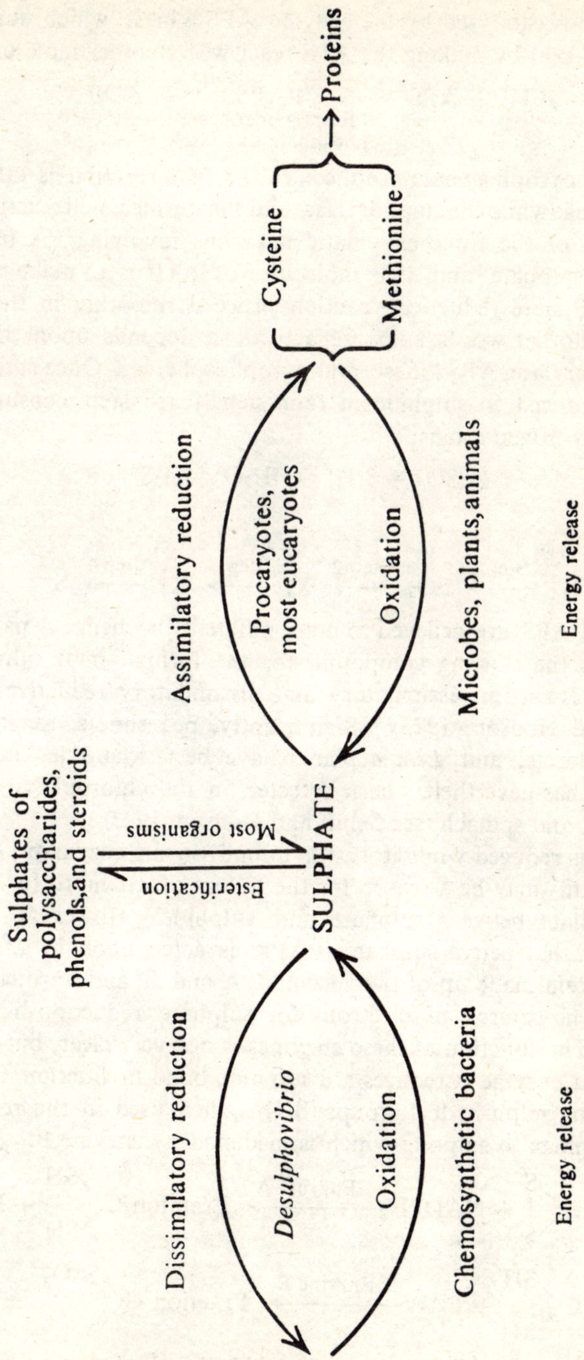

Fig. 10-4 Sulphur cycle in the biosphere.

to move in the reverse direction. This tendency is checked by another reaction which is catalyzed by the enzyme APS kinase which supplies the remaining 5–6 kcal by making the APS react with another mole of ATP:

$$\text{ATP} + \text{APS} \rightarrow \underset{\substack{\text{Phospho adenosine}\\ \text{phosphate sulphate}}}{\text{PAPS}} + \text{ADP}$$

The inorganic pyrophosphate produced in the first reaction is attacked by pyrophosphatase while the sulphurylase- and kinase-mediated reactions go on. The net effect of the three enzymatic reactions involving the use of one molecule of sulphate and two molecules of ATP is to make the highly endergonic sulphate reduction reaction proceed smoothly in the forward direction. In other words, sulphate activation depends upon the activity of ATP sulphurylase, APS kinase, and pyrophosphatase. Once activated, the sulphate is reduced to sulphide in four steps, each step consuming two electrons or hydrogen atoms:

$$H_2SO_4 + 8(H) \rightarrow 4H_2O + H_2S$$

or

$$\overset{(+6)}{SO_4^{--}} \xrightarrow{\text{Step 1}} \overset{(+4)}{SO_3} \xrightarrow{\text{Step 2}} \overset{(+2)}{X_1} \xrightarrow{\text{Step 3}} \overset{(0)}{X_2} \xrightarrow{\text{Step 4}} \overset{(-2)}{S^{--}}$$

PAPS and APS are believed to enter different biochemical pathways in that PAPS is the starting compound for the formation of sulphate ester whereas APS is used in assimilatory and dissimilatory reductive pathways (see Schiff and Hodson, 1973). Significantly, pea shoots, tomato shoots, cauliflower florets, and *Lemna* plants have been found lacking in APS kinase which has nevertheless been detected in the chloroplasts of French beans, maize, and spinach (see Schiff and Hodson, 1973).

Sulphate is reduced while it is still bound on an enzymatic or protein surface, and this may be a reason for the failure of attempts to isolate the free intermediate between sulphate and sulphide. However, the known mechanism is that active sulphate (PAPS) is acted upon by the sulphate reductase system made up of two enzymes, A and B, and a protein fraction, called C. The source of electrons for sulphate reduction is generally $NADPH_2$. The function of these enzymes is not very clear, but it may be suggested that enzyme A reduces a disulphide bond in fraction C, and one of the resulting sulph-hydryl groups (SH) is then used in the reduction of activated sulphate to sulphite which is mediated by enzyme B:

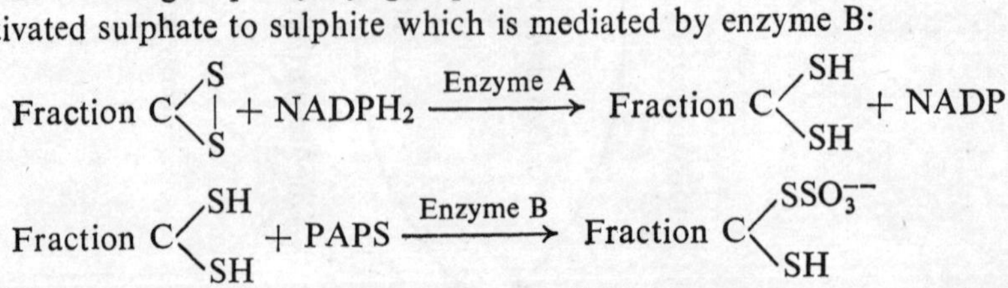

The further reduction of sulphite to sulphide involves the use of six more electrons, but the precise mechanism has not yet been determined. It is likely that the intermediates upto the thiol level of sulphate or sulphite reduction remain bound on the enzyme surface.

SULPHOTRANSFERASE REACTIONS

A number of plants, animals, and microbes have been shown to contain specific sulphotransferases which catalyze the esterification of various organic compounds with sulphate or sulphite, thus leading to the formation of specific kinds of sulphate esters and sulphonic acids. Some well-characterized sulphate esters include phenol sulphate, flavonoid sulphate, steroid sulphate, polysaccharide sulphate, and choline sulphate. In almost all analyses, PAPS has been found to be an efficient donor of the sulphate group. The occurrence of sulphated polysaccharides in the cell walls of certain marine algae and bacteria has aroused considerable speculation as to the ecological significance of these cell wall components. The general opinion is that such walls offer effective protection against desiccation.

Sulphonic acids are another class of compounds in which sulphur is in a more highly reduced state than in sulphated compounds. Examples are glucose-6-sulphonate (found in plant sulpholipids), cysteic acid, homocysteic acid, taurine, N-methyl taurine, and N-dimethyl taurine. In sulphate esters, there is an oxygen atom between the carbon and the sulphur of sulphate whereas in sulphonic acids, sulphur is directly linked to carbon. The mechanism of formation of sulphonic acid derivatives in plants or animals is not yet known, and the biological significance of such compounds also is not well understood.

SULPHUR-SELENIUM INTERRELATIONS

Both sulphur and selenium belong to the same group of the Periodic Table, show almost similar physical and chemical properties, and form an analogous series of inorganic and, to some extent, bioorganic compounds, e.g., sulphate, selenate; sulphite, selenite; methionine, selenomethionine; and cysteine, selenocysteine. An analogy between these two elements is so close that the selenium cycle, similar to the sulphur cycle, is also believed to exist in the biosphere though this has not yet been systematically studied.

Most plants and microorganisms, when confronted with selenoanalogues of sulphur compounds, first absorb them and then metabolize them by using the same enzyme system that is involved in sulphur metabolism, thereby substituting selenium for sulphur atom by atom. The biological consequence of selenium substitution is generally always lethal or fatal. Sulphur usually counteracts this toxicity in an antagonistic manner (see Shrift, 1969; Kumar and Prakash, 1971). However, some reports are available to show the essentiality of traces of selenium for animals in general and also for some selenium-accumulating plants, e.g., species of *Astragalus*. The genus *Astragalus* includes both the selenium-accumulator and a nonaccumulator species which grow side by side on seleniferous soils, and a biochemical analysis of the difference between the two groups suggests that selenium is densely stacked into selenomethyl selenocysteine in the accumulator species but not so concentrated in the nonaccumulator group (see Shrift, 1969). Since both accumulators and nonaccumulators grow on seleniferous soil without suffering any apparent damage, methylation of selenocysteine (in the accumulators) may constitute a biochemical mechanism for the detoxification of selenium. A similar mechanism of detoxification may also exist in the nonaccumulator species in which selenomethionine, instead of selenocysteine, is believed to be formed but not accumulated. Thus, there seems to exist some biochemical differentiation between the accumulator and the nonaccumulator species of *Astragalus*.

The study of sulphur-selenium antagonism or interaction is a newly-emerging field of plant physiology and carries considerable agricultural significance in those areas where seleniferous ores occur.

AMINO ACIDS

If any protein is normally hydrolyzed, it yields upto 20 different kinds of amino acids which are called protein amino acids. This however does not imply that cells form only 20 kinds of amino acids; rather they produce many more and the extra amino acids are termed the nonprotein amino acids as they are not known to form the structural components of the proteins (Tables 10-1 and 10-2).

Table 10-1 Protein amino acids

Common name	Chemical name	Structural formula
	With Aliphatic Side Chains	
Glycine	Aminoacetic acid	$^{\alpha}CH_2-COOH$ (on $^{\alpha}C$: NH_2)
Alanine	α-aminopropionic acid	$^{\beta}CH_3-{}^{\alpha}CH-COOH$ (on $^{\alpha}C$: NH_2)
Valine	α-aminoisovaleric acid	$(CH_3)_2{}^{\beta}CH-{}^{\alpha}CH-COOH$ (on $^{\alpha}C$: NH_2)
Leucine	α-aminoisocaproic acid	$(CH_3)_2{}^{\gamma}CH-{}^{\beta}CH_2-{}^{\alpha}CH-COOH$ (on $^{\alpha}C$: NH_2)
Isoleucine	α-amino-β-methyl-valeric acid	$CH_3-CH_2-{}^{\beta}CH-{}^{\alpha}CH-COOH$ (on $^{\beta}C$: CH_3; on $^{\alpha}C$: NH_2)
	With Side Chains Containing Hydroxyl Groups	
Serine	α-amino-β-hydroxy-propionic acid	$^{\beta}CH_2-{}^{\alpha}CH-COOH$ (on $^{\beta}C$: OH; on $^{\alpha}C$: NH_2)

Table 10-1 Protein amino acids (cont.)

Common name	Chemical name	Structural formula
Threonine	α-amino-β-hydroxy-butyric acid	${}^{\gamma}CH_3-{}^{\beta}CH(OH)-{}^{\alpha}CH(NH_2)-COOH$
	With Side Chains Containing Sulphur	
Cysteine	α-amino-β-mercap-topropionic acid	${}^{\beta}CH_2(SH)-{}^{\alpha}CH(NH_2)-COOH$
Methionine	α-amino-γ-methyl-thiobutyric acid	${}^{\gamma}CH_2(S-CH_3)-{}^{\beta}CH_2-{}^{\alpha}CH(NH_2)-COOH$
	With Side Chains Containing Carboxylic or Amide Groups	
Aspartic acid	α-aminosuccinic acid	$HOO{}^{\gamma}C-{}^{\beta}CH_2-{}^{\alpha}CH(NH_2)-COOH$
Asparagine	γ-amide of α-ami-nosuccinic acid	$H_2N-{}^{\gamma}C(=O)-{}^{\beta}CH_2-{}^{\alpha}CH(NH_2)-COOH$
Glutamic acid	α-aminoglutaric acid	$HOO{}^{\delta}C-{}^{\gamma}CH_2-{}^{\beta}CH_2-{}^{\alpha}CH(NH_2)-COOH$
Glutamine	δ-amide of α-ami-noglutaric acid	$H_2N-{}^{\delta}C(=O)-{}^{\gamma}CH_2-{}^{\beta}CH_2-{}^{\alpha}CH(NH_2)-COOH$
	With Side Chains Containing Basic (Amino) Groups	
Arginine	α-amino-δ-guanidi-novaleric acid	$H-N(C(=NH)-NH_2)-{}^{\delta}CH_2-{}^{\gamma}CH_2-{}^{\beta}CH_2-{}^{\alpha}CH(NH_2)-COOH$
Lysine	α, ε-diaminocaproic acid	${}^{\epsilon}CH_2(NH_2)-{}^{\delta}CH_2-{}^{\gamma}CH_2-{}^{\beta}CH_2-{}^{\alpha}CH(NH_2)-COOH$
Histidine	α-amino-β-imidazo-lepropionic acid	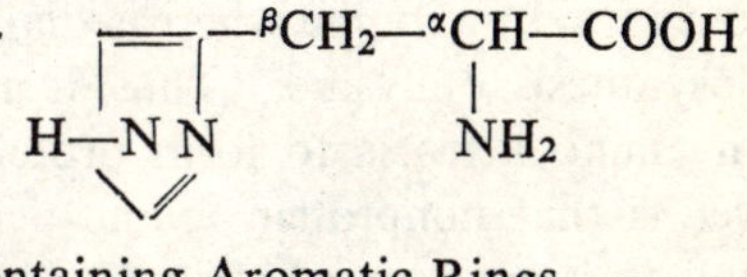
	With Side Chains Containing Aromatic Rings	
Phenylalanine	α-amino-β-phenyl-propionic acid	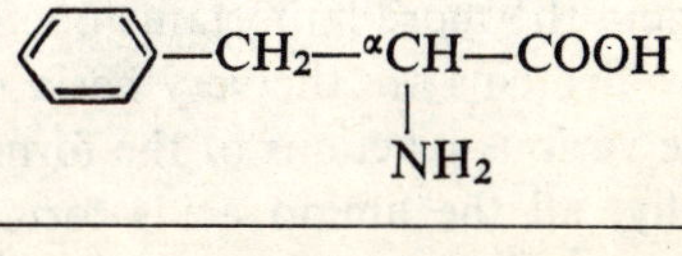

Table 10-1 Protein amino acids (cont.)

Common name	Chemical name	Structural formula
Tyrosine	α-amino-β-para-hydroxyphenylpropionic acid	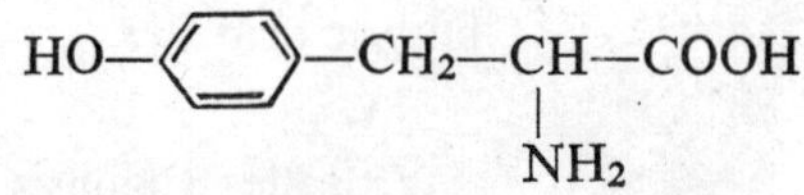
Tryptophan	α-amino-3-indolepropionic acid	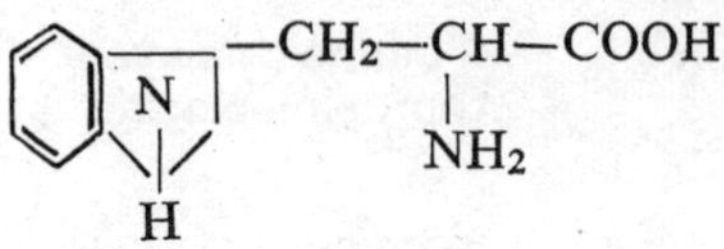
	Imino Acids	
Proline	Pyrrolidine-2-carboxylic acid	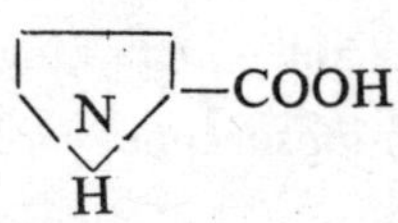
Hydroxyproline	4-hydroxy-pyrrolidine-2-carboxylic acid	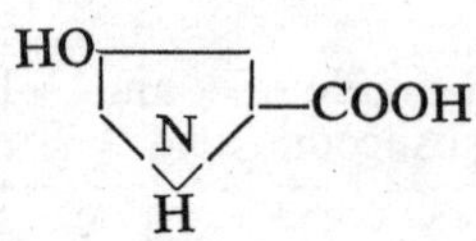

The functions of the nonprotein amino acids are not yet clear, but their concentration and number are found to vary with the plant organs and their stages of development. For instance, diaminobutyric acid and homoarginine are the predominant constituents of some seeds. Similarly, many derivatives of cysteine (S-substituted cysteines; R·S—CH_2·CH(NH_2)COOH, in which the H of SH has been replaced by R, designating any group or substance that can enter the substitution reaction) are known to accumulate in some storage organs. Since the total nitrogen in nonprotein amino acids is in considerable excess of that in protein molecules, the obvious inference is to consider them as seed nitrogen storage products which can be easily mobilized for various synthetic purposes.

Homoserine and diaminopimelic acid are other nonprotein amino acids. Homoserine is an intermediate in the biosynthesis of the protein amino acids methionine and threonine whereas diaminopimelic acid is an intermediate in the biosynthesis of lysine. Thus, it may be said that one function of nonprotein amino acids is to form protein amino acids. The general belief, however, is that nonprotein amino acids constitute a group of secondary metabolites (see Chapter 14). Further research is needed to clarify and understand the more important functions of these amino acids.

Since proteins are the very basis of all life activities, we shall describe only the various reactions of the formation of protein amino acids in plants. Generally, all the amino acids forming proteins have their amino group

Table 10-2 Nonprotein amino acids

Common name	Structural formula	Metabolic significance
α-aminoadipic acid	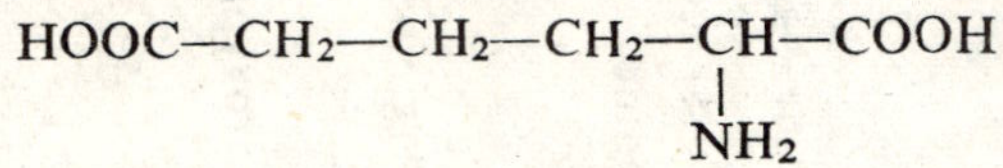	Intermediate in the synthesis of lysine
Homoserine	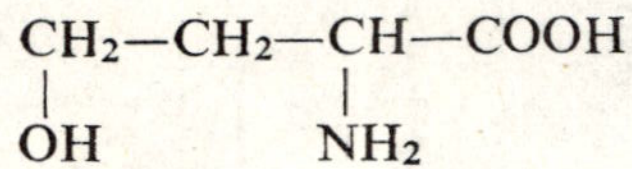	Intermediate in the synthesis of threonine, methionine, and cysteine
Homocysteine	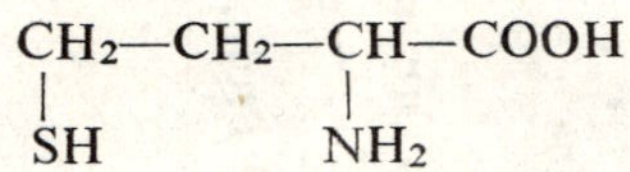	Intermediate in the synthesis of methionine and cysteine
Diaminopimelic acid	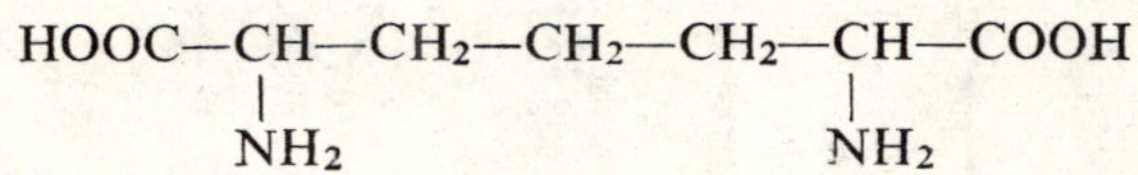	Intermediate in lysine biosynthesis; an important constituent of bacterial cell wall
Ornithine	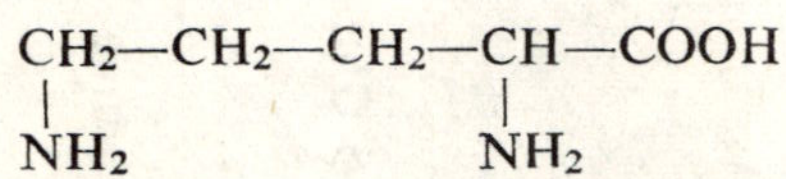	Involved in urea cycle
Citrulline	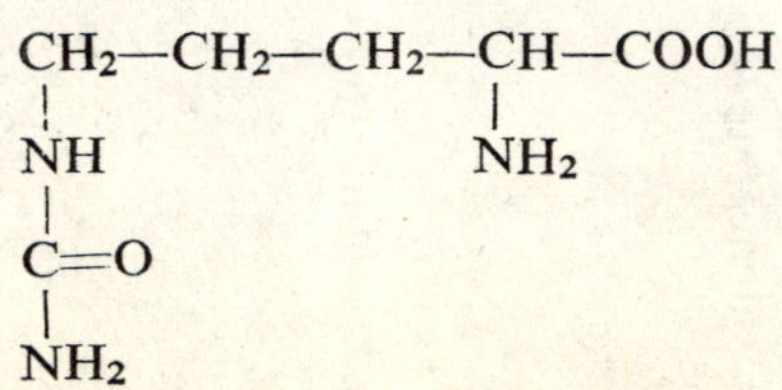	Involved in urea cycle

inserted at the α-carbon to which its carboxylic group is also attached. The general structure of an alpha-amino acid contains R as the side chain:

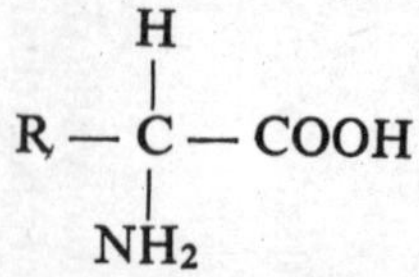

In accordance with the structure of the side chain, protein amino acids have been divided into seven groups (Table 10-1).

REACTIONS OF AMINO ACIDS

The structural requirements of an amino acid include (1) a carbon skeleton, and (2) NH_3. By using ^{14}C carbon metabolites and ^{15}N ammonia, it has been established that a number of amino acids arise from a common carbon skeleton. For example, oxaloacetate is the source of aspartic acid from further metabolism of which lysine, methionine, threonine, and isoleucine arise, i.e., these amino acids show an intimate biogenetic relationship. This has led to the concept of distinct amino acid families, e.g., the aspartate family, glutamic family, pyruvate family, and glyoxylate family (Fig. 10-5):

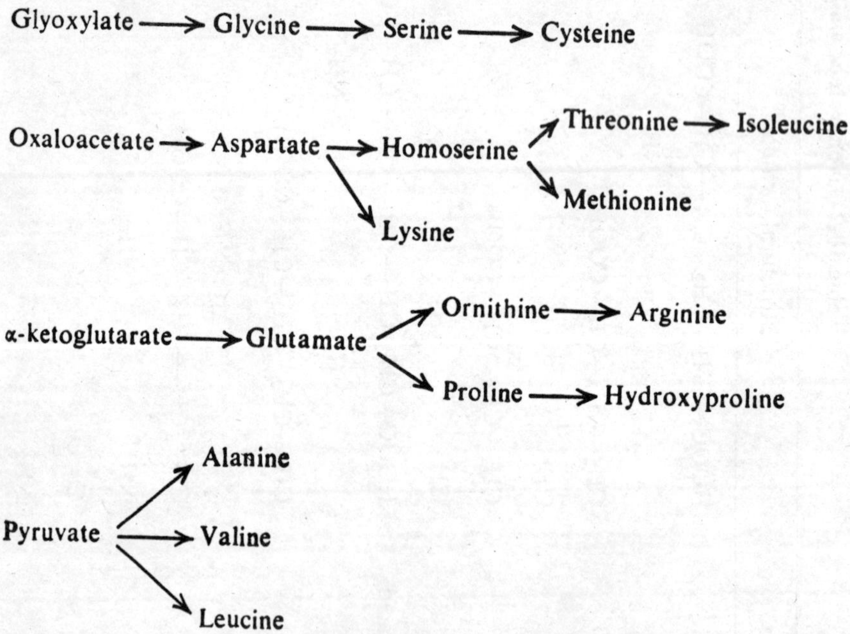

Fig. 10-5 Four distinct families of amino acids.

Glyoxylate can arise from glycollate in photosynthesis or from the operation of the glyoxylate cycle in glyoxysomes. Pyruvate, oxaloacetate, and α-ketoglutarate are the intermediates in the TCA cycle and can be produced from a variety of sources during respiration.

The formation of ammonia is perhaps the only way inorganic nitrogen (whether nitrate, nitrite, or molecular nitrogen) finds entry into amino acids. When C_4 plants (showing the dicarboxylic acid cycle of CO_2 fixation) are exposed to ammonia labelled with ^{15}N, the first amino acid to get labelled is aspartic acid. Photorespiring plants generally incorporate ammonia into glyoxylate, forming glycine, serine, and cysteine, in this order. Benson-Calvin plants use ammonia to form alanine, aspartic acid, and glutamic acid. The ammonia resulting from nitrogen fixation is utilized mainly in the formation of glutamic acid. Thus, the first organic product of ammonia utilization depends upon the type of operative metabolism and also the nature of the organism carrying out that metabolism.

We shall give a few examples to illustrate only the more well-known reactions of ammonia assimilation. One way the ammonia is assimilated is through reductive amination of alpha-ketoglutaric acid (or some other keto acid). The other method is through the formation of amides:

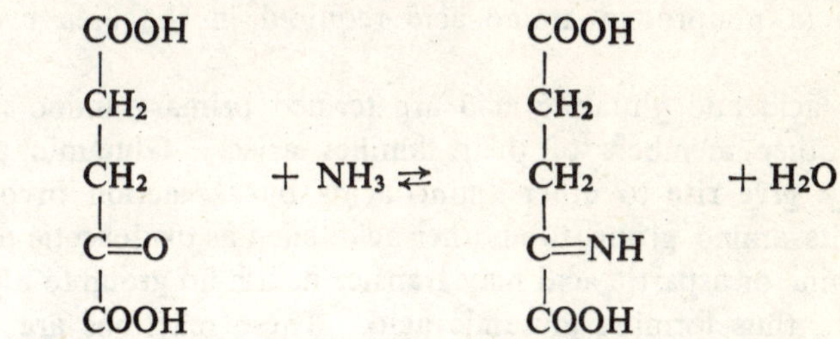

α-ketoglutaric acid α-imino glutaric acid

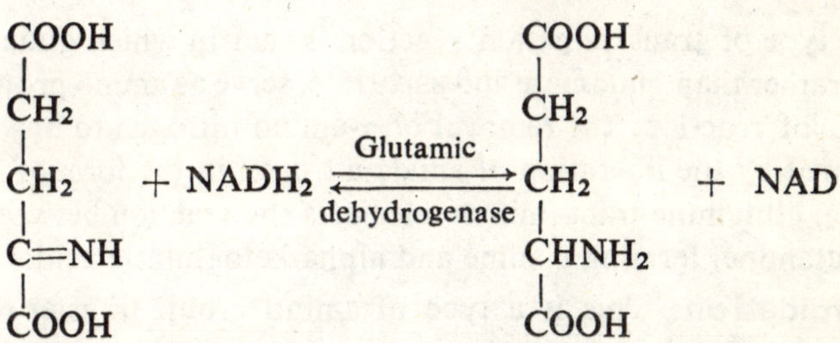

Glutamic acid

This reaction is catalyzed by the enzyme glutamic dehydrogenase, with $NADH_2$ as the hydrogen donor. In the same way, pyruvate and oxaloacetate can also undergo reductive amination in the presence of their respective dehydrogenases to give rise to alanine and aspartic acid, respectively. Similar reactions may be employed by cells to produce glycine from glyoxylate, although the mechanism of glycine formation is less certain at present.

Glutamine and asparagine are the amides of glutamic acid and aspartic acid, respectively. This reaction needs ATP as it is endergonic and is catalyzed by the enzyme synthetase (glutamic synthetase in the case of glutamine, and asparagine synthetase in the case of asparagine):

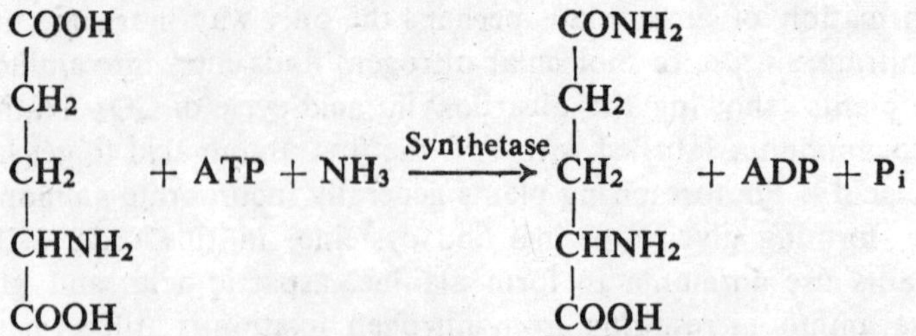

Glutamine

Two other methods of ammonia assimilation, although of relatively minor importance, are (1) the aspartase reaction, in which fumarate and ammonia react to form aspartic acid, and (2) the carbamyl phosphate synthetase reaction (see Chapter 7) where CO_2, ATP, and NH_3 react to form carbamyl phosphate ($O{=}C\langle^{NH_2}_{O\cdot P}$) which is required in the formation of citrulline (a nonprotein amino acid required in the urea cycle) from ornithine.

Aspartic acid and glutamic acid are termed primary amino acids from which the other members of their families arise. Glutamic acid once formed may give rise to other amino acids by a reaction involving the transfer of its amino group to another acid, such as oxaloacetic or pyruvic acid, or alanine or aspartic acid may transfer its amino group to alpha-keto-glutaric acid, thus forming glutamic acid. These reactions are catalyzed by specific transaminases which require pyridoxal phosphate as the prosthetic group.

Another type of transamination reaction is that in which glutamine and asparagine, rather than glutamate and aspartate, serve as amino group donors. In this kind of reaction, the removal of α-amino nitrogen to any keto acid is accompanied by the liberation of amide nitrogen in the form of ammonia. For example, glutamine transaminase mediates the reaction between pyruvic acid and glutamine, forming alanine and alpha-ketoglutaric acid.

Transamidation This is a type of amino group transfer reaction in which the amide group, and not the α-amino group, is transferred to a keto acid in exchange for the keto group. Such amide transfer reactions are called transamidation reactions and are mediated by specific transamidases. An example of such a reaction is the formation of glucosamine from glutamine and fructose-6-phosphate in the presence of the enzyme glutamine fructose phosphate transamidase.

Oxidative deamination The removal of the α-amino group of amino acids is the first step in the metabolic breakdown of amino acids. The reaction is catalyzed by amino acid dehydrogenases and proceeds as follows:

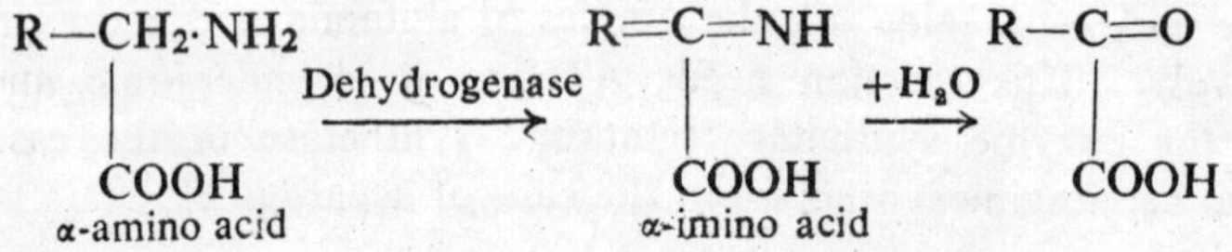

α-amino acid α-imino acid

Glutamine dehydrogenase is found in all living cells whereas other amino acid dehydrogenases occur in certain specific organs or tissues. It is because of this that the glutamic dehydrogenase reaction is the major pathway for oxidative deamination of most amino acids. This pathway begins with the removal of the amino group from an alpha-amino acid by transamination to alpha-ketoglutaric acid, thus forming glutamic acid and the corresponding keto acid. The glutamate is then deaminated to form α-ketoglutarate and ammonia. The deamination reaction is purely a catabolic reaction. We have already seen that reductive amination is the opposite of oxidative deamination and that the two reactions are catalyzed by the same enzyme, viz., glutamic dehydrogenase. Thus, the dehydrogenase reaction is reversible and, depending upon prevailing cellular conditions, can proceed either towards the synthetic or the degradative direction.

Decarboxylation Amino acids may be converted into amines by decarboxylation reaction which is catalyzed by specific decarboxylases requiring pyridoxal phosphate as the prosthetic group. Thus, glutamic acid is converted into gamma-aminobutyric acid, and aspartic acid into β-alanine, the carboxylic group attached to α-carbon being decarboxylated:

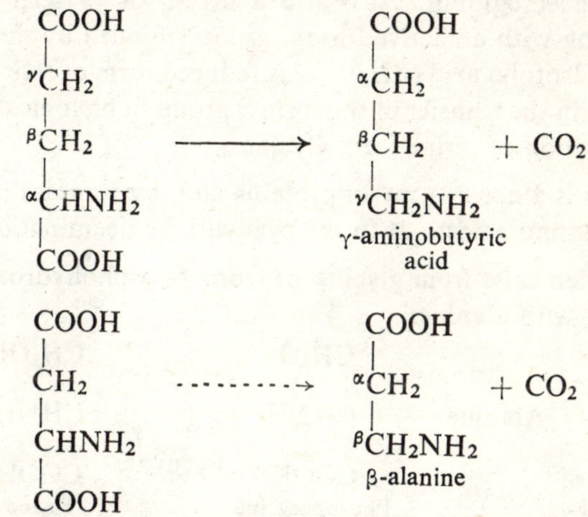

Aspartic acid

The biological significance of the amino acid decarboxylation reactions is not very clear, but the participation of β-alanine in CoA and the formation of pantothenic acid suggests the involvement of this reaction in the organization of some important biological compounds.

The various reactions of amino acids in which ammonia is liberated are of crucial importance to living beings in several respects: (1) the ammonia removed from one amino acid may be consumed to yield another amino acid; (2) the deaminated amino acids may be oxidized to water and CO_2 through the TCA cycle or may lead to glucose formation through the reversal of glycolysis, i.e., pyruvate, resulting from deamination or transamination reaction by forming phosphoenolpyruvate, is known to produce glucose; and (3) since ammonia is, even in very low concentrations, a cellular poison, there must be some means to check any rise in its intracellular concentration. In transamination reactions, the removed ammonia remains bound to the coenzyme; this is one way of detoxifying the ammonia. In animals, the urea cycle operates to remove intracellular ammonia arising through oxidative deamination of amino acids. In plants, the excess of ammonia is stored in the form of glutamine and asparagine. In other words, amide formation serves two functions, viz., trapping of free ammonia and representing a storage form of nitrogen reserve in plants.

METABOLISM AND FUNCTION OF INDIVIDUAL PROTEIN AMINO ACIDS

Glycine This can result from the transamination of glyoxylate with glutamate, glutamine, asparagine, and aspartate serving as the amino donor. When deaminated, glycine forms glyoxylate which can enter the glyoxylate cycle from where it may get completely oxidized or may enter the pathway leading to glucose formation. Besides being a constituent of proteins, it forms part of purines and the porphyrin ring of chlorophylls, cytochromes, and haemoglobin. It is also a precursor of serine which it forms by condensing with an active formyl group supplied by the formylated form of tetrahydrofolic acid (FH_4). This reduced form of folic acid is the coenzyme involved in the transfer of the formyl group in biological reactions, including the formation of serine from glycine.

Alanine This is a constituent of proteins and, by transamination, can give rise to other amino acids. It forms pyruvate by deamination.

Serine This can arise from glycine or from phosphohydroxypyruvate by transamination with alanine:

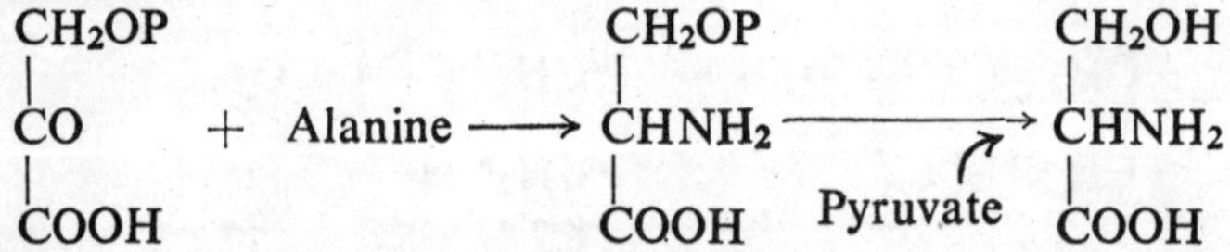

Phosphohydroxy-pyruvate — Phosphoserine — Serine

It can enter the amphibolic pathways through a reversal of its formation and participates also in purine and pyrimidine synthesis.

Threonine This is produced from phosphohomoserine ($^{\gamma}CH_2OP\cdot{}^{\beta}CH_2\cdot{}^{\alpha}CHNH_2\cdot COOH$) by the migration of —OH group from γ- to β-carbon. This reaction of intramolecular conversion is enzyme-catalyzed. Threonine is a very important functional constituent of proteins; its deamination leads to α-ketobutyrate whose further fate is not known.

Valine, Leucine, and Isoleucine The initial reaction in the formation of isoleucine is the deamination of threonine by threonine deaminase, leading to the formation of α-ketobutyrate from which isoleucine can arise in four enzymatic steps, as shown in Fig. 10-6. These details in plants,

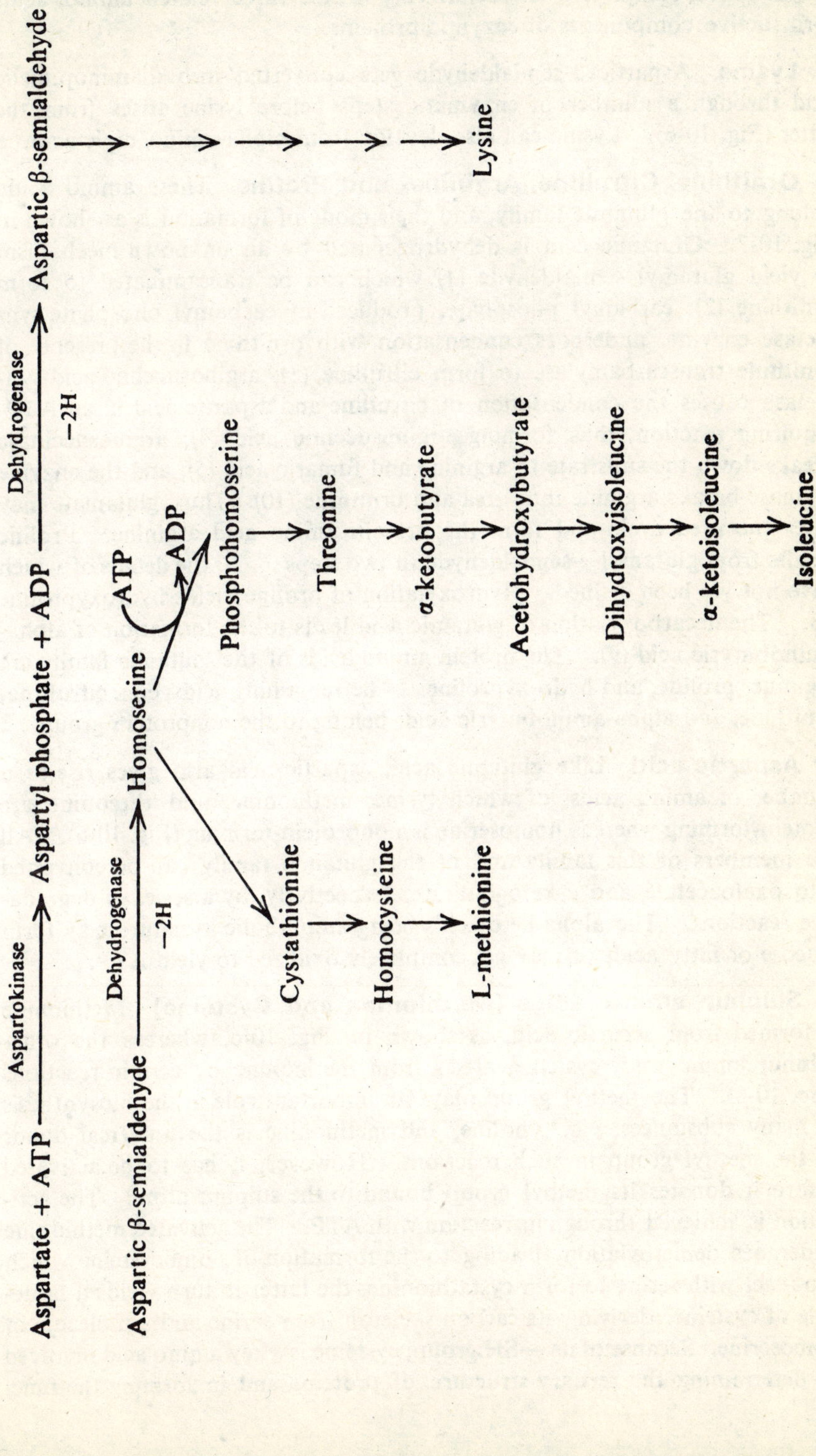

Fig. 10-6 Amino acids of aspartic acid family.

however, remain unknown. Similarly, the formation of valine and leucine is yet to be completely characterized. These three related amino acids form active components of enzymic proteins.

Lysine Aspartic β-semialdehyde gets converted into diaminopimelic acid through a number of enzymatic steps before lysine arises from the latter (Fig. 10-6). Lysine can also develop from alpha-aminoadipic acid.

Ornithine, Citrulline, Arginine, and Proline These amino acids belong to the glutamic family, and their mode of formation is as shown in Fig. 10-7. Glutamic acid is dehydrogenated by an unknown mechanism to yield glutamyl semialdehyde (1) which can be transaminated to form ornithine (2); carbamyl phosphate, produced by carbamyl phosphate synthetase enzyme, undergoes condensation with ornithine in the presence of ornithine transcarbamylase to form citrulline (3); arginosuccinic acid synthetase causes the condensation of citrulline and aspartic acid in an ATP-requiring reaction, thus forming arginosuccinic acid (4); arginosuccinase breaks down the substrate to arginine and fumaric acid (5); and the enzyme arginase breaks arginine into urea and ornithine (10). Thus, glutamate may enter the urea cycle and form the protein amino acid arginine. Proline results from glutamyl γ-semialdehyde in two steps (6, 7), the details of which have not yet been defined. Hydroxylation of proline yields hydroxyproline (8). The decarboxylation of glutamic acid leads to the formation of alpha-aminobutyric acid (9). The protein amino acids of the glutamic family are arginine, proline, and hydroxyproline. The remaining acids, e.g., citrulline, ornithine, and alpha-aminobutyric acid, belong to the nonprotein group.

Aspartic acid Like glutamic acid, aspartic acid also gives rise to a number of amino acids, of which lysine, methionine, and threonine are protein-forming whereas homoserine is nonprotein-forming (Fig. 10-6). All the members of this family and of the glutamic family can be converted into oxaloacetate and α-ketoglutarate, respectively, by a series of degradative reactions. The alpha-keto acids being amphibolic in nature may form glucose or fatty acids or may get completely oxidized to yield ATP.

Sulphur amino acids (Methionine and Cysteine) Methionine is formed from aspartic acid, as shown in Fig. 10-6, whereas the other sulphur amino acid, cysteine, arises from methionine by certain reactions (Fig. 10-8). The methyl group plays an important role in the biosynthesis of many substances, e.g., choline, and methionine is the universal donor of the methyl group in such reactions. However, it has to be activated before it donates its methyl group bound to the sulphur atom. The activation is achieved through its reaction with ATP. The activated methionine undergoes demethylation, leading to the formation of homocysteine which can react with serine to form cystathionine; the latter in turn yields a molecule of cysteine, deriving its carbon skeleton from serine and a molecule of homoserine. Because of its —SH group, cysteine is a key amino acid involved in determining the tertiary structure of proteins and in forming the func-

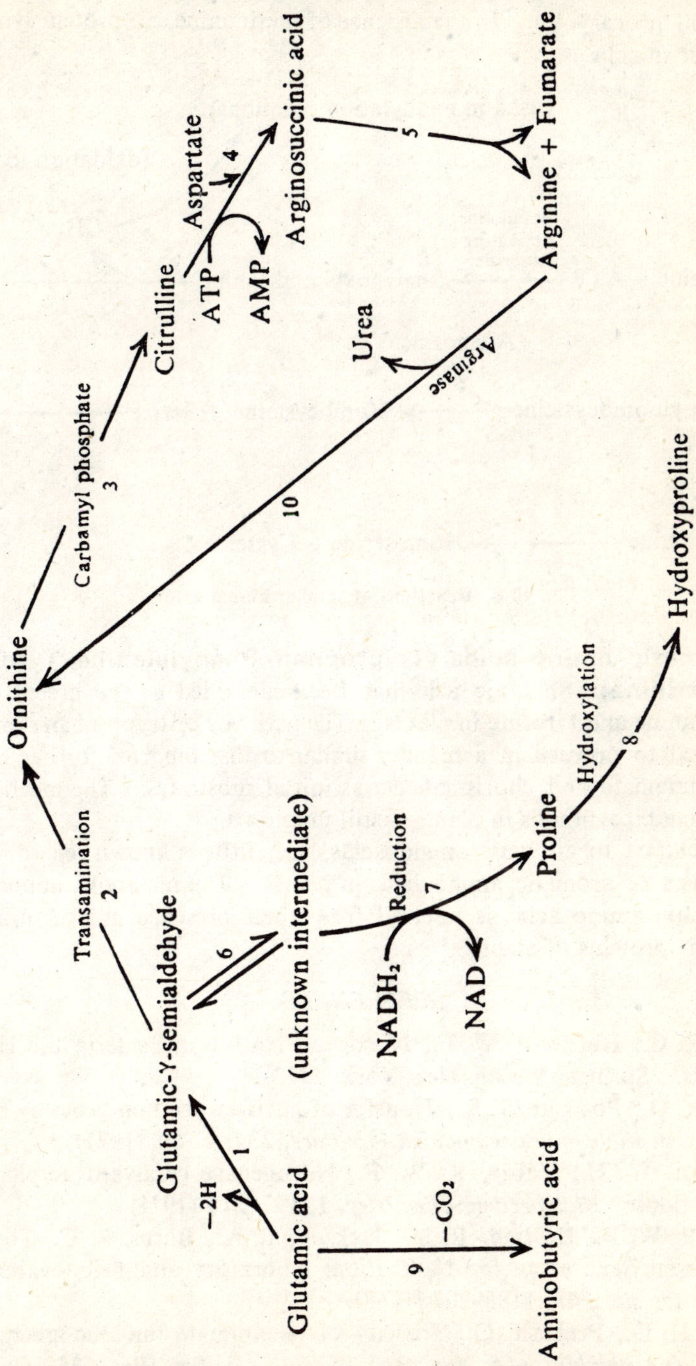

Fig. 10-7 Amino acids of glutamic acid family.

tional group (—SH) of many enzymes. As we shall see in Chapter 12, methionine is involved in the initiation of protein synthesis in both eucaryotes and procaryotes. In the absence of methionine, no protein synthesis can occur in cells.

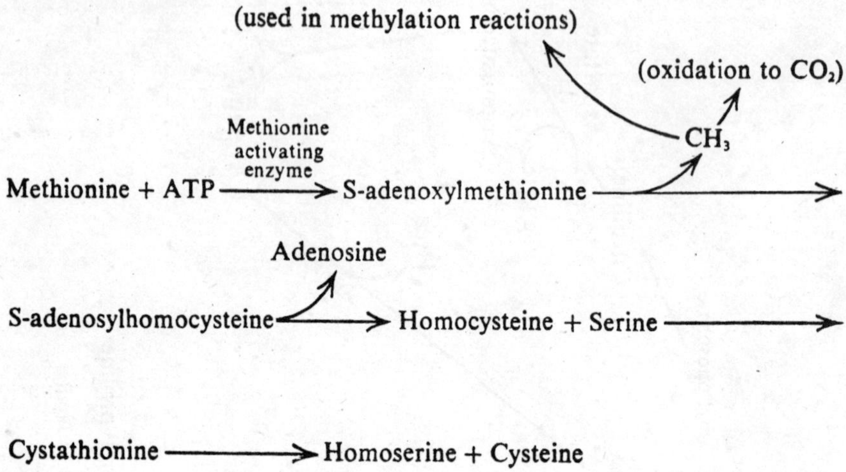

Fig. 10-8 Reactions of sulphur amino acids.

Aromatic amino acids (Tryptophan, Phenylalanine, Tyrosine, and Histidine) Shikimic acid has been identified as the precursor of phenylalanine and tyrosine in plants. The pathway of tryptophan synthesis is believed to proceed in a manner similar to that reported for bacteria in which glutamate and chorismate act as initial substrates. The mechanism of histidine biosynthesis in plants is still unknown.

In contrast to aliphatic amino acids, very little is known regarding the metabolism of aromatic amino acids in plants. The metabolic importance of aromatic amino acids is inferred from their presence in structural and enzymatic proteins of plants.

REFERENCES

Burns, R. C., Hardy, R. W. F., Nitrogen Fixation in Bacteria and Higher Plants. Springer-Verlag, New York (1973).

Dixon, R. O., Postgate, J. R., Transfer of nitrogen fixation genes by conjugation in *Klebsiella pneumoniae*. *Nature*, **234**, 47–48 (1971).

Fuchsman, W. H., Hardy, R. W. F., Nitrogenase catalyzed acrylonitrile reductions. *Bioinorganic Chemistry*, **1,** 197–215 (1971).

Hardy, R. W. F., Holsten, R. D., Jackson, E. K., Burns, R. C., The acetylene-ethylene assay for N_2-fixation: laboratory and field evaluation. *Plant Physiol.*, **43,** 1185–207 (1968).

Kumar, H. D., Prakash, G., Toxicity of selenium to the blue-green algae *Anacystis nidulans* and *Anabaena variabilis*. *Ann. Bot.*, **35**, 697–705 (1971).

LaRue, T. A., Kurz, W. G. W., Gamborg, O. L., Estimation of nitrogenase based on colorimetric determination of ethylene. *Plant Physiol.*, **49** (Suppl.), 50 (1972).

McComb, J. A., Elliott, J., Dilworth, M. J., Acetylene reduction by *Rhizobium* in pure culture. *Nature*, **256**, 209–10 (1975).

Pagan, J. D., Child, J. J., Scoweroff, W. R., Gibson, A. H., Nitrogen fixation by *Rhizobium* on a defined medium. *Nature*, **256**, 406–7 (1975).

Payne, W. J., Gas chromatographic analysis of denitrification by marine bacteria. *Sympos. Marine Bacteria*, Univ. of South Carolina (1971).

Schiff, J. A., Hodson, R. C., Metabolism of sulphate. *Ann. Rev. Plant Physiol.*, **24**, 381–414 (1973).

Shanmugam, K. T., Valentine, R. C., Molecular biology of nitrogen fixation. *Science*, **187**, 919–24 (1975).

Shrift, A., Aspects of selenium metabolism in plants. *Ann. Rev. Plant Physiol.*, **20**, 474–94 (1969).

Stewart, W. D. P., Nitrogen fixation by photosynthetic microorganisms. *Ann. Rev. Microbiol.*, **27**, 283–316 (1973).

Von Bülow, J. E. W., Dobereiner, J., Potential for nitrogen fixation in maize genotypes in Brazil. *Proc. Natl. Acad. Sci.*, USA, **72**, 2389–93 (1975).

phosphate (CMP), and uridine monophosphate (UMP) are also known as adenylic acid, guanylic acid, cytidylic acid, and uridylic acid, respectively. The 5′-monophosphates of deoxyribonucleotides and ribonucleotides are the most common forms of mononucleotides encountered in biological systems and they are converted into their respective triphosphates by further phosphorylation at the carbon-5 position in the presence of specific kinases and ATP. Thus, we have AMP, ADP, ATP, CMP, cytidine diphosphate (CDP), cytidine triphosphate (CTP), GMP, GDP, GTP and UMP, uridine diphosphate (UDP) and uridine triphosphate (UTP). The prefix 'd' before these signifies the corresponding deoxyribonucleotide.

Recently, adenosine tetraphosphate too has been described. Cyclic monophosphates of nucleosides are also known and serve to discharge some of the most important biological functions. Cyclization involves the carbon-2 and carbon-3, or carbon-3 and carbon-5 positions of the sugar moiety. One of the biologically most important compounds is adenosine-3,5-cyclic monophosphate (known as cyclic-AMP or c-AMP), also sometimes designated as the second messenger; it participates in many hormone-controlled processes, including the transcription of the lactose operon in *Escherichia coli.* Nucleoside triphosphates are considered to be the normal substrates for nucleic-acid-polymerizing enzymes, such as RNA polymerase and DNA polymerase. The chemical structure of the nucleosides and nucleotides is as shown in Fig. 11-2.

In addition to serving as substrates for RNA and DNA synthesis, nucleotides play an indispensable role in intermediary metabolism (e.g., NAD^+ and $NADP^+$), in protein synthesis (e.g., GTP), and in phospholipid synthesis (e.g., CTP).

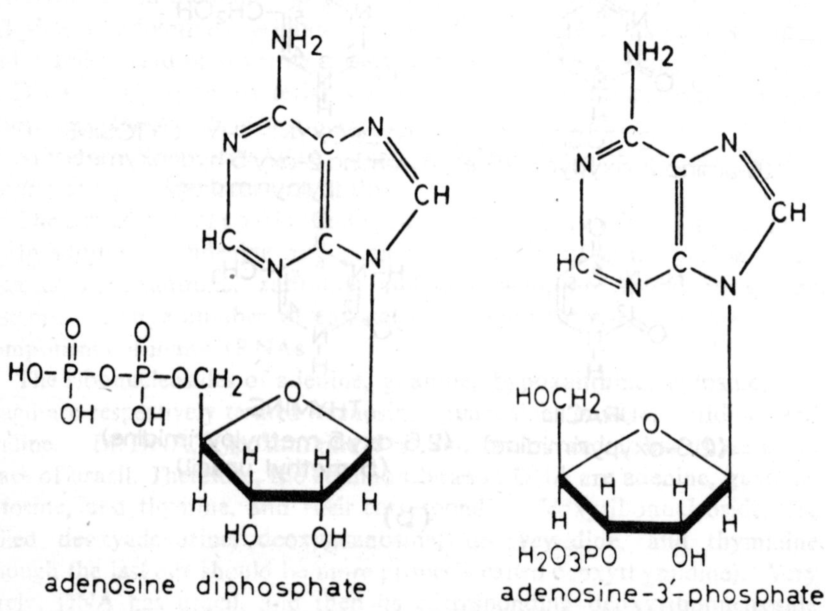

Fig. 11-2 Chemical structure of some nucleosides and nucleotides (cont.).

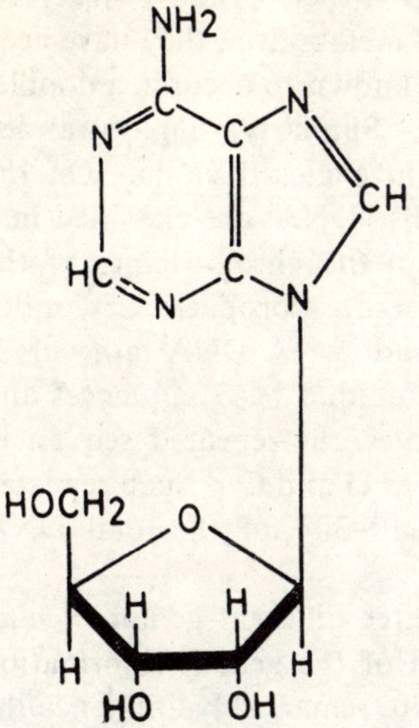

adenosine
9-β-D-ribofuranosido-adenine

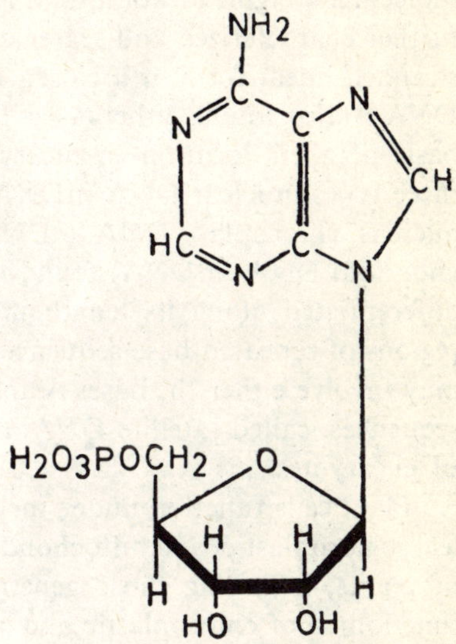

adenosine-5′-phosphate

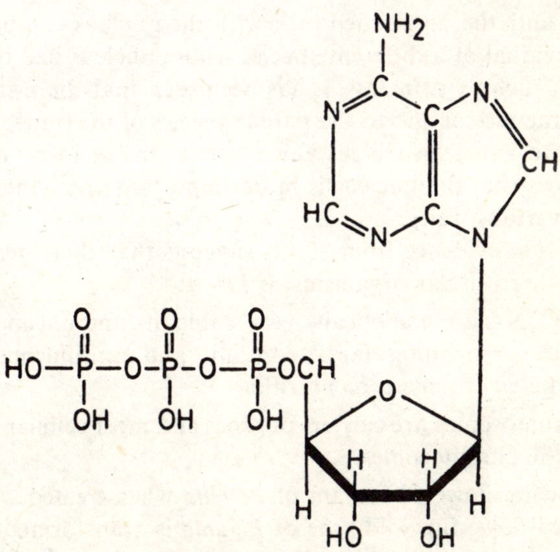

adenosine triphosphate

Fig. 11-2 Chemical structure of some nucleosides and nucleotides.

KINDS

Two main types of nucleic acid, RNA and DNA, are defined on the basis of their chemical components. Depending upon their primary structure, molecular weight, location, and function in cell metabolism, they have been further characterized and segregated. DNA is known to occur in a double-stranded linear form or a closed circular form. Similarly, single-stranded DNA is also found either as a linear rod or in a closed circle. On the basis of their location in eucaryotic plant cells, DNAs are classified into three types: nuclear DNA (nDNA), occurring in the chromosomes of the nucleus, chloroplast DNA (cDNA) present in the chloroplast, and mitochondrial DNA (mDNA) found in the mitochondria. A DNA molecule is differentiated along its length into regions of unique base sequences and regions of repeated base sequences. In eucaryotes, the repeated sequences may involve either the bases A and T, or the bases G and C. Such repeated sequences, called satellite DNA, may constitute 10–30% of the total DNA of eucaryotes.

Plant cells function under the influence of three different genomes: nuclear, chloroplastic, and mitochondrial. But most of the genetic information of plants, including those genetic elements that remarkably influence the functioning of chloroplastic and mitochondrial genomes, is thought to be located within the nuclei on the chromosomes. Nuclear transplantation and grafting experiments involving various species of the unicellular green alga *Acetabularia* have yielded valuable information regarding the relative role of the nucleus and the cytoplasm. The nucleus can be easily extracted by amputating the nucleus-containing rhizoidal lobe from a young individual of this alga, and the amputated lobe with the nucleus can be transplanted on to an individual of a different species whose nucleus had been previously removed. In such experiments, it has been seen that the transplanted alga develops characters specific to the parent species of the transplanted nucleus and not to the enucleated alga which acts as the recipient of the nucleus. This also shows that the nucleus is more important than the cytoplasm in determining various traits.

The following evidence from plants suggests that their genetic material, like that of other cellular organisms, is DNA:

(a) The DNA is metabolically very stable, its amount in any particular species remaining fairly constant and uninfluenced by changes in the environment or nutrition.

(b) Chromosomes are carriers of genes and most cellular DNA is found in the chromosomes.

(c) A white-flowered mutant of *Petunia* when treated with DNA from a red-flowering wild type of *Petunia* is transformed into the red-flowering type which subsequently breeds according to the red-flower character.

(d) The proportion of the four bases found in DNA is constant for all the tissues and cells of the same species but varies for different species.

Three main types of RNA are found in all cellular organisms: (1) mRNA, which carries information from DNA to the site of protein synthesis; (2) tRNA, which transports a specific amino acid to the site of protein synthesis and translates the specificity of the base sequence into an amino acid sequence; and (3) rRNA, whose function is not yet certain. Cells produce as many types of mRNA as there are functional genes (cistrons). Also, there occur about as many tRNAs as there are amino acids, with at least one specific tRNA for one protein-forming amino acid. Further details about tRNA and rRNA are given in Chapter 12.

STRUCTURE

The primary structure of both DNA and RNA is a long polynucleotide linear chain. Both alkaline and enzymatic digestion of polyribonucleotides and polydeoxyribonucleotides suggest that phosphodiester linkages connecting carbon-3 of one nucleotide with carbon-5 of another constitute the internucleotide bonds. For convenience, the formulae of polynucleotide chains are represented in graphic form or by short notations. In Fig. 11-3, the vertical line represents the pentose sugar with its upper end denoting carbon-1′, and attached to the purine or pyrimidine base, its lower end denoting the carbon-5′ position, and its middle point the carbon-3′ position. The two diagonal lines, one at the middle point and the other at the lower end of the vertical line, represent phosphodiester linkage.

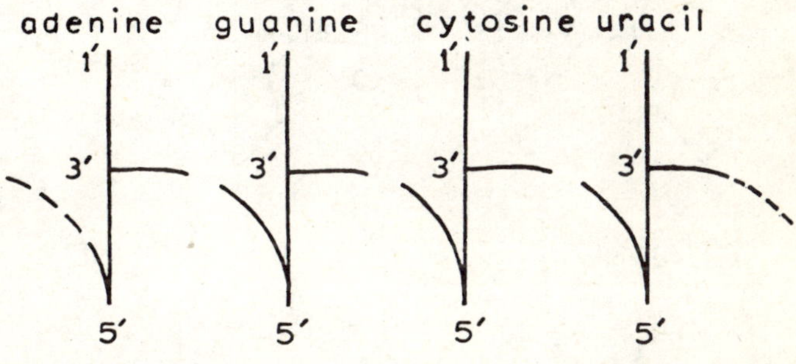

Fig. 11-3 Diagrammatic representation of nucleotide interlinks (3′ to 5′).

In short notation, letters AGTUC represent nucleosides. The phosphate group, denoted by P, when placed to the right of the nucleoside signifies esterification at the carbon-3′ position of the pentose sugar, and when placed to the left of the nucleoside it indicates esterification at the carbon-5′ position. For instance, GPGP is a dinucleotide with a diester bond between carbon-3′ of the first guanosine and carbon-5′ of the second guanosine, and with a phosphate monoesterified at carbon-3′ of the second guanosine. The letter P between nucleoside residues may be replaced by a hyphen (-); thus, GPGP could be written also as G-GP.

Chargaff (1950) examined the base composition of DNA from a wide range of organisms and found that, in any one organism, the amount of adenine is always equal to that of thymine, and the amount of guanine equal to that of cytosine. The X-ray diffraction studies on purified DNA, carried out by Astbury (1947) and Rosalind and Gosling (1953), revealed that DNA is a double-stranded helix with a 20-Å diameter and with the two chains wound around each other every 34 Å. The bases lie flat, 3.4 Å apart along each chain at right angles to the sugar phosphate backbone. These observations prompted Watson and Crick (1953) to propose a model of DNA structure in which (1) the two polynucleotide chains are antiparallel (i.e., of opposite polarity in that the internucleotide linkage in one chain is from 3′ to 5′ whereas in the other it is from 5′ to 3′) and held together by weak hydrogen bonds; (2) steric reasons permit the pairing of a purine base with a pyrimidine base through hydrogen bonds; and (3) the specificity of hydrogen bonding is reason for adenine to pair with thymine, and guanine with cytosine, i.e., base pairing is of complementary nature, and is determined mainly by the availability of hydrogen bonds (Fig. 11-4). This pairing pattern, and the antiparallel nature of the two chains, led the researchers to suggest that one chain is complementary to the other in that the sequence of the bases in one chain will determine the sequence of the bases in the other. This reasoning also prompted them to predict that DNA replication

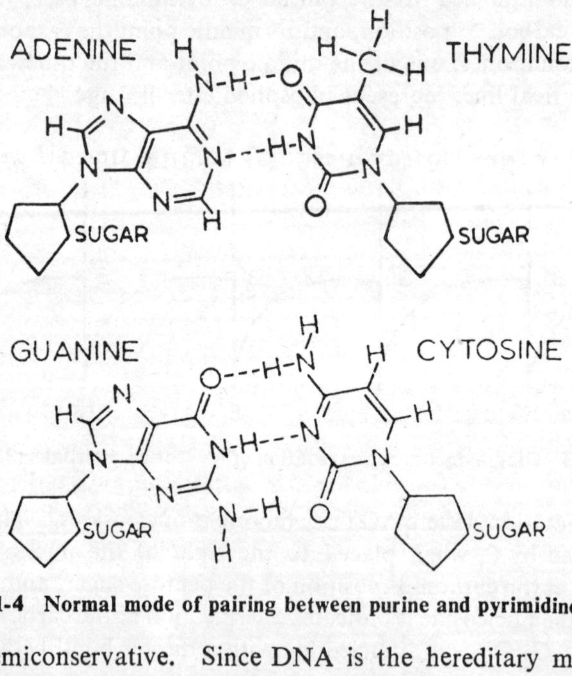

Fig. 11-4 Normal mode of pairing between purine and pyrimidine bases.

should be semiconservative. Since DNA is the hereditary molecule containing information for all the characters of an organism, the problem was to suggest the mode and mechanism of information storage in this molecule. The solution they provided was the proposal that the four bases can occur in any arbitrary sequence along one chain of the DNA double helix, thus allowing DNA to maximize its capacity for information storage. The obvious implication of this proposal is that the sequence of bases distinguishes one DNA from another, or one gene from another gene, or one organism from another organism. Thus, any change in the base sequence of DNA, either through mutation or recombination, will result in a change in the quality or function of its information. Since the two chains of a DNA molecule are complementary, the base sequence of one chain will automatically determine the base sequence of its complementary chain, and vice versa. This also indicates that hereditary information is stored in two complete sets, each corresponding to one chain of the DNA molecule. The obvious advantage of this to the organism is that it can easily escape from the genetic damage inflicted on any one chain of the double helix.

Subsequent research by Kornberg *et al.* (1967), Meselson and Stahl (1958), and others have confirmed the accuracy of the Watson and Crick model of DNA structure. Further, it is now becoming clear that the stability of DNA molecules is not due to hydrogen bonding, but due to base-stacking forces or apolar bonds, and that the main role of hydrogen bonding is in the specificity of base pairing. The work of Wilkins *et al.* (1965) further indicates that, under differing humidity and ionic conditions, DNA fibres show three possible kinds of structure.

Methods have been established to distinguish between single-stranded and double-stranded nucleic acids. By one method the relative amounts of nucleic acid bases are determined. If A=T and G=C in any given nucleic acid, it may be concluded that the nucleic acid has a double-stranded structure. Heating breaks the hydrogen bonding between the bases and results in the separation of the two strands. Such separation of the two chains of DNA, after the application of heat to its solution (by boiling) and then rapid cooling of it, is termed helix-coil transition. This transition causes a drastic change in the properties of DNA. One consequence is the lowering of viscosity. The second is its reactivity with formaldehyde which is known to react with the amino group of the bases. Since the amino groups remain engaged in hydrogen bonding in the double-stranded configuration, their reaction with a DNA solution should be indicative of the single-strandedness of DNA. Heat-denatured double-stranded DNA solution shows about 40% more absorption at 260 nm than its corresponding normal solution. Thus, a third consequence of heat denaturation is an increase in UV absorption by the nucleic acid solution.

Results obtained from the application of such methods have suggested that various types of RNA molecules too have structures in which some regions are linear and others are helically coiled with typical hydrogen bonding and complementary base pairing. The advantage of such secondary structures is that they provide more stability and, hence, longer life to the RNA molecules in general. The RNA viruses, in which RNA directly serves as mRNA, have been studied in detail in this respect, and it has been observed that each RNA molecule contains many hairpin loops which are double-stranded regions with complementary base pairing (Fig. 11-5). In comparison to other types of RNA, tRNAs involve a considerable part

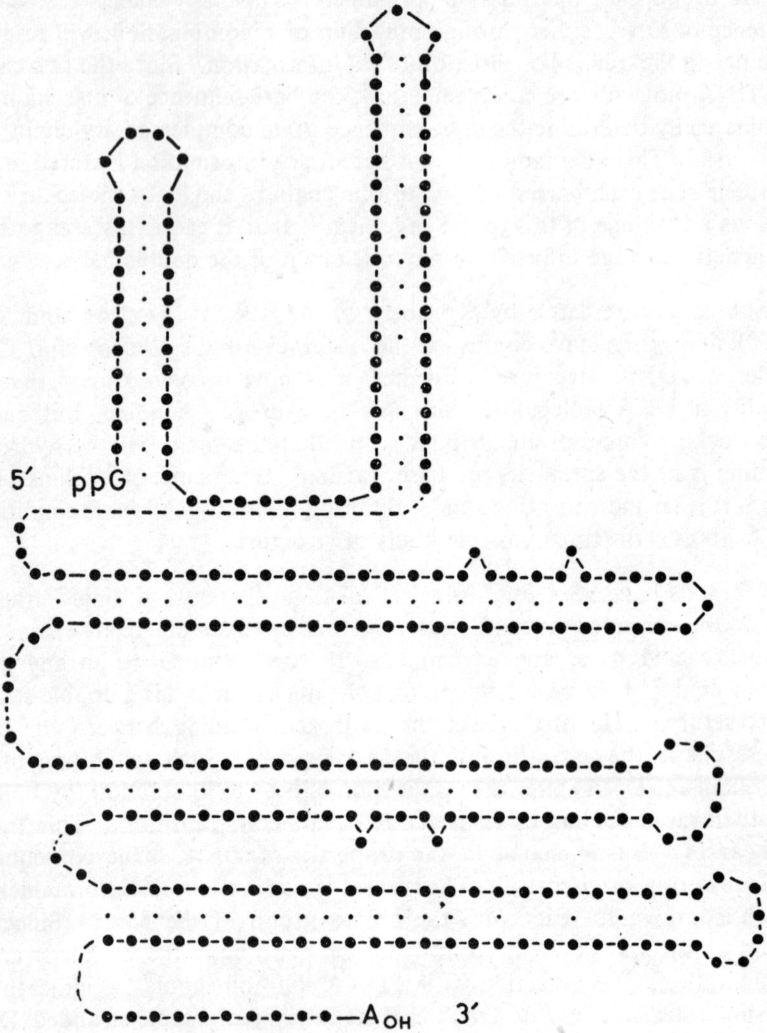

Fig. 11-5 Structure of RNA of bacteriophage R17 showing fine 'hairpin' loops. (After Davidson, 1972.)

of their molecule in the formation of secondary and tertiary structures (Fig. 11-6).

DNA SYNTHESIS

DNA synthesis consists of four major stages: (1) formation of purine and pyrimidine ribonucleotides; (2) reduction of ribonucleotides to their respective deoxy forms; (3) conversion of deoxyribonucleotide monophosphates into their respective 5′-triphosphate forms; and (4) polymerization of the triphosphates produced into the DNA chain.

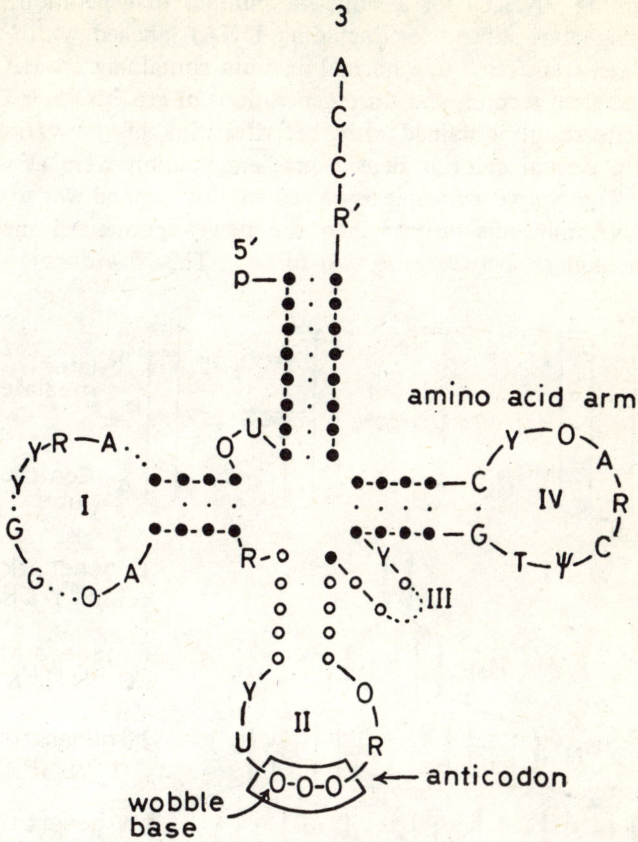

Fig. 11-6 Generalized clover-leaf structure of tRNA. (After Davidson, 1972.)

Since DNA is hereditary material, its synthetic mechanism must produce a replica of the parental DNA molecule. This may be achieved in three ways:

(a) The newly produced DNA double helix includes an old parental chain and a newly-synthesized chain, using either of the two strands of the old parental chain as a template; this mode of replication is called semi-conservative and is one of the predictions of the Watson and Crick model of DNA structure.

(b) Both chains of the daughter DNA double helix are newly-synthesized, using the entire parental DNA duplex as a template; this mode is termed conservative.

(c) The parental DNA double helix is disintegrated so that all the progeny receives only newly-made DNA; this method is termed nonconservative.

The findings of Kornberg and his coworkers and of Meselson and Stahl have provided sufficient evidence in support of the semiconservative mode of DNA replication. Meselson and Stahl (1958) grew *E. coli* in a medium that, containing $^{15}NH_4Cl$ for a sufficient number of generations, had all cellular nitrogenous substances (including DNA) labelled with ^{15}N. The cells were then transferred to a normal medium containing $^{14}NH_4Cl$; subsequent to the first, second, and third generations of growth, their DNA was isolated. The results obtained after centrifugation of the various DNA samples in a cesium chloride density-gradient column were as shown in Fig. 11-7. The simple principle involved in this method was to make the parental DNA molecule heavier than the newly-synthesized molecule in order to distinguish between the two forms. This was done by growing

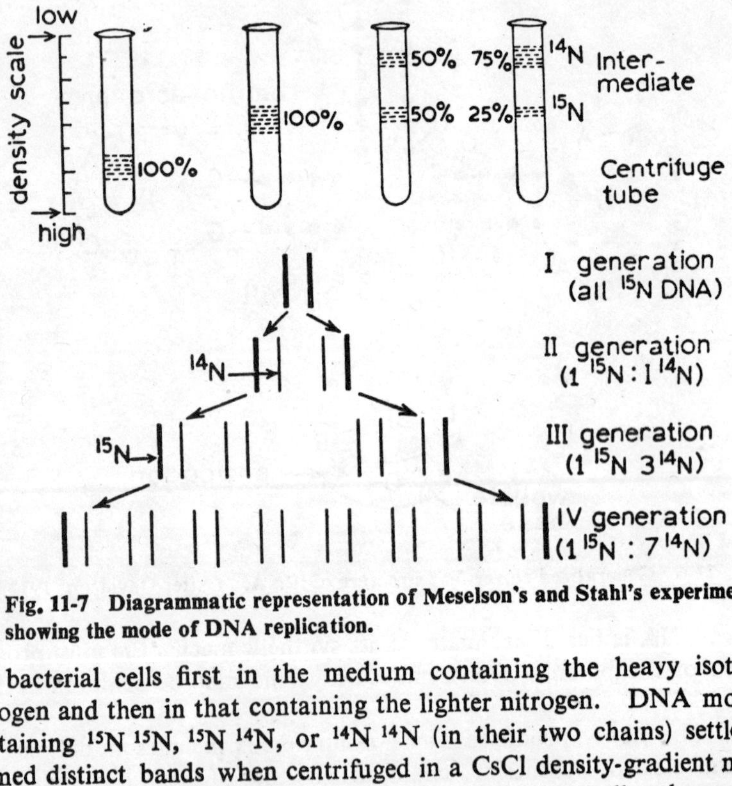

Fig. 11-7 Diagrammatic representation of Meselson's and Stahl's experiment showing the mode of DNA replication.

the bacterial cells first in the medium containing the heavy isotope of nitrogen and then in that containing the lighter nitrogen. DNA molecules containing ^{15}N ^{15}N, ^{15}N ^{14}N, or ^{14}N ^{14}N (in their two chains) settled and formed distinct bands when centrifuged in a CsCl density-gradient medium in an ultracentrifuge. In this experiment, it was actually observed that all double-stranded DNA molecules extracted from zero generation cells were of the ^{15}N type; those of first generation cells had one chain of the ^{14}N type and the other of the ^{15}N type; and those of second generation cells had two kinds of double-stranded DNA molecules, viz., one in which both strands contained ^{14}N, and the other in which one strand was ^{14}N and the other ^{15}N. Thus, the DNA of second generation cells included 50% molecules containing only ^{14}N whereas the other 50%, being hybrids of the two isotopic forms, contained both ^{15}N and ^{14}N in equal proportions. Thus, the conclusion was that DNA replication is semiconservative in this bacterium. Simultaneously, Taylor (1959), using tritiated thymidine, showed that nuclear DNA also replicates semiconservatively in plants. Now it is almost certain that DNA replication is universally semiconservative, be it a single- or double-stranded linear or circular DNA. Further, cDNA and mDNA have also been shown to replicate semiconservatively.

Kornberg and his coworkers (1967) purified an enzyme from cells of *E. coli*, called DNA polymerase I or DNA nucleotidyl transferase or DNA-dependent DNA polymerase, and incubated this preparation at a suitable temperature in a mixture containing a template DNA chain (obtained from calf thymus DNA after denaturation), deoxycytidine-5′-triphosphate, deoxyadenosine-5′-triphosphate, deoxyguanosine-5′-triphosphate, and deoxythymidine-5′-triphosphate, along with magnesium ions. The result was the synthesis of a new DNA chain which was found to be complementary to the template DNA chain. The omission of any of the components, such as dATP, dGTP, deoxythymidine triphosphate (dTTP), template DNA, and Mg^{2+} ions, inhibited the reaction and the production of the new DNA chain. This again supports the semiconservative mode of DNA replication.

Further investigations with this enzyme *in vitro* have established that the newly-synthesized chain is complementary to the template on the basis of (1) equivalence between A and T, and G and C, and (2) heat denaturation, leading to the transition of the helix coil, a reaction with formaldehyde, and an increase in UV absorption. That the Kornberg polymerase enzyme is the enzyme involved in DNA synthesis through a semiconservative mechanism was further substantiated by Kornberg *et al.* (1967) when they were able to produce an *in vitro* biologically-infective species of DNA of $\phi\times 174$ virus which infects *E. coli*. Their reaction mixture included labelled 5′-triphosphates of all four deoxynucleotides, magnesium ions, DNA polymerase I, a primer polynucleotide ligase, and the virus DNA as a template. This virus contains a single-stranded cyclic DNA termed the +DNA circle. During replication, the +DNA circle becomes a double-stranded closed circle, called the replicative form (RF), in which the newly-synthesized cyclic strand is termed the −DNA circle. Thus, RF included both +DNA and −DNA circles, held together by hydrogen bonding. DNA synthesis proceeds linearly in all systems, but the infectivity of such linearly-synthesized DNA in $\phi\times 174$ depends upon its conversion into the cyclic form; the polynucleotide ligase enzyme catalyzes this cyclization in which NAD^+ is a requirement.

In the foregoing reaction, DNase is the enzyme that produces a break in either of the chains of the duplex DNA molecule and, by controlling the conditions, the enzyme can be made to introduce a nick (break) in the −DNA circle or the +DNA circle. Subsequent denaturation of the enzyme-treated, circular, duplex DNA leads to the separation of +DNA and −DNA circles. Since the −DNA circle is as infective as the +DNA circle (which serves as the template for the synthesis of the −DNA circle), there is no doubt regarding the role of DNA polymerase in the *in vivo* replication of at least single-stranded DNA molecules. The primer is always a short piece of oligonucleotides and its function is to provide a fixed point with a free 3′—OH end for the addition of an incoming mononucleotide by DNA polymerase I. The primer achieves this function by making available a base sequence complementary to certain regions of the template DNA (Fig. 11-8).

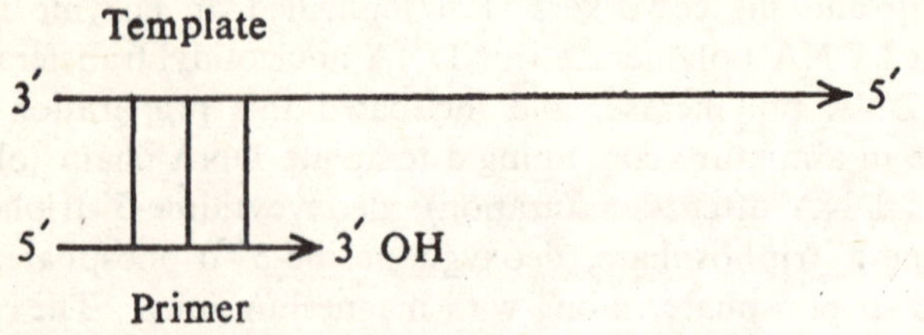

Fig. 11-8 Diagram showing DNA-polymerase-I-catalyzed elongation of polynucleotide chain in 5′ to 3′ direction.

The enzymes associated with DNA synthesis include DNA polymerases, endonucleases, exonucleases, and the polynucleotide ligase. There are three types of DNA polymerase, termed polymerase I (the Kornberg enzyme), polymerase II, and polymerase III. Polymerase I can bind to single-stranded and double-stranded DNA, but only in regions containing single-stranded breaks or nicks. This is true for the *E. coli* enzyme; polymerase I from T4 bacteriophage binds only to single-stranded DNA.

Kornberg (1969) has outlined the functions of the enzyme polymerase I which include (1) DNA synthesis only in the 5′ to 3′ direction in the presence of a template and a primer containing a free 3′—OH group to which a nucleotide triphosphate is added; (2) hydrolysis of the DNA chain containing the free 3′—OH group in the 3′ to 5′ direction (exonuclease II activity); (3) hydrolysis of the DNA chain having 5′-phosphate or 5′—OH as the terminus in the 5′ to 3′ direction; (4) pyrophosphorolysis of the DNA chain from its 3′ end in which inorganic pyrophosphate is used up to generate nucleoside triphosphate from the DNA chain; and (5) exchange of inorganic pyrophosphate with the pyrophosphate of the deoxyribonucleoside triphosphate.

Kornberg has proposed a model to explain the enzyme's multiple action (Fig. 11-9). In this model, there is (1) a template binding site; (2) a primer binding site; (3) a substrate binding site which seems to be the same for all the four types of nucleotides; (4) a site recognizing the free 3′—OH group of the primer which is believed to correspond to its 3′ to 5′ hydrolytic activity; and (5) a site for the 5′ to 3′ hydrolytic activity. According to this model, the enzyme binds to the template in the region containing a base-paired primer with its free 3′—OH group terminus. An incoming complementary deoxyribonucleotide triphosphate is permitted to bind to the substrate site on the right of the 3′—OH group. This leads to a conformational change in the enzyme which now catalyzes the formation of a phosphodiester bond with the elimination of an inorganic pyrophosphate molecule, resulting in the growth of the primer chain by one deoxyribonucleotide. The enzyme then moves to the right by one nucleotide so that 3′—OH of the newly-added nucleotide now occupies the primer terminus site, vacating the substrate site for another complementary nucleotide triphosphate. A repetition of this process results in the formation of a completely new DNA chain.

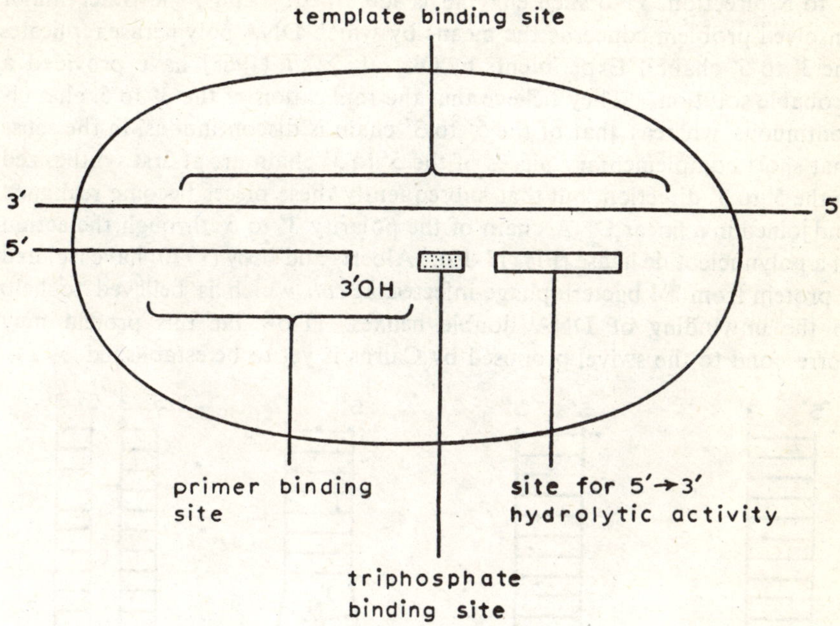

Fig. 11-9 Model of DNA polymerase I showing various sites of action. (After Kornberg, 1969.)

There seems no doubt that polymerase I is the enzyme involved in the replication of single-stranded DNA molecules and the repair synthesis of damaged regions in either of the DNA double chains. RNA viruses contain a single-stranded RNA which, however, becomes double-stranded just before replication inside the host. In fact, the double-strandedness of DNA or RNA is a prerequisite for their replication. But there are certain major unsolved problems concerning the mechanism of replication *in vivo* of helically coiled double-stranded DNA molecules; the separation of the two chains during replication is necessary, and so far there is no definite information on how the twisted chains become untwisted during replication.

Cairns (1963) made an autoradiographic study of DNA replication in *E. coli*. He suggested that the replication starts from a fixed point, and the replicating fork which envisages simultaneous synthesis of two new chains of opposite polarity moves sequentially along the circular DNA duplex from the starting point. He proposed a swivel mechanism for unwinding the DNA duplex, but the nature of the swivel is not yet defined although there is a suggestion that the swivel might correspond to an action of the endonuclease alternating with an action of the ligase which alternately breaks and rejoins the template DNA chains at specific points along their twists. Polymerase I is the enzyme that always synthesizes new DNA chains in the

5′ to 3′ direction. Since the two chains of a DNA duplex are of opposite polarity, one running in the 3′ to 5′ direction and the other in the 5′ to 3′ direction, their simultaneous replication will need an enzyme working in the 3′ to 5′ direction. No such enzyme is identified. Thus, the other major unsolved problem concerns the means by which DNA polymerase replicates the 3′ to 5′ chain. Experiments by Okazaki *et al.* (1968) have provided a probable solution. They believe that the replication of the 3′ to 5′ chain is continuous whereas that of the 5′ to 3′ chain is discontinuous, in the sense that short complementary pieces of the 5′ to 3′ chain are at first synthesized in the 5′ to 3′ direction, but that subsequently these pieces become realigned and joined in a linear DNA chain of the polarity 3′ to 5′ through the action of a polynucleotide ligase (Fig. 11-10). Alberts and Frey (1970) have isolated a protein from T4 bacteriophage-infected *E. coli* which is believed to help in the unwinding of DNA double helixes. How far this protein may correspond to the swivel proposed by Cairns is yet to be established.

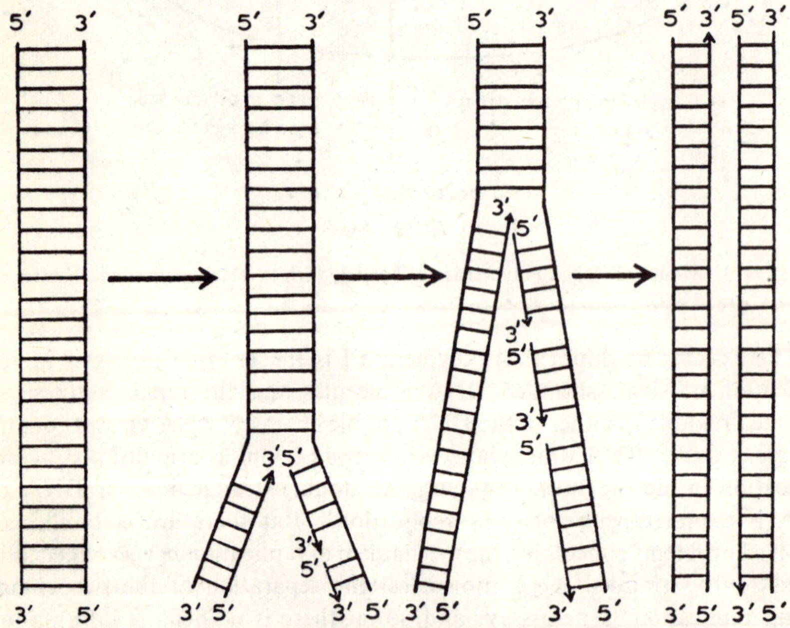

Fig. 11-10 Discontinuous synthesis of DNA in the 5′ to 3′ direction.

Polymerases II and III seem to catalyze replication in certain mutants of *E. coli* which lack polymerase I. Unlike polymerase I which is found in cell cytoplasm, polymerases II and III are membrane-bound and are thought to be associated with the membrane machinery of DNA replication. The 5′-triphosphates are natural substrates of all the DNA polymerase enzymes. By purely chemical methods, Hargobind Khorana and his coworkers have successfully synthesized short chain polydeoxyribonucleotides of repeating sequences which were converted into long polynucleotide chains of a known sequence with the help of the enzyme polynucleotide ligase (Gupta *et al.*,

1968). Using the same technique, they achieved the unique feat of synthesizing a gene that codes for alanine tRNAs.

RNA SYNTHESIS

There are roughly four steps in the synthesis of RNA: (1) template binding; (2) chain initiation; (3) chain elongation; and (4) chain termination. RNA is a polynucleotide, normally synthesized from four kinds of 5′-ribonucleosides, viz., CTP, UTP, GTP, and ATP. At present, three kinds of enzyme are known to participate in polyribonucleotide synthesis, viz., DNA-dependent RNA polymerase, also called transcriptase; RNA-dependent RNA polymerase, also known as RNA replicase; and polynucleotide phosphorylase.

The chain growth of polyribonucleotides is always in the 5′ to 3′ direction, and the 5′-terminal nucleotide of nearly every transcribed RNA is a purine nucleotide triphosphate. Other kinds of RNA (mRNA, tRNA, and rRNA) are produced from specific regions of the DNA template through transcription, and evidence indicates that nuclei, chloroplasts, and mitochondria have their own separate transcription machineries. Much of our information about transcription is based on studies conducted on bacteriophages and bacteria. The mechanism of transcription in eucaryotic systems is comparatively poorly understood. The nuclear chromosomes in particular remain complexed with histones and nonhistone proteins, making eucaryotic systems over-complicated for analyses of the transcription process. In procaryotes, both transcription and translation are tightly coupled, taking place almost simultaneously and at the same site. In eucaryotes, nuclear transcripts of the nature of mRNA move out from the nucleus to the cytoplasm and are translated there. Procaryotes are known to contain only one type of DNA-dependent RNA polymerase which, in association with other different protein factors, is able to transcribe different mRNAs, tRNAs, and rRNAs. On the other hand, eucaryotes seem to possess multiple forms of RNA polymerases and perhaps transcribe different kinds of RNAs. For instance, RNA polymerase I is located in the nucleolus and transcribes precursors of rRNAs; RNA polymerases II and III are located in the nucleoplasm and function in transcribing heterogeneous RNAs. An RNA polymerase IV is also known in eucaryotes and is suspected to be a component of the mitochondrial transcription machinery. In eucaryotic plant cells, the chloroplast has its own transcriptional machinery and its RNA polymerase may represent a polymerase that is different from the RNA polymerases already identified.

The DNA-dependent RNA polymerase of *E. coli* has been examined in detail and its composition and properties are as follows: the enzyme consists of one beta chain (β chain), one β prime chain (β' chain), two alpha chains (α chains), and one molecule of sigma factor (σ factor); this may be briefly represented as $\alpha 2\beta\beta'\sigma$ which indicates the composition of the holoenzyme. The deletion of σ leaves the core enzyme, designated as $\alpha 2\beta\beta'$. The enzyme is bound to the template in the absence of triphosphates of ribonucleosides. The core enzyme is known to bind at random, but the holoenzyme is found to show greater specific affinity for binding to the promoter site of the template. This is taken as evidence to suggest that specific binding of the enzyme is affected by the σ factor. The β subunit of the enzyme is known to be involved in template binding and the σ factor is found to interact with this subunit like an allosteric effector. This leads to specific and stable binding of the enzyme with the promoter region of the template. In bacterial systems, the promoter is a component of the transcriptional unit, called the operon (Fig. 11-11). The other components of this operon include operator genes and structural genes. The operator is the site of binding of repressor protein produced by a regulator gene. Cells contain as many operator and regulator genes as there are operons.

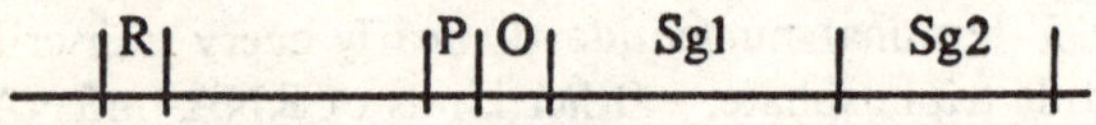

Fig. 11-11 The operon model. (R, regulator gene; P, promoter gene; O, operator gene; Sg1 and Sg2, structural genes 1 and 2.)

The repressor is thought to block transcription by preventing the movement of RNA polymerase beyond the promoter region by simply acting as a physical block in the way of the enzyme. This is termed the *negative* control of transcription. Since chain initiation begins with the formation of a dinucleotide in which the 5′ terminus is always a purine nucleotide triphosphate, it is believed that this is estimated on the base sequence in the promoter gene. Since this is true of RNA initiation in every case so far examined, it has been suggested that various promoter genes may be almost identical with regard to the codons they contain for RNA chain initiation. The reaction of RNA chain initiation takes place in the β subunit of the enzyme in the presence of the σ factor. But the σ factor is thought to be released from union with the enzyme before chain elongation. In respect of specificity of enzyme binding and RNA chain initiation, the σ factor is a mechanism of positive control of transcription unlike that of repressor protein. Other mechanisms that exercise positive control of transcription are (1) the M factor, which causes a general increase in the transcription of most templates; (2) catabolite gene activator protein (CAP) which, in conjunction with cyclic AMP, acts on the promoter gene by stimulating the chain initiation event; and (3) the psi factor (ψ factor), which is known to be specifically involved in the transcription of *E. coli* DNA.

The elongation of the RNA chain proceeds step by step by the addition of an incoming 5′-triphosphate at the free 3′—OH group of the growing chain in the manner already described for DNA polymerase I. The growing RNA chain finally stops at a place on the template that is unable to be transcribed. The newly-synthesized RNA chain then gets released from the template by the interaction of another protein factor, called the rho factor (ρ factor), with the transcription assembly at the stop codons of the template.

Both double- and single-stranded DNAs can be transcribed, but the unique *in vivo* feature of transcription seems to be its asymmetric nature, i.e., only one of the two chains of DNA undergoes transcription whereas *in vitro* both the chains are transcribed. Furthermore, denatured DNA is a very poor template for transcription in comparison to native DNA, and is unable to promote a significant synthesis of proteins. This means that denatured DNA, though transcribed, produces a biologically-inactive species of RNA. How the enzyme selects the proper DNA strand for the transcription and production of biologically-active RNA is not yet clear.

Nuclear RNA polymerases of eucaryotes have been studied in some detail with regard to their composition and ability to transcribe phage DNA and bacterial DNA. It seems that each is made up of a number of components. Nuclear polymerases produce (1) 45S RNA, the precursor of RNAs of 80S ribosomes; (2) the precursor of tRNA; (3) heterodisperse nuclear RNA, which is highly unstable, but which may ultimately give rise to mRNA; (4) mRNA associated with polysomes in the cytoplasm; and (5) RNA types whose functions are still not clearly defined.

Most of the RNA species, such as mRNA, rRNA, and precursors of tRNA, undergo post-transcriptional changes which are recognized as a mechanism of their stable functioning. A bacterial chromosome contains distinct cistrons for three kinds of RNA, viz., 5S, 16S, and 23S. The 45S RNA formed in the nucleolus is a precursor of the 80S ribosomal RNA; it is first cleaved and then methylated to yield 18S, 7S, and 28S rRNA. Another species of rRNA, termed 5S, is synthesized independent of the 45S RNA. Similarly, in procaryotes containing 70S ribosomes, precursors are known to be involved in the synthesis of the 16S and 23S rRNAs in a tandem fashion, i.e., the precursor rRNA is first produced and subsequently cleaved and trimmed to give the actual size of rRNAs.

Procaryotic mRNAs are more unstable than nuclear mRNAs of eucaryotes. Recent studies suggest that eucaryotic mRNAs normally contain a repeated sequence of nearly 200 AMP residues at their 3′ end. This polyribonucleotide segment consisting of only AMP originates in the nucleus and is added to the mRNA after the mRNA has been transcribed. The enzymes capable of adding polyadenylates to the mRNA are located in the nucleus. The role played by the sequence of the adenylate in the biology of nuclear mRNAs of eucaryotes remains unknown.

Just as there are precursor transcripts of rRNAs, so too there are precursor transcripts of tRNAs both in procaryotes and eucaryotes; these precursors are subsequently trimmed and methylated to reach the proper size and the required secondary and tertiary structures of the tRNAs. This is followed by the addition of two CMP moieties sequentially at the 3′ end of the tRNA and, finally, the terminal CMP residue at the 3′ end accepts an AMP residue, giving rise to the sequence PCPCPA at the 3′ end of the tRNA molecule. It is noteworthy that all tRNAs have a PCPCPA sequence at their 3′ end.

The enzymatic changes brought about in various transcript RNAs subsequent to their transcription are termed post-transcriptional processing or maturation and their relevance to the overall regulation of RNA transport, stability, and function is not yet defined.

In eucaryotes, the nuclear DNA is complexed with histone and nonhistone proteins. Normally, a eucaryotic chromosome shows 5–10% transcription activity in comparison to its protein-free chromosome. Deproteinized chromosomes, when exposed to histone proteins, show a considerably reduced transcription ability, an observation interpreted to mean that the histones in eucaryotes, like repressors in procaryotes, regulate the activity of genes. Subsequent analyses of the relative amounts of histone and nonhistone proteins of eucaryotic chromosomes in their transcribing regions suggest no change in the amount of histones, but indicate a relatively large increase in the amount of nonhistone proteins. This has led to the proposition that histones mask the template ability of the chromosome whereas nonhistone proteins unmask this effect. Chemically, the unmasking seems to involve the phosphorylation, methylation, and acetylation of histones by nonhistone proteins, all of which appear to be enzymatic reactions.

That mRNA exists in higher plants has been proved. Research on the subject covers antimetabolites which block the transcription of all kinds of RNA (e.g., neomycin) or of only tRNA and rRNA (e.g., 5-fluorouracil), leaving the formation of mRNA to proceed normally. This feature has been noticed in isolated sections of soybean hypocotyl. In another approach, DNA-RNA hybridization studies were made in germinating groundnut cotyledons from which DNA and RNA were extracted and purified; the purified DNA was then denatured and hybridized with various fractions of cotyledonary RNA. One such fraction, believed to be mRNA, was found to hybridize more readily with the cotyledonary DNA than classes of RNA, such as tRNA and rRNA.

In 1958, Crick put forward the theory of Central Dogma which stipulates that a specific protein is produced when information from a gene flows from DNA to RNA and from RNA to protein, through the processes of transcription and translation, respectively. In 1970, Temin and Mizutani demonstrated that certain oncogenic RNA viruses of mammals possess the enzyme RNA-dependent DNA polymerase. This enzyme used an RNA template for the synthesis of new DNA molecules which were complementary to the template, and such RNA-DNA hybrids were successfully isolated. This observation constituted the reverse of the transcription discussed earlier in this chapter and, therefore, the process was termed *reverse transcription* and the enzyme involved was named *reverse transcriptase*. So far there has been no report of any reverse transcriptase from plants or their viruses. Reverse transcriptase has been demonstrated in normal mouse and human cells and in virus-free tumour cells of human beings. The implication of the presence of reverse transcriptase in normal cells is that DNA can be made from RNA in the same way as RNA is made from DNA; in other words, the biological importance of DNA and RNA may be fairly equal, and the flow of information is from nucleic acids to proteins.

REFERENCES

Alberts, B. M., Frey, L., T4 bacteriophage gene 32: A structural protein in the replication and recombination of DNA. *Nature*, **227**, 1313–18 (1970).

Astbury, W. T., X-ray studies of nucleic acids. S. E. B. Symposium, No. 1, Cambridge Univ. Press, London (1947).

Baltimore, D., Viral RNA-dependent DNA polymerase. *Nature*, **226**, 1209–11 (1970).

Cairns, J., Bacterial chromosome and its manner of replication as seen by autoradiography. *J. Mol. Biol.*, **6**, 208–13 (1963).

Chargaff, E., Chemical specificity of nucleic acids and mechanism of their enzymatic degradation. *Experientia*, **6**, 201–9 (1950).

Crick, F. H. C., On protein synthesis. S. E. B. Symposium, No. 12, Cambridge Univ. Press, London (1958).

Davidson, J. N., The Biochemistry of the Nucleic Acids, 7th edn. Methuen, London (1972).

Gupta, N. K., Ohtsuka, E., Buchi, H., Kumar, A., Weber, H., Khorana, H. G., Studies on polynucleotides. Enzymatic joining of chemically synthesized segments corresponding to the gene for alanine-tRNA. *Proc. Natl. Acad. Sci.*, USA, **60**, 1338–44 (1968).

Kornberg, A., Biological synthesis of deoxyribonucleic acid. *Science*, **131**, 1503–8 (1960).

Kornberg, A., Active control of DNA polymerase. *Science*, **163**, 1410–18 (1969).

Kornberg, A., Gorlian, M., Sinsheimer, R. L., Enzymatic synthesis of DNA. 24. Synthesis of infectious phage $\phi\times174$ DNA. *Proc. Natl. Acad. Sci.*, USA, **58**, 2321–28 (1967).

Meselson, M., Stahl, F. W., The replication of DNA in *Escherichia coli*. *Proc. Natl. Acad. Sci.*, USA, **44**, 671–82 (1958).

Okazaki, R., Okazaki, T., Sakabe, K., Sugimoto, K., Kainuma, R., Sugino, A., Iwatsuki, N., *In vivo* mechanism of DNA chain growth. *Cold Spring Harbor Sympos. Quant. Biol.*, **33**, 129–43 (1968).

Rosalind, F., Gosling, R. G., Molecular configuration in sodium thymonucleate. *Nature*, **171**, 740–41 (1953).

Taylor, J. H., The organization and duplication of genetic material. *Proc. 10th Internat. Congr. Genetics*, Montreal, **1**, 63–78 (1959).

Temin, H., Mizutani, S., RNA-dependent DNA polymerase in virions of Rous Sarcoma virus. *Nature*, **226**, 1211–13 (1970).

Watson, J. D., Crick, F. H. C., Genetical implications of the structure of deoxyribonucleic acid. *Nature*, **177**, 964–67 (1953).

Wilkins, M. H. F., Fuller, W., Wilson, H. R., Hamilton, L. D., The molecular configuration of deoxyribonucleic acid. IV. X-ray diffraction study of the A form. *J. Mol. Biol.*, **12**, 60–75 (1965).

12: Protein Metabolism

INTRODUCTION

Proteins, whether viral, bacterial, plant, or animal, are polymers of twenty different kinds of amino acids. But despite this fundamental gross similarity, no two proteins of a plant or any other organism are structurally and functionally alike. The basis of this difference lies at the level of the amino acid sequence characteristic of each protein. The differences among proteins are ultimately reflected in character variations between any two organisms or even between different developmental stages of the same organism. For example, a pea seed germinates and produces a pea plant with a characteristic organization into roots, stem, leaves, and flowers, each associated with a particular function. The biochemical causes of such phenotypic diversities among different organs of the same plant are the proteins, i.e., phenotypic differences are biochemical differences in proteins. Proteins determine these phenotypic variations, the precise sequence of amino acids within a protein distinguishing it from other proteins.

The genotype of an organism determines its phenotypes, and the currently accepted mode of flow of information in this process is as shown in Fig. 12-1.

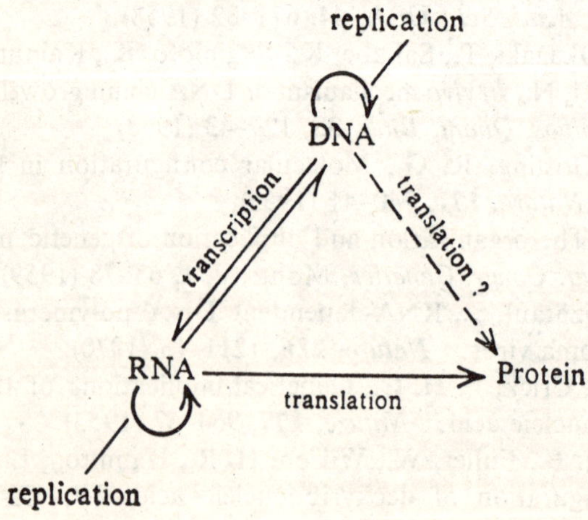

Fig. 12-1 Proposed mode of biological information transfer. (Crick's Central Dogma.)

Since the discovery of the enzyme *reverse transcriptase* (Temin, 1971), it is commonly agreed that information transfer between DNA and RNA can be bidirectional whereas that between RNA and protein is always unidirectional—from mRNA to protein. This is true of most cases; however, in one exception reported, bacteriophage DNA was observed to be translated into protein in the presence of the drug neomycin sulphate without the involvement of RNA as an intermediate. The model of information transfer in Fig. 12-1 is based on studies carried out with bacteria, bacterial viruses, and oncogenic (tumour-inducing) animal viruses. How far this model is applicable to eucaryotic plant cells is not known, but the original Central Dogma of Crick (1958), as shown in Fig. 12-1, is valid today for almost all cellular systems, including plants. According to Crick's theory, the autocatalytic function of DNA is confined to replication whereas its heterocatalytic function includes transcription and translation. Transcription involves the conversion of one stretch of base sequences into another set that is of a complementary nature. In contrast, translation is much more complex as it concerns the conversion of the nucleotide language of a nucleic acid into the amino acid alphabet of a protein.

DNA is segmented into several regions, called genes, which differ in respect of the number of base pairs and their order or sequence. The functional, mutational, and recombinational units of a gene are called cistron, muton, and recon, respectively. A cistron contains data for the formation of a single polypeptide. A protein may be composed of one, two, or more than two polypeptides. Allosteric proteins in general have more than one polypeptide chain. Thus, an organism will have fewer kinds of proteins than the total number of cistrons. *In vivo* transcription is asymmetric, involving only one of the two chains of a cistron at any one time. What determines the strand selection in transcription is not known. The chain transcribed is termed 'code complement' to differentiate it from the complementary chain called 'code copy' which is not transcribed. The transcription RNA (mRNA) is labelled the 'code' and contains at least as many codons as there are amino acids in the polypeptide chain synthesized on the protein-synthesizing machinery. The translation apparatus consists of ribosomes (70S or 80S), mRNA, nearly 40 kinds of tRNA, and almost the same number of aminoacyl tRNA synthetases (ligases), more than a dozen protein factors, 20 kinds of amino acids, and GTP, ATP, and Mg ions. If any of these substances is lacking, the process of translation will either be inhibited or become abnormal.

The RNA of RNA viruses acts directly as mRNA, and the satellite tobacco mosaic virus (STMV) has been found to contain just sufficient RNA to code for one protein (the coat protein), i.e., it contains a single cistron for its coat protein. Much of our knowledge about the mechanism and control of translation has emerged from research on RNA viruses (R17, Qβ) of bacteria. The unit of transcription is an operon. It involves at least four kinds of genes, called regulator, promoter, operator, and structural. Only the mRNA of the regulator gene and the portion of mRNA corresponding to the structural gene or genes of the operon are translated. How

Table 12-1 Genetic code

First base (at 5′—OH end)	Middle base U	C	A	G	Third base (at 3′—OH end)
U	Phe	Ser	Tyr	Cys	U
	Phe	Ser	Tyr	Cys	C
	Leu	Ser	**	**	A
	Leu	Ser	**	Trp	G
C	Leu	Pro	His	Arg	U
	Leu	Pro	His	Arg	C
	Leu	Pro	Gln	Arg	A
	Leu	Pro	Gln	Arg	G
A	Ileuc	Thr	Asn	Ser	U
	Ileuc	Thr	Asn	Ser	C
	Ileuc	Thr	Lys	Arg	A
	Met*	Thr	Lys	Arg	G
G	Val	Ala	Asp	Gly	U
	Val	Ala	Asp	Gly	C
	Val	Ala	Glu	Gly	A
	Val	Ala	Glu	Gly	G

* chain-initiating codon
** chain-terminating (nonsense) codon

For instance, an mRNA of repeating sequence (U-C) produced two kinds of homopolymer, viz., one polyserine and the other polyleucine. The repeating sequence U-C-U-C-U-C- produces two kinds of codons, viz., UCU and CUC. This finding, when compared with the results of the triplet binding test, at once suggested UCU as the codon for serine, and CUC as the codon for leucine. It also indicated that the protein-synthesizing machinery selected at random a nucleotide residue of the mRNA as a starting point for reading the genetic code; this in turn fixed the reading frame which continued being read as a group of three consecutive nucleotides of the mRNA at a time, in the 5′ to 3′ direction. However, such cursory selection of the nucleotide residue as a starting point for message reading in biosynthetic mRNA does not occur *in vivo*. It has been found that all biosynthetic mRNAs have a fixed point which initiates protein synthesis and that this point corresponds to the codon AUG which codes for formylmethionine (f-met) in case it lies at the 5′ terminus of mRNA and codes for methionine (met) if it is in an intercalary position along the message-reading frame. Depending on the size of the operon, the mRNA may be monocistronic or polycistronic. Both mono- and polycistronic mRNAs are believed to contain codons which act as terminating signals for protein synthesis. In polycistronic mRNA, there are thought to be as many initiation and termination codons as the number of resulting polypeptides produced. That nonsense codons are responsible for the termination of protein synthesis (see Davidson, 1972) is validated by several findings.

TRANSFER RNA

The structure and composition of tRNAs as studied from various sources have a relatively common and uniform secondary structure, often called clover-leaf structure (see Holley, 1966) (Fig. 11-6). The general features include (1) the presence of GMP at the 5′ terminus; (2) a sequence of cytidylic cytidylic-adenylic acids (CCA) at the 3′ terminus to which an amino acid gets attached by the hydroxyl group of adenosine moiety under the influence of an appropriate aminoacyl tRNA synthetase enzyme; and (3) four base-paired regions and four loops of unpaired regions containing most of the unusual bases, the anticodon always being located in the second or middle loop. Among the many unusual nucleotides that tRNAs are known to contain, the more prominent ones are ribothymidine, methylguanosine, dimethylguanosine, methylinosine, pseudouridine, and dihydrouridine. Evidence suggests that unusual bases result from the modification of the common bases subsequent to the synthesis of the tRNA polynucleotide chain. The significance of the unusual nucleotides is not yet certain, but it has been conjectured that their presence may prevent RNAs from being used as templates in protein synthesis. The tRNAs are in general very stable perhaps because of the involvement of their 5′ terminal residues in complementary base pairing (Fig. 11-6) in the amino acid arm which terminates in the CCA sequence. The most obvious difference between various tRNAs lies in their anticodons and the number and kind of nucleotide residues comprising the third unpaired arm. The paired regions, loops, and bands differ one from the other in respect of nucleotide composition and the number of nucleotide residues (see Davidson, 1972).

The mRNA codons are recognized by the anticodons of corresponding tRNAs and their identity is established by complementary base pairing between a triplet codon and a triplet anticodon. This codon-anticodon hypothesis theoretically demands the occurrence of as many tRNA types as there are codons coding for the protein-forming amino acids. Experimental findings report around 40 types of tRNA which is more than the different kinds of amino acids involved in protein formation but less than the total number of codons available for these amino acids. Five species of tRNA accepting leucine, three species of tRNA for serine and valine, and two for tyrosine and phenylalanine have been observed. When several tRNA species accept the same amino acid, e.g., leucine, they are called isoacceptors.

Figure 12-2 lists the anticodons of known tRNAs with their corresponding codons. The anticodon-codon pairing is antiparallel with regard to the polarity of the two polynucleotide chains. It is also seen that the alanine anticodon CGI can recognize three codons, viz., GCU, GCC, and GCA. The only difference among the three codons is with regard to their third position. Obviously, the pairing in the first and second positions is quite strictly followed whereas this is not so for the third position. This kind of data on codon-anticodon pairing led to the development of the Wobble hypothesis by Crick (1966) who explained the degeneracy of the code.

According to this hypothesis, strict complementary base pairing and codon specificity depend upon only the first two bases of the codon. This also explains why a single species of tRNA can recognize more than one codon and why cells should not possess as many species of tRNA molecules as there are codons. None of the common species of tRNA can recognize the three codons UAA, UAG, and UGA, and because of this feature protein synthesis terminates in the region of the mRNA that contains any of these three codons.

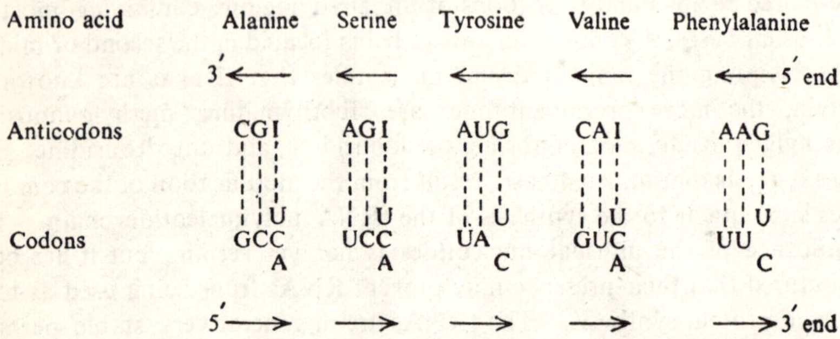

Fig. 12-2 Wobble hypothesis illustrating variations in third nucleotide of codon, and base pairings in an antiparallel direction between codons and anticodons. (After Davidson, 1972.)

The analysis of the sequence of some tRNAs has revealed that they are 75–77 nucleotides long. It has been visualized that each tRNA molecule has four specific sites: (1) the anticodon site, for specific interaction with the corresponding codon on mRNA; (2) the site that interacts with the ribosome, thus making the tRNA attachment possible and stable on the ribosome surface; (3) the site with which the aminoacyl tRNA synthetase associates; and (4) the amino-acid-accepting site, which consists of the CCA sequence. Apparently, the specificity of tRNAs in protein synthesis seems to reside in their anticodon in the sense that the anticodon of a tRNA will, by its specificity for a particular codon of mRNA, always be accepted by that codon, irrespective of the amino acid that the tRNA is artificially bluffed to carry. This also indicates that the rRNA-charging enzyme may control the specificity in terms of charging a particular tRNA with an appropriate amino acid.

Aminoacyl tRNA Synthetases

The formation of a charged tRNA or aminoacyl tRNA from an amino acid and tRNA involves two major reaction steps, both mediated by a specific enzyme called aminoacyl tRNA synthetase or aminoacyl tRNA ligase. The enzyme first activates the amino acid by the formation of an aminoacyl-adenylate-enzyme complex as follows:

$$\underset{\text{Amino acid}}{AA_1} + ATP \xrightarrow{\text{Enzyme } E_1} AA_1\text{-}AMP\text{-}E_1 + PP_i$$

This complex is then used for the aminoacylation of the appropriate tRNA as follows:

$$AA_1\text{-}AMP\text{-}E_1 + tRNA_1 \rightarrow AA_1\text{-}tRNA_1 + E_1 + AMP$$

The aminoacylation of tRNA involves the formation of an ester bond of the amino acid with the 2′—OH or 3′—OH of the terminal nucleotide of the tRNA. These two reactions, catalyzed by the enzyme aminoacyl tRNA synthetase, must be highly selective for a given amino acid and for the corresponding tRNA so that the correct peptide sequence is synthesized. Since the genetic code is universal, the various tRNA species and aminoacyl tRNA synthetases of one organism should be almost similar in activity and structure to those of any other organism. There should also be as many species of synthetases as there are tRNA species. Firm experimental evidence in support of these speculations is, however, lacking. If, in the formation of aminoacyl tRNA, the synthetase enzyme from one organism and the tRNA from another are used, the combination is known as heterologous; if both are from the same organism, the combination is termed isologous. The use of heterologous combinations has led to the important discovery of multiple forms of tRNA, all called isoacceptors because they accept the same amino acid. Subsequent studies on isologous and heterologous combinations have revealed that conditions for charging isologous tRNA are entirely different from those for charging heterologous tRNA. It has further been reported that all the five tRNAs for alanine of the human spleen can be aminoacylated by the corresponding synthetase extracted from rabbit liver, but that only three tRNAs show aminoacylation when the synthetase is derived from *Escherichia coli.* This indicates some difference in the tRNA binding site of the *E. coli* and that of rabbit liver synthetases. The *E. coli* synthetase seems comparatively more conservative than liver synthetase in this regard. The evolutionary significance of such observations will depend greatly upon the extent of how general and widespread is such heterologous specificity in the biological world.

In the entire aminoacylation reaction of tRNA, the enzyme reacts first with ATP, forming an enzyme-ATP complex. This leads to a change in the conformation of the enzyme in such a way that the complex becomes ready to accept the proper amino acid, thus resulting in the formation of an enzyme-AMP-amino acid complex and the elimination of pyrophosphate. The AMP-amino acid complex is the activated form of the amino acid. The arrival of the proper tRNA close to the enzyme-AMP-amino acid complex induces a transfer reaction in which the activated amino acid gets covalent bonded to the tRNA molecule at its 3′ terminus. The specificity of the enzyme for the amino acid activation is less stringent than for the amino acid transfer reaction. For example, the enzyme complex can activate a wrong amino acid, but the arrival of the right tRNA, though resulting in a transfer reaction, brings about the transfer of the wrong amino acid to water instead of to tRNA. It is conjectured that such specificity of the transfer reaction might be one of the chief ways of charging a tRNA with the correct amino acid (see Jacobson, 1971).

RIBOSOMES

Ribosomes fall into two main classes according to their sedimentation coefficients, viz., 70S ribosomes characteristic of procaryotes and such organelles as chloroplasts and mitochondria, and 80S ribosomes characteristic of eucaryotic cytoplasm and found free of or associated with the endoplasmic reticulum. 70S ribosome is composed of two subunits, called 30S and 50S whereas 80S ribosome is made up of the subunits 40S and 60S. 70S ribosome has 60–65% RNA and 35–40% protein whereas 80S ribosome seems to have almost equal proportions of RNA and proteins. The structure and function of 70S bacterial ribosomes are better defined than those of the 80S ribosomes of eucaryotes. The 50S subunit of 70S ribosome contains two species of ribosomal RNA having sedimentation coefficients of 23S and 5S. Its protein part is made up of nearly 34 kinds of protein subunits. The 30S subunit consists of 16S rRNA and 20 kinds of distinct proteins. Since one of the established functions of the ribosome is to orient the mRNA and charged tRNA in such a way as to enable the proper translation of the base sequence into an amino acid sequence, it must possess sites for the attachment of mRNA and tRNA. The mRNA is bound to the 30S subunit which contains also the decoding site where correct codon-anticodon pairing occurs. The 50S subunit has two distinct binding sites, one for the binding of the aminoacyl tRNA, called site A, and the other for the binding of the peptidyl donor tRNA, termed site B.

Not much is known about the protein types of the 60S and 40S subunits of 80S ribosomes. However, these subunits play the same role in protein synthesis as do the 50S and 30S subunits. The 60S subunit contains two rRNAs, 25S (or 28S) and 5S whereas the 40S includes only 18S rRNA.

When protein synthesis is not active, 70S ribosomes exist as 30S and 50S subunits, but when active protein synthesis takes place, the two subunits join to form 70S ribosome which helps in reading the genetic code and translating it into a polypeptide. Once a ribosome has completed the synthesis of a polypeptide, it dissociates from the translation apparatus and gets cleaved into 30S and 50S subunits; these subunits can then reassociate to start the next round of protein synthesis. Such a merger of subunits into a ribosome, followed by a breaking down of the ribosome into subunits, is termed the *ribosome cycle*. The structural and functional dynamism of 70S ribosomes suggests that these ribosomes are probably the most active component of the procaryotic cell. Before the commencement of polypeptide synthesis, the 30S and 50S subunits associate with each other through a complex process and the resulting 70S ribosome becomes ready to read the mRNA in the 5′ to 3′ direction, reading one codon at a time or permitting the formation of one peptide bond at a time. As soon as this ribosome has moved a distance of a few codons, another ribosome gets attached to the initiation codon (AUG) and repeats the behaviour of the first ribosome. In this way, a single mRNA is loaded with a number of ribosomes, each synthesizing a new growing polypeptide chain. Such a cluster of ribosomes strung along the mRNA is called a *polysome* or *polyribosome*; a polysome will often aid the synthesis of a number of copies of the polypeptide corresponding to the number of ribosomes composing the polysome. In other words, a polysome of eight ribosomes will produce eight copies of the polypeptide.

The functions of various ribosomal proteins and ribosomal RNAs in protein synthesis are not yet very clear, but studies on the reconstitution of functional 30S and 50S subunits from their components have elucidated this aspect (see Revel *et al.*, 1972). In protein synthesis, the ribosome helps in correct recognition of the anticodon by the relevant codon since a synthetic trinucleotide acting as mRNA can bind to any region of the corresponding tRNA, but the addition of the ribosome restricts this binding to the anticodon region. As stated earlier, the decoding site lies in the region of the 30S subunit, and evidence suggests that the proteins of this subunit participate in influencing the codon-anticodon pairing. Streptomycin kills procaryotic organisms by inactivating the 30S subunit of 70S ribosomes, and this ultimately leads to the inhibition of protein synthesis. It has been found that a protein component of the 30S subunit, called P10, binds the streptomycin, and the resulting protein-streptomycin complex is responsible for the inhibition of protein synthesis. Another effect of the drug includes the misreading of codons by the streptomycin-affected 70S ribosomes, thereby leading to faulty synthesis of polypeptide. Streptomycin-resistant mutants have their P10 protein which, however, does not bind streptomycin and, hence, normal proteins are formed in these mutants. This obviously means that P10 protein is an essential component of the ribosomal machinery involved in the specific and correct decoding process. Similarly, another protein component of 70S ribosomes, called P7, has been found to restrict the degree of ribosomal misreading of codons.

The 50S subunit contains the enzyme peptidyl transferase and perhaps also other enzymatic proteins which catalyze the formation of the peptide bond and the transfer of peptidyl moiety from peptidyl tRNA to aminoacyl tRNA. There are specific and distinct sites on 16S rRNA where the various proteins of the 30S subunit bind in a way that produces the functional 30S subunit. Proteins of the 50S subunit do not associate with 16S rRNA to produce the functional subunit. Even a small fragmentation of the 16S rRNA results in complete loss of the functional reconstitution of the particle. All these features clearly imply a definite structural role of rRNA in appropriate orientation and organization of ribosomal proteins which ultimately leads to the formation of functionally viable particles.

PROTEIN FACTORS

In addition to ribosomes, mRNA, 20 kinds of amino acid, ATP, and various aminoacyl tRNA synthetases and tRNAs, protein synthesis requires the participation of a number of proteins called (1) initiation factors; (2) elongation factors; (3) translocation factors; and (4) release factors. Precise functions of these factors have not been resolved. Three kinds of proteins

by tRNA but are recognized by protein factors called release factors. Three such factors, designated R1, R2, and R3, have been detected in the bacterial cell that is engaged in protein synthesis. R1 recognizes UAG and UAA, and R2 identifies UAA and UGA (Capecchi and Klein, 1969). It is proposed that the release factors R1 and R2, under the direction of the nonsense codon, bind to the ribosome in the presence of the other release factor, R3, and GTP. These reactions trigger the transfer of the peptide from tRNA to water and are thus freed from the ribosome.

MODEL OF PROTEIN SYNTHESIS

A polypeptide chain grows step by step by the addition of individual amino acids, beginning at the amino end and terminating at the growing carboxylic end. The information for a particular polypeptide synthesis is encoded in the corresponding mRNA which contains a codon (or codons) for the initiation of the chain and codons for the termination of the chain, in addition to codons coding for amino acid residues of the polypeptide chain. The translation machinery reads the codons in a 5′ to 3′ direction. The synthesis of the polypeptide chain includes three steps: (1) chain initiation, which occurs only once; (2) chain elongation, which is repeated as many times as there are peptide bonds in the chain; and (3) chain termination, resulting in the release of a completed polypeptide from the mRNA-ribosome-tRNA complex.

CHAIN INITIATION

In procaryotes and viruses, chain initiation involves the formation of an initiation complex consisting of mRNA, the 30S ribosomal subunit, f-met-tRNA, initiation factors F1, F2, and F3, and GTP. This complex soon permits the association of the 50S subunit, giving rise to an initiation complex on the 70S monosome. There is good reason to believe that the signal for the initiation of translation in most procaryotes is AUG or GUG which is recognized in procaryotes by a special aminoacyl tRNA complex, called f-met-tRNA. But in eucaryotes, met-tRNA seems to replace f-met-tRNA. Certain procaryotes contain two species of methionine tRNA, one of which undergoes formylation at the α-amino group of its methionyl moiety by the enzyme formylase whereas the other is not formylated. It is known that peptide bond formation involves a reaction between the α-amino group of one amino acid with the carboxylic group of the preceding amino acid. Accordingly, the ideal initiator of protein synthesis would be the amino acid that has its α-amino group blocked, thus rendering it nonavailable for peptide bond formation with any preceding amino acid. The involvement of f-met-tRNA in all kinds of polypeptide initiation in bacterial and viral systems supports this concept, but when considered in the light of met-tRNA initiating the proteins, this theory is open to question. Protein initiation in eucaryotes is attributed to nonformylated methionyl tRNA perhaps because of the lack of the formylase enzyme in eucaryotes. However, the details of protein synthesis in eucaryotes have yet to be worked out.

All kinds of proteins in procaryotes are synthesized, beginning with f-methionine at their amino terminal end, but none of them retains the formyl group after the completion of their synthesis. Even methionine is often found missing in some of these proteins. Procaryotic cells seem to have special enzymes which recognize and hydrolyze the formyl group or methionine together with a few more amino acids from the amino terminal end of the completed polypeptide.

Proteins contain methionine residues which generally lie in the internal region of the molecules. The only codon for methionine is AUG and this is recognized by both f-met-tRNA and met-tRNA. The function of the f-met-tRNA is to initiate the polypeptide chain synthesis, but that of met-tRNA is to decode the internal codon of methionine, leading to proper placement of the methionyl residue in the intercalary position of the growing polypeptide chain. Another initiator codon, GUG, which normally codes for valine, is also recognized by f-met-tRNA, but not by met-tRNA. Since the initiator codon identified in Qβ and R17 viruses is AUG, which can be read by both f-met-tRNA and met-tRNA, it may be concluded that there is nothing in the codon specificity to explain the observed behavioural difference between the two kinds of met-tRNAs. Thus, available information indicates that initiator factors together with 70S monosomes may have the answer to why only formylmethionyl tRNA acts as the chain initiator. The RNA of the R17 virus contains three cistrons, one for its coat protein, another for its maturation protein, and the third for its RNA replicase. Its proteins contain methionine in their intercalary position. The intercistronic regions are found to contain AUG. Analyses of the secondary structure of its RNA suggest that most of the RNA is organized in the form of hydrogen-bonded hairpin loops, leaving only short interloop regions in the linear, single-stranded form in which initiator AUG codons are believed to be located. It seems that the secondary structure of translatable RNA also plays a role in the selection and expression of cistrons.

Another significant relevance of these observations is that the organizational pattern of mRNA permits the translating machinery to initiate protein synthesis on the AUG-containing region of mRNA which is exposed and not buried within the hydrogen-bonded loops. That is, the organizational pattern of the mRNA molecule determines whether AUG will act as a chain initiator or as a codon for intercalary methionine residue, at least in the case of the R17 virus.

CHAIN ELONGATION

Chain elongation starts after the successful completion of the initiation complex which consumes a molecule of GTP during the process. As shown in Fig. 12-3, the elongation phase consists of three steps. The first step involves aminoacyl tRNA binding to site A, as detected by the codon adjacent to the 3′ side of the initiator codon. The elongation factors TS and TU together with GTP are required for the transport of the proper

of bacteria, especially *E. coli,* but in general it may be applicable to any level of biological organization ranging from viruses to higher plants and mammals.

Inducible enzymes are a class of proteins produced only when their substrates or closely related substances are available. The other group of enzymes, qualified as repressible, have their production blocked in the presence of excess end products. We shall give some examples of inducible and repressible enzyme synthesis to clarify this concept.

In *E. coli,* the disaccharide β-galactoside, called lactose, is enzymatically cleaved into glucose and galactose. There are two enzymes functioning in sequence in the entire operation. β-galactoside permease or lactose permease transports the lactose from the outside to the inside of the cell. The enzyme β-galactosidase then hydrolyzes the lactose into glucose and galactose. Both these enzymes are almost completely absent from cells of *E. coli* that are not fed with lactose. A supply of lactose to the cells immediately leads to the appearance and subsequently to the increase in the level of the two enzymes. Genetical and biochemical studies have established the existence of an operon, called the lactose operon, consisting of a promoter gene which controls the rate of transcription, and an operator gene which is involved in the switching off and on of the expression of structural genes or cistrons; these cistrons, numbering three, are conveniently symbolized as Z, X, and Y (in this order) with reference to the operator gene. The Z cistron codes for mRNA involved in the synthesis of the enzyme β-galactosidase; the X cistron in a similar way specifies lactose permease; and the Y cistron controls the structure of the thiogalactoside transacetylase enzyme. All the three cistrons are either made to function together or prevented from so functioning; that is, their expression is coordinated. The molecule that participates, together with the operator gene, in the coordinated control of expression of the three cistrons is a protein called the repressor. It is a product of another gene, called the regulator, which is not linked to the operon. It is important to note here that most of bacterial operons have their structural genes closely linked or clustered. For example, in the lactose operon, the three cistrons Z, X, and Y are closely linked to operator genes. The repressor protein interacts only with the operator and this specificity must be due to some unique base sequence in the operator gene which is recognized only by the repressor molecule. The result of the repressor-operator interaction is the termination of the transcription process, probably in the following manner: the DNA-dependent RNA polymerase binds to the promoter gene and then moves along the template DNA chain in the operator-to-cistron direction; the operator, forming a complex with the repressor, prevents further progress of the polymerase enzyme, thus leading to the stoppage of transcription of the three cistrons; since polycistronic mRNA of the lactose operon is not produced, the corresponding enzymes fail to appear.

We have already noted that the lactose operon enzymes permease and galactosidase, along with thiogalactoside transacetylase, are not produced in the absence of lactose. The addition of lactose leads to their immediate production. The molecular basis for this is that each repressor protein molecule which remains in a bacterial cell possesses two specific sites, one interacting with the operator, and the other (in the case we quote) with lactose. Since the repressor is an allosteric protein, its configuration and activity depend upon its interaction with either of the two effectors, the operator or the lactose. The presence of lactose inactivates it in the sense that it is no longer able to recognize the operator which is thus left open for the RNA polymerase to proceed for transcription of the entire operon. The absence of lactose allows the repressor to block transcription by forming a complex with the operator gene. In other words, the repressor is active in the absence of the inducer with the result that transcription is blocked. This sort of transcriptional control is termed negative control because the activity of the repressor protein bars the transcription of its relevant operon. Figure 12-4 illustrates the features of the operon model.

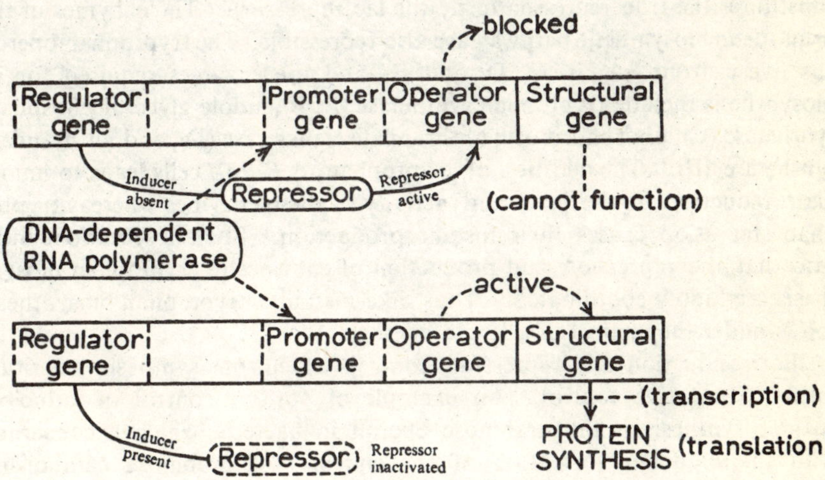

Fig. 12-4 Operon model of gene expression and regulation of enzyme synthesis.

Repression control is the reverse of induction control in that the repressor protein is unable to switch off transcription on its own and operates only when biosynthetic enzymes, such as those synthesizing tryptophan or histidine, occur in a metabolic pathway. This is in contrast to inducible enzymes, all of which are catabolic in nature and function mainly in supplying the raw material and energy source for the biosynthesis of essential cellular metabolites. In repression control, the excess of end products of a multienzyme biochemical pathway signals the stoppage of formation of all the enzymes concerned in the pathway. This may result from exogenous supply or from less biosynthetic consumption. The end product commonly forms a complex with the repressor and the resulting complex becomes active in the sense that it alone is able to recognize the operator of the biosynthetic pathway, thereby switching off the transcription process in the way the lac repressor does in the lac operon.

More recent findings on the mechanism of repression of enzymes of amino acid biosynthesis have led to the unexpected discovery that amino acids are able to repress the formation of their biosynthetic enzymes, only if they are first activated by the corresponding aminoacyl tRNA synthetase, perhaps also involving the relevant tRNA molecules. This condition applies at least in the repression of enzymes of the histidine operon by histidine when the production of histidine exceeds its consumption or when it is exogenously supplied. The histidine operon has nine cistrons corresponding to the nine enzymes involved in its biosynthesis. The expression of all the nine cistrons is coordinated by the interaction of the operator and the repressor-activated histidine complex in exactly the same way as in the lac operon. In repression control, the repressor by itself is not active and, hence, is called the aporepressor which jointly with the activated histidine constitutes the true repressor (e.g., the lac repressor). The enzymes of the tryptophan biosynthetic pathway are also repressible. The tryptophan operon has five cistrons—A, B, C, D, and E—and the enzymes required for its biosynthesis include tryptophan synthetase (A, B), indole glycerol phosphate synthetase (C), phosphoribosyl anthranilate transferase (D), and anthranilate synthetase (E). The addition of tryptophan to *E. coli* cells leads to immediate reduction in the level and activity of these enzymes whereas tryptophan starvation causes their instant production. Such observations indicate that the repression and production of enzymes of tryptophan biosynthesis are highly coordinated. Thus, like histidine, tryptophan biosynthesis too is under the control of a single operon.

Both induction and repression controls of enzyme synthesis constitute forms of negative control. An example of positive control of inducible enzyme synthesis is the arabinose operon in bacteria which is concerned with the inducible formation of the enzymes of arabinose catabolism. Findings suggest that the product of the regulatory gene of the arabinose operon is inactive by itself in the sense that it cannot interact with the arabinose operon. The availability of arabinose to the cells activates the protein that signals the initiation of transcription of the arabinose operon. That is, the complex of regulatory protein and arabinose acts as the initiator of the arabinose operon.

Yet another important mode of regulation of bacterial genes is termed catabolite repression or glucose effect. The functioning of operons in *E. coli*, each controlling the breakdown of a specific sugar, is stopped by glucose. For example, bacterial cells growing in a medium containing glucose and lactose stop synthesizing the enzymes of the lactose operon. This is so also in galactose and arabinose operons. Recent biochemical studies explain catabolite or glucose repression as follows: cyclic AMP is involved in the expression of almost all catabolite operons; the products of glucose catabolism lead to a reduction in the level of c-AMP whose nonavailability in turn blocks the transcription of catabolite operons; in the absence of glucose, the intracellular level of c-AMP rises and so does the rate of transcription of catabolite operons in the presence of their respective effectors.

Not all cistrons controlling the structure of enzymes of a linked biochemical pathway necessarily occur in clusters. Sometimes the cistrons may be scattered over one chromosome, as in procaryotes, or may be found in different chromosomes, as in eucaryotes. Let us assume, for instance, that six cistrons of a certain linked biosynthetic pathway are clustered in three groups of two cistrons each, located in three regions of one chromosome or in three distinct chromosomes. In either case, the result is the formation of three operons, all of which are under the control of a single regulator gene. It is believed that the scattering of the cistrons of a linked pathway is a very common feature of eucaryotes whereas in procaryotes it is much less widespread.

The regulation of gene expression in eucaryotes is further complicated by the presence of chromosomal proteins, called histones. The artificial removal of histones from such chromosomes is known to considerably increase their transcriptional ability. This suggests that the simple model proposed by Jacob and Monod for bacterial gene regulation may not find direct applicability to eucaryotes. Various modifications of this model are now being worked out in different laboratories in attempts to apply them to eucaryotic systems, but without any apparent success so far. One view regarding the action of plant hormones in higher plants is that the hormones activate genes in the same way as the effectors of bacterial systems. For instance, gibberellin has been shown to specifically activate those genes in barley (during seed germination) that code for amylases (Johri and Varner, 1968); then, again in the presence of gibberellin, molecules have been formed by the nuclei of cells in the aleurone layer and, finally, translated into amylase proteins. Furthermore, the chromatin of pea buds, unable to synthesize the pea seed globulin protein *in vivo*, failed to synthesize this protein even *in vitro*. However, the removal of histone from the chromatin led to the production of mRNA that was capable of synthesizing globulin protein (Bonner *et al.*, 1964). These findings again endorse the role of histones in the regulation of gene activity in eucaryotes. Very little is known about the relative role of histone and nonhistone proteins in chromosomes in the regulation of activity of eucaryotic genes, but the fact remains that both these classes of proteins participate in the control of protein synthesis in eucaryotes.

P-PROTEINS

P-proteins, a newly-discovered class of proteins, is so named mainly because of its specific location in the phloem of higher plants where they occur as aggregates of tubular, fibrous, crystalline, or amorphous material. P-proteins are very similar to actin-like contractile microfilaments and, therefore, it has been suggested that they may function in the movement of assimilates in sieve elements. However, there are remarkable differences also between P-proteins and actin; for example, unlike actin, P-proteins do not show an associated ATPase activity, and do not respond to treatments with agents such as colchicine and cytochalasin-B. Some workers (see Helper and Palevitz, 1974) assume that P-proteins, like actin and tubulin, are a class of structural proteins in the phloem of higher plants.

REFERENCES

Bonner, J., Huang, R. C. C., Murray, K., Physical and biological properties of soluble nucleohistones. *J. Mol. Biol.*, **8**, 54–64 (1964).

Capecchi, M., Klein, H., Characterization of three proteins involved in polypeptide chain termination. *Cold Spring Harbor Sympos. Quant. Biol.*, **34**, 469–77 (1969).

Carlson, P. S., The use of chloroplast in genetic research. *Proc. Natl. Acad. Sci.*, USA, **70**, 598–602 (1973).

Crick, F. H. C., On protein synthesis. S. E. B. Symposium, No. 12, Cambridge Univ. Press, London (1958).

Crick, F. H. C., Codon-anticodon pairing: The Wobble hypothesis. *J. Mol. Biol.*, **19**, 548–55 (1966).

Davidson, J. N., The Biochemistry of the Nucleic Acids, 7th edn. Methuen, London (1972).

Helper, P. K., Palevitz, B. A., Microtubules and microfilaments. *Ann. Rev. Plant Physiol.*, **25**, 309–62 (1974).

Holley, R. W., The nucleotide sequence of a nucleic acid. In Haynes, R. H., Hanawalt, P. C. (eds.), The Molecular Basis of Life, 72–81, W. H. Freeman, San Francisco (1966).

Jacob, F., Monod, J., Genetic regulatory mechanisms in the synthesis of proteins. *J. Mol. Biol.*, **3**, 318–56 (1961).

Jacobson, K. B., Reaction of aminoacyl-tRNA synthetases with heterologous tRNA. In Davidson, J. N., Cohn, W. E. (eds.), Progress in Nucleic Acid Research and Molecular Biology, Vol. 11, 423–54, Academic Press, New York (1971).

Johri, M. M., Varner, J. E., Enhancement of RNA synthesis in isolated pea nuclei by gibberellic acid. *Proc. Natl. Acad. Sci.*, USA, **59**, 269–76 (1968).

Khorana, H. G., Synthesis in the study of nucleic acid. *Biochem. J.*, **109**, 709–25 (1968).

Kuwano, M., Schlessinger, D., Binding of 3′:5′-cyclic phosphate to G factor of *E. coli*, and its effects on GTPase, RNase and protein synthesis. *Proc. Natl. Acad. Sci.*, USA, **66**, 146–52 (1970).

Nirenberg, M., Leder, P., RNA code words and protein synthesis. *Science*, **145**, 1399–407 (1964).

Nirenberg, M., Matthaie, J. H., The dependence of cell-free protein synthesis in *E. coli* upon naturally occurring or synthetic polyribonucleotides. *Proc. Natl. Acad. Sci.*, USA, **47**, 1588–602 (1961).

Ochoa, S., Enzymes in polynucleotide metabolism. *Ann. N. Y. Acad. Sci.*, **81**, 511–804 (1959).

Revel, M., Groner, Y., Pollack, Y., Scheps, R., Berissi, H., Protein synthesis machinery and the regulation of messenger RNA formation. In Polymerization in Biological Systems, 69–85, Ciba Foundation Symposium, Associated Scientific Publishers, London (1972).

Rudland, P., Whybrow, W., Clark, B. F. G., Recognition of bacterial initiator tRNA by an initiation factor. *Nature New Biol.*, **231**, 76–78 (1971).

Temin, H. M., Mechanism of cell transformation by RNA tumor viruses. *Ann. Rev. Microbiol.*, **25**, 609–48 (1971).

13: Lipid Metabolism

INTRODUCTION

Though usually not present in as large quantities as carbohydrates and proteins, lipids are nevertheless important structural components of living cells, such as cell membranes. Lipids often equal the amounts of carbohydrates and proteins in organelles and exceed them in many seeds. If the waxes on many tropical leaves are also included, the total amount of lipids is probably equal to at least all the proteins. Lipids, which include fats and oils, are generally found in a combined form with either proteins or carbohydrates or both, the two exceptions being seeds and osmiophilic globules in plastids. The term 'lipid' is broadly used to cover a wide range of heterogeneous, fat-like, organic substances, all of which are insoluble or sparingly soluble in water but freely soluble in nonpolar solvents, such as alcohol, ether, and benzene. Examples of such heterogeneous substances are fats and oils, phospholipids, sphingolipids, sterols, waxes, etc. Most lipids are highly complex molecules, but the natural fats and oils which are the most abundant and widely distributed forms of lipids, and which are of central importance in our discussion, are relatively simpler in structure and chemistry. Unlike proteins and carbohydrates which are polymeric macromolecules formed by respectively the condensation of numerous amino acid and sugar moieties, lipids do not form any such polymers. In fact, lipids owe much of their biological usefulness to the fact that, by virtue of their nonpolymeric form coupled with their soluble character, they can bridge the gap between the water-soluble and water-insoluble phases without a sharp demarcation line. Recent findings indicate that the nature and functions of various cell membranes (including those of subcellular organelles) which are simply organizations of macromolecules (including some smaller, nonpolymeric lipid molecules) may best be appreciated in terms of the unique solubility properties of lipids.

Fats are important as fuel and for the formation of a concentrated food source, especially in animals; in plant leaves, the majority of food reserves are commonly stored in the form of starch and other substances, but most seeds and fruits are rich in fats. Contrary to the popular belief that fatty deposits in animals are consumed only during periods of starvation, analyses of labelled fatty acids have demonstrated that fats are constantly being used and reformed.

The simpler fats and oils (the former are normally solid at ambient temperatures and the latter liquid) are similar to carbohydrates in that they contain only carbon, hydrogen, and oxygen. Unlike carbohydrates, however, they have less oxygen than carbon and hydrogen. Upon hydrolysis, fats and oils yield glycerol and fatty acids, and the general formula for these simpler lipids (fats and oils) is:

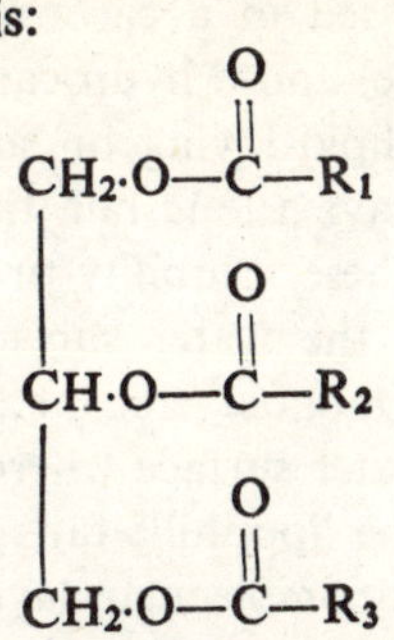

(R radical may be identical or may vary as shown.) Mono-, di-, and triglycerides are esters of one, two, and three fatty acids with glycerol. Particular species of plants and animals seem to possess distinctive glyceride patterns or spectra.

In the distant past, it used to be assumed that natural fats were mixtures of simple monoglycerides. The adherence to this belief was sustained by the lack of suitable techniques for complete separation and analysis of fat components. The development of highly sophisticated and automatic instruments and analytical techniques, such as a combination of argentation and thin layer chromatography, countercurrent distribution, lipase hydrolysis, gas-liquid chromatography, and reversed-phase chromatography, has been responsible for most of the current precision analyses of natural fats. The use of such modern methods and techniques has disclosed the interesting fact that most natural glycerides possess a stereospecific distribution of fatty acids rather than a random distribution (Hübscher, 1970; Coleman, 1963; James and Morris, 1964).

FATTY ACIDS AND THEIR SYNTHESIS

Fatty acids are straight chain (rarely, branched chain) organic acids containing a carboxylic acid group (—COOH) and an even number of carbon atoms. They may be represented by the general formula $CH_3(CH_2)_nCOOH$, where *n* is an even number. Fatty acids may be saturated (i.e., with no double bonds), or unsaturated (i.e., having one or more double bonds). Common saturated fatty acids are palmitic ($C_{153}H_1COOH$), stearic ($C_{17}H_{35}COOH$), and butyric (C_3H_7COOH, a short chain acid). Unsaturated acids are crotonic ($CH_3{\cdot}CH{=}CH{\cdot}COOH$), oleic [$CH_3{\cdot}(CH_2)_7CH{=}CH(CH_2)_7COOH$], and linoleic [$CH_3{\cdot}CH_2{\cdot}CH{=}CH{\cdot}CH_2{\cdot}CH{=}CH{\cdot}CH_2CH{=}CH{\cdot}(CH_2)_7COOH$].

Except for many bacteria, organisms are capable of introducing two or more double bonds into the fatty acid chain (Gurr and James, 1971). Such double bonds are always interrupted by a methylene group. It has also been noticed that animals and plants synthesize different families of poly-

with a fatty acid. Acetyl CoA is found in all aerobic cells (see Krebs cycle and Respiration in Chapter 8), and malonyl CoA can be formed from unsaturated fatty acids whereas bacteria do not synthesize these.

Each fatty acid is composed of a carboxyl group which is hydrophilic (water-loving or water-soluble) and a hydrocarbon chain, or tail, of variable length, which is lipophilic (lipid-loving or soluble in nonpolar solvents). In higher fatty acids which have a long tail, the solubility in water is greatly decreased. As a result of these solubility properties, any contact between a long fatty acid chain and the water surface leads to the appearance of what are known as 'spread monolayers', i.e., a one-molecule thick film of fatty acid is formed on the water surface where the hydrophilic heads come in contact with water and the lipophilic tails remain projecting away from the water surface. Such monomolecular layers constitute good working models to explain the organization of biological cell membranes.

The first step in fatty acid metabolism usually involves the activation of the fatty acid and its conversion into the thiol ester of the acid using either CoA or a relatively small protein, called acyl carrier protein (ACP) (see Fig. 13-1). Such activation results in two situations: (1) the molecule acquires a high energy thiol ester bond between the carboxyl group of the fatty acid and the sulphhydryl (—SH) group of the coenzyme, and (2) the acyl chains are rendered water-soluble. The activation of fatty acids is catalyzed by the enzyme thiokinase or acyl CoA synthetase, and magnesium and ATP (or GTP) are essential requirements for the action of this enzyme.

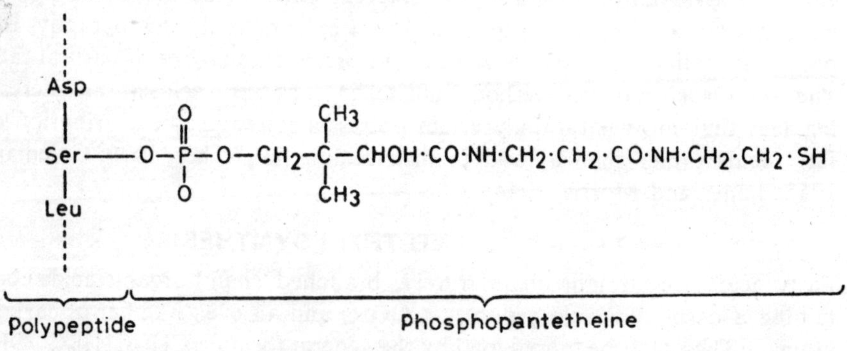

Fig. 13-1 Acyl carrier protein.

Fatty acids may be synthesized within a cell through any one or both of two independent pathways. One pathway is located within the mitochondria and the other in the cytoplasm. In the mitochondrial pathway, an existing short fatty acid chain condenses repeatedly with acetyl CoA, and thus gets progressively converted into a long fatty acid chain. The cytoplasmic pathway, on the other hand, represents the total synthesis of fatty acids, instead of merely leading to the elongation of the chain of a pre-existing acid. The cytoplasmic pathway, which is accepted as the major pathway (Awapara, 1968), is thought to involve a series of enzyme-catalyzed condensations, starting with acetyl CoA and malonyl CoA and ending

acetyl CoA as follows:

$$CH_3{\cdot}CO{\cdot}S{-}CoA + CO_2 \text{ (Acetyl CoA)} + ATP \xrightarrow{\text{Carboxylase}} COOH{\cdot}CH_2{\cdot}CO{\cdot}S{-}CoA \text{ (Malonyl CoA)} + ADP + P_i$$

In addition to this carboxylase, the cytoplasm contains the enzyme ACP which catalyzes the binding of acyl intermediates during the development of long fatty acid chains. The reduced ACP (ACP—SH) combines with malonyl CoA to form the malonyl derivative of ACP:

$$\text{Malonyl—S·CoA} + \text{ACP—SH} \rightarrow \text{Malonyl—S—ACP} + \text{CoA·SH}$$

The malonyl derivative then condenses with a primer fatty acid molecule joined to ACP, and CO_2 is released:

$$\text{Malonyl—S—ACP} + \text{Acetyl—S—ACP} \rightarrow \text{Acetoacetyl—S—ACP} + CO_2 + \text{ACP—SH}$$

Through further reactions, the acetoacetyl compound is changed to the butyryl form:

$$\text{Acetoacetyl—S—ACP} + \text{NADPH} + H^+ \rightarrow \alpha\text{-hydroxybutyryl—S—ACP} + \text{NADP}^+$$

$$\alpha\text{-hydroxybutyryl—S—ACP} \rightarrow \text{Crotonyl—S—ACP} + H_2O$$

$$\text{Crotonyl—S—ACP} + \text{NADPH} + H^+ \rightarrow \text{Butyryl—S—ACP} + \text{NADP}^+$$

The chain length increases still further when butyryl—S—ACP likewise condenses with the malonyl derivative and (through intermediate reactions analogous to those already stated) gives rise to the propionyl derivative. A new molecule of malonyl CoA thus enters into the sequence of reactions and recommences the process of chain elongation, with two carbons being added to the growing fatty acid molecule. Each complete turn of this system results in the net addition of two carbon atoms to the original (primer) fatty acid till such time as a long fatty acid chain is formed. Such a process of elongation continues until the final product is released from the synthetase enzyme as an acyl CoA derivative:

$$\text{R·CO—S—ACP} + \text{HS—CoA} \rightleftharpoons \text{R·CO·S—CoA} + \text{HS—ACP}$$

The whole series of reactions involved in the synthesis of fatty acids is summarized in Fig. 13-2.

Unsaturated fatty acids are synthesized by the direct introduction of double bonds into saturated acids through the enzymatic action of desaturases which require oxygen and a reduced coenzyme (I or II) (Mazliak, 1973). However, although the existence of desaturases is well-documented in animal tissues, these enzymes have not so far been isolated from higher plants.

Polyunsaturated fatty acids, such as linolenic acid, which are more abundant in higher plants than monounsaturated acids, are synthesized rapidly during seed maturation. Linolenic acid which is concentrated in chloroplasts may well represent as much as 70% of the weight of all fatty acids in chloroplasts.

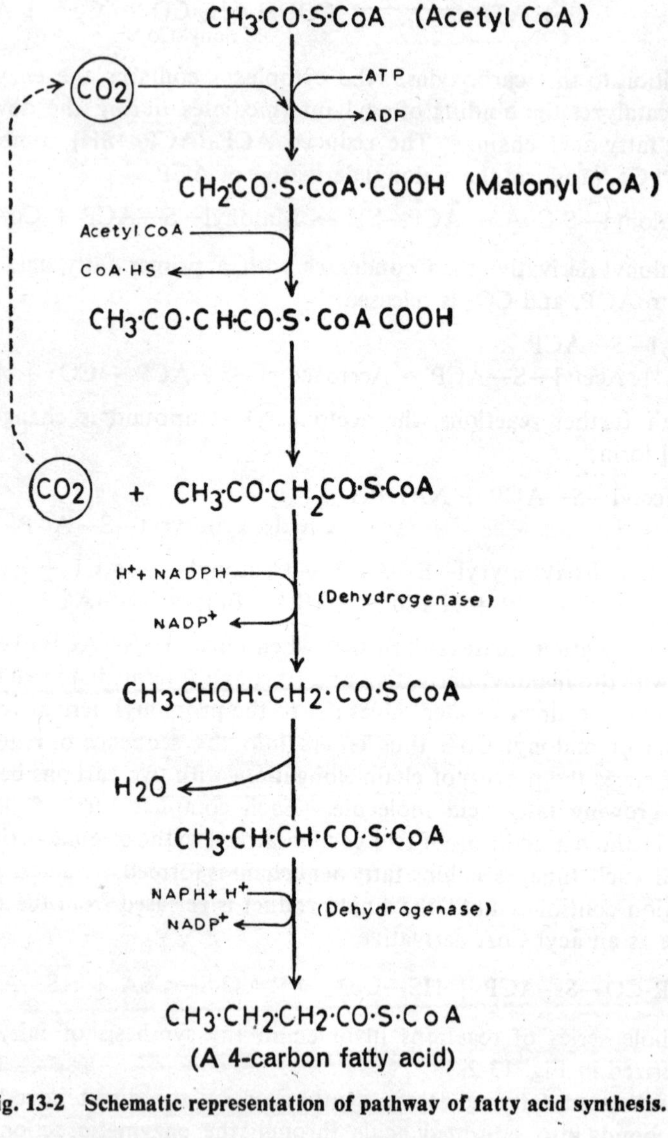

Fig. 13-2 Schematic representation of pathway of fatty acid synthesis.

GLYCEROL SYNTHESIS

Since a fat is formed by the condensation of three fatty acid molecules with one molecule of glycerol, the latter constitutes the second requirement for fat synthesis. It results from the reduction of dihydroxyacetone phosphate, which is an intermediate in glycolysis, through particular reactions as follows:

$$\begin{array}{c} CH_2OH \\ | \\ C{=}O \\ | \\ CH_2\cdot O\cdot H_2PO_3 \\ \text{Dihydroxyacetone} \\ \text{phosphate} \end{array} + NADH + H^+ \underset{}{\overset{\text{Dehydrogenase}}{\rightleftharpoons}} \begin{array}{c} CH_2OH \\ | \\ CHOH \\ | \\ CH_2\cdot O\cdot H_2PO_3 \\ \text{Glycerol} \\ \text{phosphate} \end{array}$$

$$\text{Glycerol phosphate} \underset{+H_2O}{\overset{\text{Phosphatase}\; -H_2O}{\rightleftharpoons}} \begin{array}{c} CH_2OH \\ | \\ CHOH \\ | \\ CH_2HO \\ \text{Glycerol} \end{array} + H_3PO_4$$

FAT SYNTHESIS

Three fatty acid molecules, which may or may not be identical, condense with one molecule of glycerol to form a fat. Each of the three fatty acid molecules forms an ester linkage with an —OH group of glycerol.

$$\begin{array}{l} H_2C{-}OH \\ \;\;| \\ HC{-}OH \\ \;\;| \\ H_2C{-}OH \end{array} + \underset{\text{Stearic acid}}{3HOOC\cdot(CH_2)_{16}\cdot CH_3} \overset{\text{Lipase}}{\rightleftharpoons} \begin{array}{l} H_2C{-}O{-}C{=}O{-}(CH_2)_{16}\cdot CH_3 \\ \;\;| \\ HC{-}O{-}C{=}O\cdot(CH_2)_{16}\cdot CH_3 \\ \;\;| \\ H_2C{-}O{-}C{=}O\cdot(CH_2)_{16}\cdot CH_3 \\ \qquad\qquad \text{Stearin} \end{array}$$

Lipases are widespread in microorganisms, plants, and animals. Enzymes of glyceride synthesis are thought to be located in the endoplasmic reticulum.

FATTY ACID CATABOLISM

β-OXIDATION

The energy stored in long fatty acid chains can be released through their oxidation which generally occurs in the mitochondria. The oxidation proceeds through a number of steps which finally results in acetyl CoA. (Acetyl CoA can in turn yield ATP.) The oxidation process involves a sequential elimination of acetyl CoA molecules from the carboxyl end of the fatty acid molecule. After such elimination, a new CoA-activated —COOH group is regenerated at the end of the remaining part of the fatty acid molecule. Since during this series of reactions it is the β-carbon atom (i.e., carbon-3 from the —COOH end) that gets oxidized to give rise to the newly-generated —COOH group, the process is referred to as β-oxidation. Some important enzymes of the β-oxidation cycle are acyl CoA dehydrogenase, enoyl CoA hydratase, 3-hydroxyacyl dehydrogenase, thiokinase, thiolase, and acetyl-acyl CoA transacetylase.

Apart from the straight chain saturated acids, fatty acids containing branched chains, double bonds, and an odd number of carbons can be oxidized.

The β-oxidation pathway is believed to operate very actively in germinating seeds and its enzymes are thought to be located extramitochondrially in glyoxysomes and peroxisomes. The degradation of unsaturated fatty acids through this pathway has been reported in the slime mould *Dictyostelium discoideum* (see Mazliak, 1973).

α-OXIDATION

Although the β-oxidation pathway is the most important means whereby fatty acids are degraded, especially in animal tissues, it is certainly not the only medium. An alternative pathway is α-oxidation in which the α-carbon (carbon-2), rather than the β-carbon, is oxidized and converted into the new —COOH group. This pathway is probably widespread in plants, especially in young leaves.

PEROXIDATION

When exposed to oxygen, lipids characteristically form peroxides either through autooxidation or enzyme-catalyzed oxidation. Peroxidation is economically significant since it can lead to tissue deterioration and food spoilage. It can be prevented or retarded through the use of certain antioxidants, such as tocopherols (see Gurr and James, 1971).

CONVERSION OF FATS INTO SUGARS

Fats or oils constitute the chief food reserves of many seeds. Several workers have observed that the utilization of reserve fat during the germination of seeds such as castor (*Ricinus*) is often not associated with a direct conversion of the fat into CO_2 and water, but with an initial transformation of fat into carbohydrates; it is the carbohydrates that then serve as respiratory substrates for the developing embryo. Significant increases in the sugar content of germinating seeds, coupled with reduction in the oil content, have frequently been recorded.

In living cells, carbohydrates may be converted into fats, or vice versa, depending upon cellular conditions.

Kornberg and Beevers (1957) have defined the operation of the glyoxylate cycle (Fig. 8-7) in castor oil endosperm and it is believed that this cycle, which starts with acetyl CoA, probably accounts for the net synthesis of a sucrose precursor, such as oxaloacetate. The two key enzymes involved in the glyoxylate cycle, viz., isocitritase and malate synthetase, are widespread in diverse germinating seeds. It is also believed that such transformations of fat into sugar occur in the glyoxysomes.

OTHER PLANT LIPIDS

Phospholipids These differ from fats in having a complex phosphorylated moiety substituted for one of the three fatty acids. Unlike fats, phospholipids are highly dispersed within the cell instead of being deposited as food reserves. Phospholipids form an integral part of the cell membranes in which they are intimately bound to other substances, such as proteins and carbohydrates, with their polar heads associating with proteins, and their nonpolar tails extending freely and perpendicularly to the axis of the protein layers.

Phospholipids have been shown to exhibit a high metabolic turnover in the cell thus invalidating the theory of their inertness (Ansell and Hawthorne, 1964).

Glycolipids and Sulpholipids These are lipids containing one or more molecules of sugar or sulphur. Most glycolipids are structurally analogous to phospholipids with sugar being attached to position 3 of glycerol and fatty acids to positions 1 and 2. Plant chloroplasts have a fairly high content of galactosyl diglycerides and sulphoquinovosyl diglyceride (a sulpholipid in which sulphur is found in a sulphonic acid group and not as a sulphate ester). Recently, certain sulpholipids containing organic chlorine have been located in some green algae (see Gurr and James, 1971).

Though the functions of glycolipids are not yet established, some workers claim that these may act as carriers of sugars across cell membranes.

Isoprenoid lipids These include different substances, such as cholesterol, bile acids, some vitamins, carotene, and sex hormones, all of which have the basic isoprenoid structure and resemble fats as regards their solubility properties:

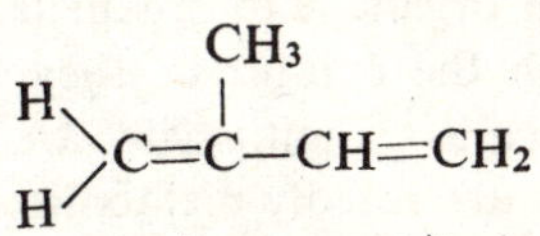

(C_5H_8).

An isoprene unit

Cholesterol ($C_{27}H_{45}OH$) is much more widely distributed in animals than in plants which have certain other closely related sterols, e.g., ergosterol, stigmasterol and β-sitosterol. Ergosterol is especially important in that upon exposure to UV light it is converted into calciferol (vitamin D).

Animal bile contains certain amides of bile acids which help in the emulsification of fats during digestion. The occurrence of such bile salts in plants is not yet determined.

Carotenoids are reddish or yellowish pigments found in most plants. They function in phototropism and serve as accessory pigments in photosynthesis. Carotene is a molecule with a 6-carbon ring at each end of a long chain of carbon atoms which is connected by means of alternating single and double bonds. Vitamin A is derived from a carotene molecule split into half.

Waxes These are esters of long fatty acid chains with long chain saturated monohydroxyl alcohols, such as cetyl alcohol ($C_{16}H_{33}OH$). Waxes generally occur as coatings on the surfaces of leaves and other organs of plants and serve to minimize water loss through transpiration. They are also evident on animal surfaces.

According to some researchers, waxes include also certain odd-numbered, long chain alkanes (formed by the condensation of normal, long fatty acid chains with 2-carbon units, followed by decarboxylation at the C_{30} level) (Kolattukudy, 1968).

Suberin and Cutin These, like waxes, are of great importance in plants, especially tropical plants. Their chemistry is still open to conjecture, but they are regarded as lipids because of their solubility characteristics. Suberin is observed mostly in the walls of cork cells whereas cutin is prevalent on the outer surfaces of epidermal cell walls. Both substances act as water-repellent films or coatings, minimizing the loss of water through transpiration.

LIPID TRANSFERS AND TURNOVER *IN VITRO*

Lipid biosynthesis is not a continuous process; rather it exhibits a sort of rhythmicity in maturing seeds and fruits. The major part of fat accumulation occurs in a few days in the course of seed or fruit development.

Mazliak (1973) and his coworkers have suggested that an exchange or transfer of lipids amongst different subcellular organelles or protoplasmic constituents is possible. For instance, labelled fatty acids have been shown to be transferred *in vitro* from one kind of organelle to another without any apparent metabolic change.

The relative rates of metabolic turnover of different fatty acids may also vary from acid to acid and from organelle to organelle. Thus, though the turnover rates are quite high in the cytoplasm, they are rather low in the mitochondria. Some acids, such as palmitic acid, are metabolically stable whereas others, e.g., oleic acid, are rapidly metabolized. Cell membranes rich in polyunsaturated fatty acids exhibit a slow turnover of these acids. Such observations have led to the conclusion that cellular lipids may be divided into several metabolic pools, some showing fast turnover and others being metabolically inert but significant as structural components.

REFERENCES

Ansell, G. B., Hawthorne, J. N., Phospholipids. Elsevier, Amsterdam (1964).

Awapara, J., Introduction to Biological Chemistry. Prentice-Hall, Englewood Cliffs, New Jersey (1968).

Coleman, M. H., The structural investigation of natural fats. *Adv. Lipid Res.*, **1**, 2 (1963).

Gurr, M. I., James, A. T., Lipid Biochemistry: An Introduction. Chapman and Hall, London (1971).

Hübscher, G. H., Glyceride metabolism. In Wakil, S. J. (ed.), Lipid Metabolism, Academic Press, New York (1970).

James, A. T., Morris, L. J. (eds.), New Biochemical Separations. Van Nostrand, London (1964).

Kolattukudy, P. E., Biosynthesis of surface lipids. *Science*, **159**, 498–505 (1968).

Kornberg, H. L., Beevers, H., The glyoxylate cycle as a stage in the conversion of fat to carbohydrate in castor beans. *Biochim. Biophys. Acta*, **26**, 531–37 (1957).

Mazliak, P., Lipid metabolism in plants. *Ann. Rev. Plant Physiol.*, **24**, 287–310 (1973).

14: Secondary Plant Products

INTRODUCTION

Progress in plant sciences is intimately linked with man's constant need for better sources of food, feed, fodder, wood, medicine, drug, and fibre. Phytochemists have already isolated, identified, and characterized some 70,000 plant chemicals and this number is expected to increase in the future. How these chemicals are made by plants and what their significance is for the plants that produce them and for those (whether man and/or animal) who consume them are questions that have aroused the attention of specialists in plant sciences: plant physiologists are keen to understand the functions of plant chemicals in the life and survival of plant species; plant biochemists are concerned with the mechanisms of the biosynthetic and biodegradative pathways of plant chemicals; agriculturists are engaged in increasing the production of economically important plants and chemicals; pharmacologists analyze those plant chemicals that can serve as drugs and medicine; nutritionists are absorbed in estimating the nutritional values of different foods; industrialists search for plants that may serve as raw material in industry; and chemotaxonomists are attracted by the importance of secondary plant products in the phyletic relations of certain plant groups involved in their biosynthesis.

Plant chemicals, such as carbohydrates, fats, amino acids, proteins, nucleic acids, and chlorophylls are called primary metabolic products because they are (1) essential for the growth and survival of plants, and (2) common to all plant species. Substances such as terpenoids, phenols, and alkaloids are called secondary metabolic products because (1) they are not so far known to be involved in the biosynthesis of primary plant metabolites; (2) their biosynthesis and accumulation are generally confined to a few plant species; and (3) they are functionally accessory to primary plant metabolites even though they are produced in plants from simple components of carbohydrates, fats, and amino acids. Such a classification of plant chemicals suggests that they are functionally unimportant to the producer plants and are a kind of metabolic waste product, but this may not be true in the light of recent findings on the functional importance of some of them.

Figure 14-1 shows the metabolic links between primary and secondary plant metabolites. Functionally, some secondary metabolites are as vital to plants as primary metabolites. For instance, both abscisic acid and gibberellins belong to the class of secondary metabolites called terpenoids, but both are well-established phytohormones involved in controlling the development, differentiation, and dormancy in plants. Phytosterols, which are components of plant cell membranes, are also classified as terpenoids. Lignin, the polymeric woody substance of higher plants, is a derivative of phenols, i.e., phenylpropane units; without lignin, the transition of plant life from water to land might not have been possible. Volatile oils, another kind of secondary plant metabolite, are engaged in the control of pollination by insects; they also seem to protect plants against phytophagous insects and animals.

On the other hand, there are secondary metabolites whose precise functional significance to their producer plants is still uncertain. These include alkaloids, calcium oxalate crystals, resins, and tannins—all chemicals generally considered plant metabolic wastes deposited within cells, vacuoles, or in the barks of trees. But it is difficult to comprehend why plants should produce such substances on a large scale—secondary metabolites are undoubtedly produced on a fairly large scale in most cases—when these find no functional use in their biochemical economy. Perhaps they had some functional relevance to the plants in the remote past and they might therefore represent a kind of physiological relic. Another possibility is that they still serve some useful purpose which though unknown today may be clarified in the future. The notion that secondary plant substances are unimportant to their producer plants is questionable. In fact, these substances are secondary only biosynthetically and not functionally (see Fig. 14-1). For the present, therefore, we can define secondary plant metabolites as substances biosynthetically derived from carbohydrates, fats, and amino acids. This definition may be far from complete, but it does suffice to separate the whole range of plant chemicals into two distinct groups, viz., the primary metabolic products and the secondary metabolic products.

Secondary substances such as rubber and some alkaloids are produced by only a few plants. Other secondary products, though yielded by many species, are created in much greater quantities in some plants than in others. This may be due to (1) differences in the efficiency of metabolic regulation; (2) the presence or the absence of enzyme systems that degrade them; and (3) deposition of the secondary substances in places such as vacuoles and barks which are inaccessible to degradative enzyme systems.

The formation and accumulation of secondary substances vary from plant to plant and are also dependent on the plant's age. Even in a single plant species, the formation and accumulation of secondary products may show seasonal variations, or a secondary metabolite that is abundant in one part of a plant may be scanty in another part of the same plant. Besides, day-to-day variations in light, temperature, hydration, etc., are equally important as control factors in the biogenesis of secondary products. The wide variety of secondary substances in plants is in sharp contrast to their absence from animals. It may be that the secondary substances of animals are excreted as waste products whereas in plants they are retained and deposited in such structures as vacuoles, resin canals, and cell walls.

Although the sites and mechanisms of the biosynthesis of terpenoids,

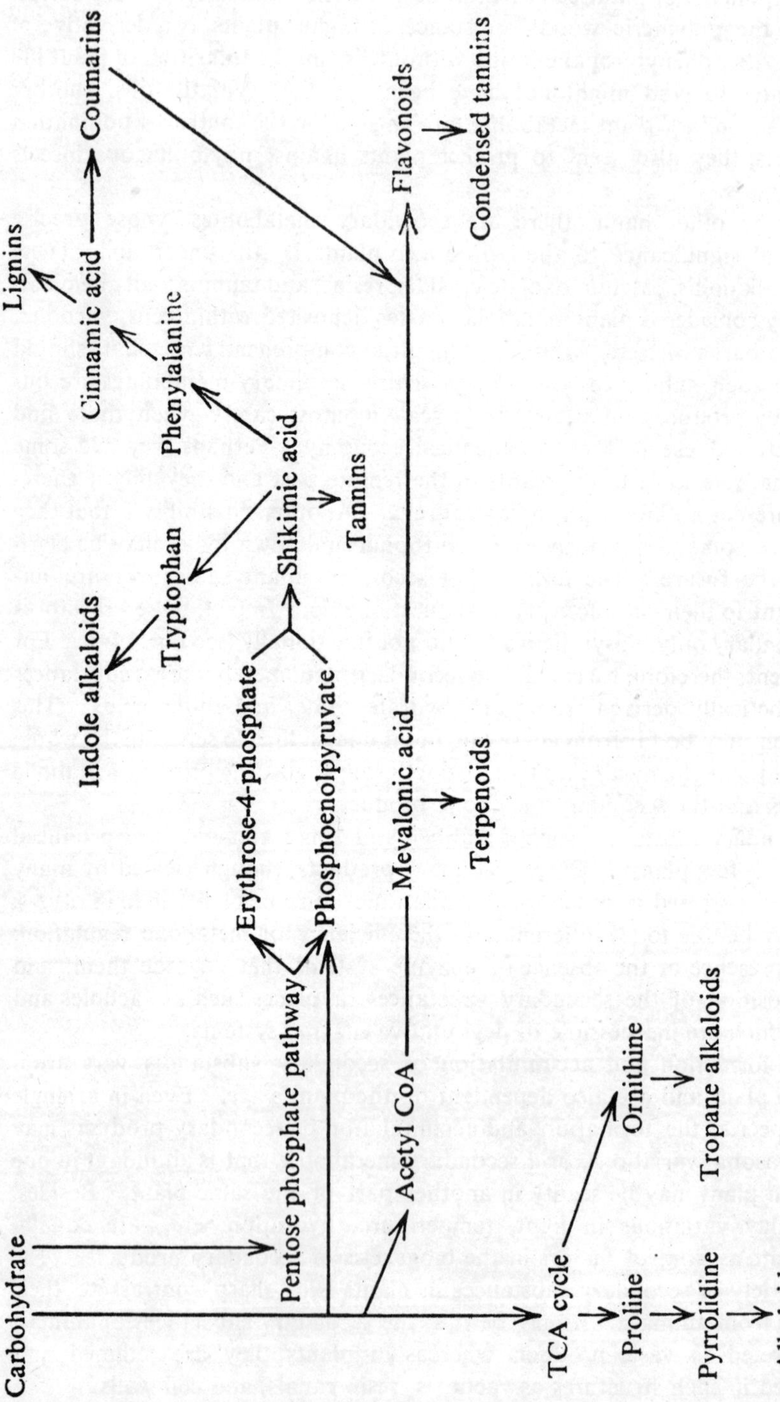

Fig. 14-1 Outline of biosynthetic routes of secondary plant products.

phenols, and alkaloids have been investigated, there is still considerable ignorance in respect of their enzymology, metabolic controls, and areas of synthesis. However, some secondary substances, such as flavonoids and other phenolic compounds, are toxic to both the producer and nonproducer plants when allowed to remain mobile within them. But the producer plants have developed a mechanism to detoxify them. They do this by individually linking the secondary metabolites to one or more sugar molecules, thus giving rise to a class of compounds called glycosides. The sugar molecule may be a glucose, fructose, or rhamnose; the nonsugar molecule, known as aglycone, includes one of the varied kinds of secondary plant substances. Anthocyanins are the best example of glycosides in which the aglycone component is one of the flavonoids and the most common sugar may be either glucose, fructose, or rhamnose. The formation of glycosides is also a mechanism for rendering water-insoluble aglycone substances water-soluble. The sugar molecule released when glycosides are hydrolyzed may be utilized in respiration for the generation of ATP and carbon source whenever the photosynthetic supply of sugar or other carbohydrates diminishes. Certain glycosides are known to be involved in protecting the producer plants against fungal or bacterial disease because of their toxicity to the invading pathogen.

We shall now give brief accounts of the biosynthesis and functions of some of the more important groups of terpenoids, phenols, flavonoids, and alkaloids. Although these alone are generally considered the major groups of secondary metabolic products, nonprotein-forming amino acids and waxes are also members of this class. The considerable research on nonprotein-forming amino acids not withstanding our knowledge of the biochemistry of waxes is still very meagre.

TERPENOIDS

Terpenoids are a class of compounds constructed from 5-carbon building blocks, called isoprene units. The structural formula of an isoprene unit is:

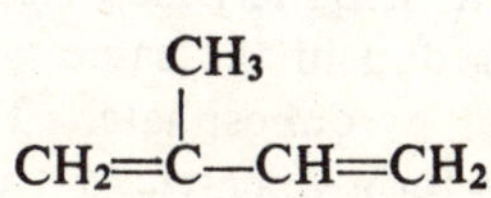

Isoprene units are biologically inert since the terpenoid biosynthetic enzymes are unable to recognize and use them for the synthesis of terpenoids. The biologically functional form of the isoprene unit is isopentenyl pyrophosphate:

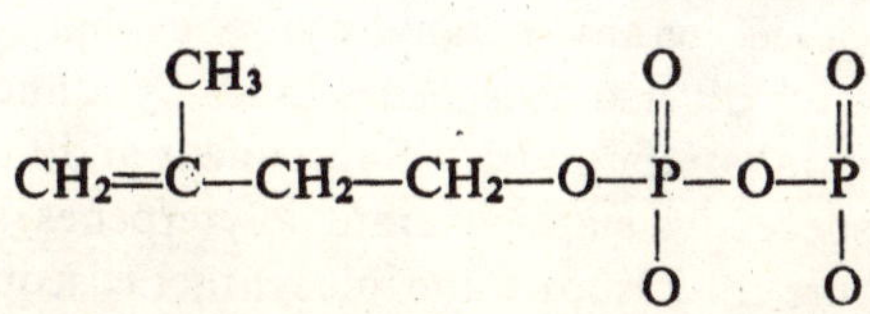

Terpenoids, also known as isoprenoids because of the involvement of isoprenes in their synthesis, are classified according to the number of isoprene units they contain: hemiterpenes, monoterpenes, sesquiterpenes, diterpenes, triterpenes, tetraterpenes, and polyterpenes are respectively made up of one, two, three, four, six, eight, and more than eight 5-carbon units. Hemiterpenes are always found as components of mixed terpenoids which have a terpenoid component and a nonterpenoid component; certain quinones, for example, are mixed terpenoids in which hemiterpenes are the terpenoid component. Monoterpenes, sesquiterpenes, and diterpenes may be open-chain or cyclic compounds having one or several ring systems whereas polyterpenes are exclusively open-chain compounds of 5-carbon units linked in series. Terpenoids in general can undergo various kinds of modifications, such as oxidation, reduction, addition, and subtraction of carbon atoms, thus leading to the possible occurrence of a wide variety of terpenoids in plants.

GENERAL BIOSYNTHESIS

Isopentenyl pyrophosphate is derived from acetyl CoA with mevalonic acid acting as an intermediate. First, two acetyl CoA units are joined to form acetoacetyl CoA which combines with another molecule of acetyl CoA to produce the 6-carbon compound β-hydroxy-β-methylglutaryl CoA. This undergoes reduction, leads to the formation of mevalonic acid, and CoA gets released during this reaction. Mevalonic acid is then phosphorylated and oxidatively decarboxylated to form isopentenyl pyrophosphate which isomerizes to dimethyl allyl pyrophosphate, the enzyme catalyzing this reaction being an isomerase.

The formation of monoterpenes (10-carbon compounds) results from the condensation of one molecule of isopentenyl pyrophosphate with one molecule of dimethyl allyl pyrophosphate. The resulting parent monoterpenoid is geranyl pyrophosphate (10-carbon) which can give rise to open-chain or cyclic monoterpenes after the liberation of the pyrophosphate moiety.

Farnesyl pyrophosphate (15-carbon) is produced by the addition of a molecule of isopentenyl pyrophosphate to geranyl pyrophosphate and this can generate open or cyclic sesquiterpenes (15-carbon compounds) after losing the pyrophosphate group in the same way as in the formation of monoterpenes from geranyl pyrophosphate. Two molecules of farnesyl pyrophosphate can unite to form open-chain triterpenes (30-carbon) from which cyclic triterpenes are obtained; the latter include steroids which are known to be essential for all living organisms, including plants.

Diterpenes (20-carbon) are produced by the addition of a molecule of isopentenyl pyrophosphate to farnesyl pyrophosphate. The resulting parent diterpene, called geranyl-geranyl-pyrophosphate, by further modification gives rise to diverse diterpenes and by dimerization produces carotenoids (40-carbon compounds). Subsequent additions of isopentenyl pyrophosphate in successive steps generate polyterpenes, such as rubber and gutta percha. Figure 14-2 shows the biosynthetic sequence of terpenoid formation.

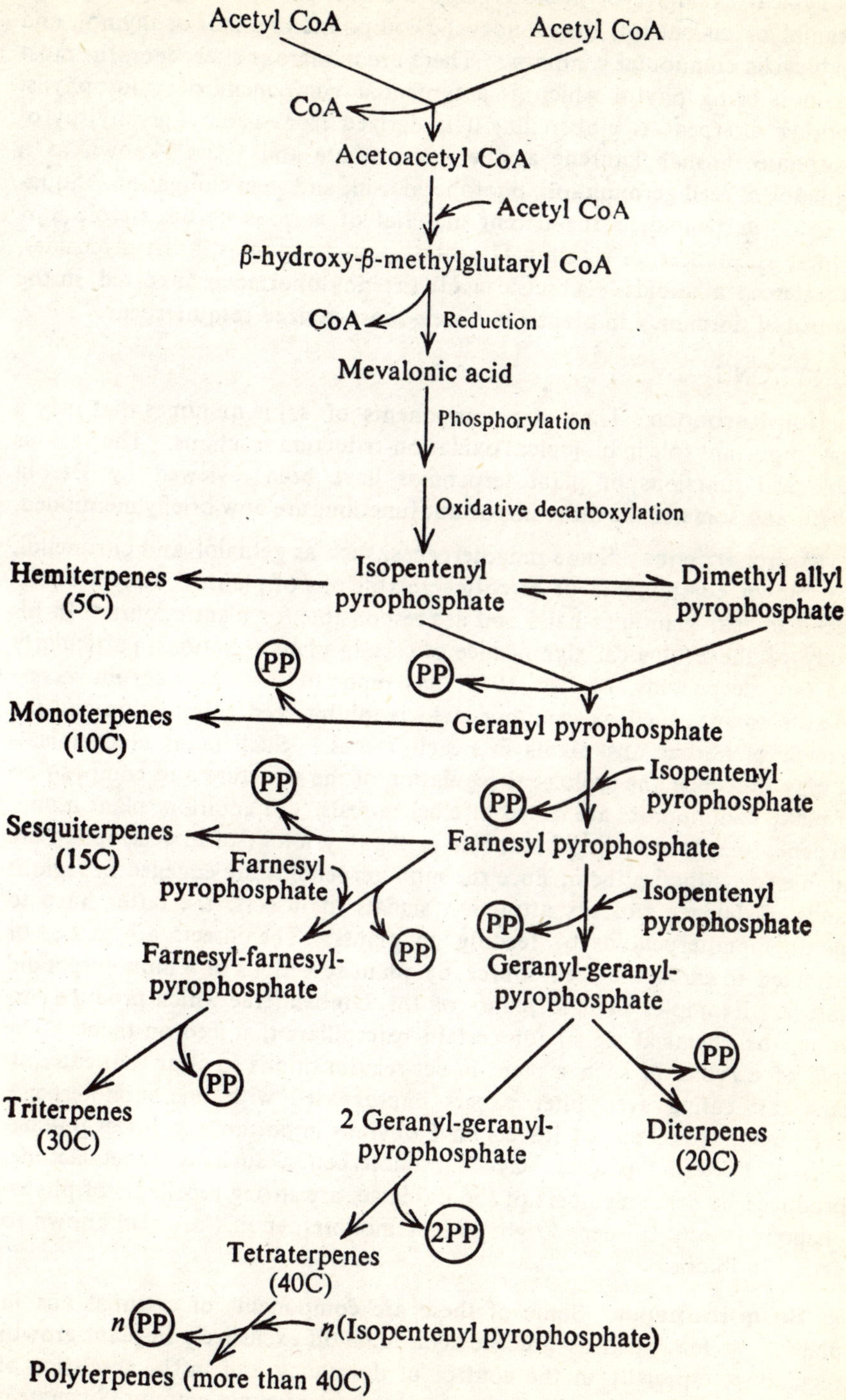

Fig. 14-2 Biosynthetic sequence of terpenoid formation.

Typical examples of monoterpenes include the open-chain compound geraniol or citronellol, the monocyclic compound menthol or thymol, and the bicyclic compound camphor. There are numerous diterpenes, the most common being phytol which is a terpenoid component of chlorophylls. Another diterpene is gibberellin; it is derived from geranyl-geranyl-pyrophosphate through kaurene as the intermediate and is well-known as a regulator of seed germination, photoperiodism, and stem elongation. Squalene, a triterpenoid, is the parent material of various sterols, steroid hormones, substances of vitamin D group, steroid saponins, heart glycosides, and steroid alkaloids. Abscisic acid, the phytohormone involved in the control of dormancy in plants, is a well-characterized sesquiterpene.

FUNCTIONS

Hemiterpenes These are components of some quinones that play a very important role in biological oxidation-reduction reactions. The various biological functions of plant terpenoids have been reviewed by Herout (1970) and some of the more important functions are now briefly mentioned.

Monoterpenes Some monoterpenes, such as geraniol and citronellol, are known components of essential volatile oils of plants. They are produced in leaf glandular hairs and are responsible for plant odours. In his study of the ecological significance of certain plant secretions, particularly the monoterpenoids, Muller (1969) has reported that some shrubs secrete volatile toxins, such as camphor, which inhibit seed germination and the growth of herbaceous plants in nearby areas. Such plant biochemicals as may influence the ecological regulation of the structure and composition of plant communities are called allelochemicals. In addition, plant monoterpenes play a very important role in the physiology of insects, which are unable to synthesize them; since the monoterpenoids are engaged in various kinds of defence and sex attraction signals in insects, the latter have to obtain monoterpenoids by feeding on plants. The insects are guided or attracted towards the feed source by plant secretions of a monoterpenoid nature. Examples include plants of the Umbelliferae which produce carvone, the chemical signal for certain caterpillars that feed on them. The role of carvone in such a plant-insect relationship is so clear that caterpillars start eating even filter papers impregnated with this monoterpene. Such plant attractants of insects have obvious importance in insect pollination of flowering plants. Cyclic monoterpenes, such as nepetalactone, produced by some members of the Labiatae, are strong repellents of phytophagous insects (Eisner, 1964). Most monoterpenoids are also known to act as antibiotics.

Sesquiterpenes Some of these are components of essential oils in plants. A few, such as abscisic acid, function exclusively in plant growth regulation, especially in the control of dormancy and in the abscission of leaves and fruits. Xanthinin is an antagonist of auxin activity (Silverstein *et al.*, 1966). Farnesol and its derivatives are traced in a wide variety of insects where they function as juvenile hormones, defence devices, and sex attractants. It is significant to remember that insects in general lack the ability to synthesize terpenoids from acetate.

Diterpenes Gibberellins are diterpenes and serve in almost all aspects of plant development and differentiation, either independently or in association with indole acetic acid, cytokinins, and abscisic acid.

Triterpenes The triterpene meliantriol, traced in a few species of the Meliaceae, is a well-known repellent of phytophagous insects. An important group of triterpenes called steroids is produced from the simple triterpene squalene in plants. Phytosterols constitute essential components of the membranes of cells, the chloroplast, mitochondria, nucleus, the endoplasmic reticulum, and the Golgi apparatus. The most important plant sterols are β-sitosterol and stigmasterol. Steroid saponins are glycosides in which the aglycone part is sapogenin. The saponins in general are found in members of the Compositae and Rutaceae and are poisonous to fish, man, and animal. In this respect, they certainly constitute a kind of defence mechanism against herbivorous animals. Another group of steroids, called heart glycosides because of their application in the treatment of heart diseases, is found in abundance in *Digitalis purpurea.* These glycosides have cardenolides or bufadienolides as their aglycone components. Some cardenolides, such as calactin, calotropin, and calotoxin, are produced exclusively in members of the Asclepiadaceae and are used in the defence of the feeder insects against their predators. Precursors of the insect-moulting hormone ecdysone are actually produced by plants (pteridophytes, gymnosperms, and angiosperms) and then elaborated into the actual ecdysone within the insects themselves. The functional significance of the ecdysone precursor to the producer plant remains largely unknown, but it has been suggested that it may act as a deterrent to phytophagous insects (Johnson *et al.*, 1963). Steroid alkaloids are biosynthetically related to triterpenes. An example is demissidine, found in the genus *Solanum*, which acts as a deterrent to Colorado beetle.

Tetraterpenes These include carotenes and their oxygenated derivatives xanthophylls. These pigments are used in photosynthesis and tend to protect plants against solar oxidation of chlorophyll molecules.

Polyterpenes These are components of the latex produced in a few members of the Apocyanaceae and Asclepiadaceae. Rubber is a polyterpene and its main source is *Hevea brasiliensis*. It is made up of 500–5000 isoprene units. As compared to rubber, gutta percha is a shorter chain molecule. In rubber, the double bonds are in the *cis* position whe-eas in gutta percha they occur in the *trans* position. The utility of these compounds for the producer plants remains unknown but, of course, both rubber and gutta percha have great application in industry.

PHENOLS

Phenols are found in a wide variety of plants, from algae to angiosperms

(Walker, 1975). Chemically, the basic skeleton of phenols includes an aromatic ring containing at least one hydroxyl group or its functional derivative. The aromatic ring can be substituted by other organic groups, thus giving rise to a vast range of phenolic compounds. On the basis of their structural complexity, four groups of phenolic compounds are generally recognized: (1) simple phenols, with one or more hydroxyl groups attached to the ring system which may contain other substituents as well; (2) phenol carboxylic acids, which are simple phenols having a carboxylic group as a substituent; (3) phenylpropanes, which are simple phenols with a 3-carbon side chain; and (4) flavan derivatives, with a flavan skeleton which consists of two aromatic rings, A and B, flanking the central oxygen-containing heterocycle. Phenols are found in plants in the form of their glycosides in cell vacuoles.

BIOSYNTHESIS

Tracer techniques have been very useful in tracking the path of phenol biosynthesis in plants, and although phenol biogenesis seems to proceed along three different metabolic pathways, namely, the shikimic acid pathway, the acetate-malonate pathway, and the acetate-mevalonate pathway, the most common and significant pathway in both higher plants and microorganisms seems to be that involving shikimic acid. This pathway is a source of phenolic substances and also of aromatic amino acids, such as phenylalanine, tryptophan, and tyrosine. The mode of generation of shikimic acid and the reactions through which it is transformed into coumaric acid and cinnamic acid are illustrated in Fig. 14-3. Cinnamic acid is the simple parent phenylpropane from which the phenylpropane derivatives are obtained. Plastoquinones and ubiquinones are derived from *p*-hydroxyphenylpyruvate (see Fig. 14-3) by a complex biosynthetic route. Cinnamic acids in plants are normally found in a bound form in glycosides or as esters.

Biosynthesis of Phenol Carboxylic Acid and Simple Phenols

In higher plants, phenol carboxylic acids and simple phenols are produced from *p*-coumaric acid which undergoes β-oxidation (in the same way as fatty acids). As a consequence, the side chain of *p*-coumaric acid is shortened by two carbon atoms, and this in turn leads to the formation of aromatic acid, called *p*-hydroxybenzoic acid, which is an example of a phenol carboxylic acid. Similarly, caffeic acid and ferulic acid can generate protocatechuic acid and vanillic acid, respectively. Gallic acid is another important phenol carboxylic acid. The simple phenol hydroquinone results from the decarboxylation of *p*-hydroxybenzoic acid.

Biosynthesis of Coumarins

The introduction of an hydroxyl group from the *ortho* position to the side chain of *trans*-cinnamic acid forms *o*-hydroxyl-*trans*-cinnamic acid (or, *o*-coumaric acid) which is glucosylated to *o*-coumaric acid β-glucoside.

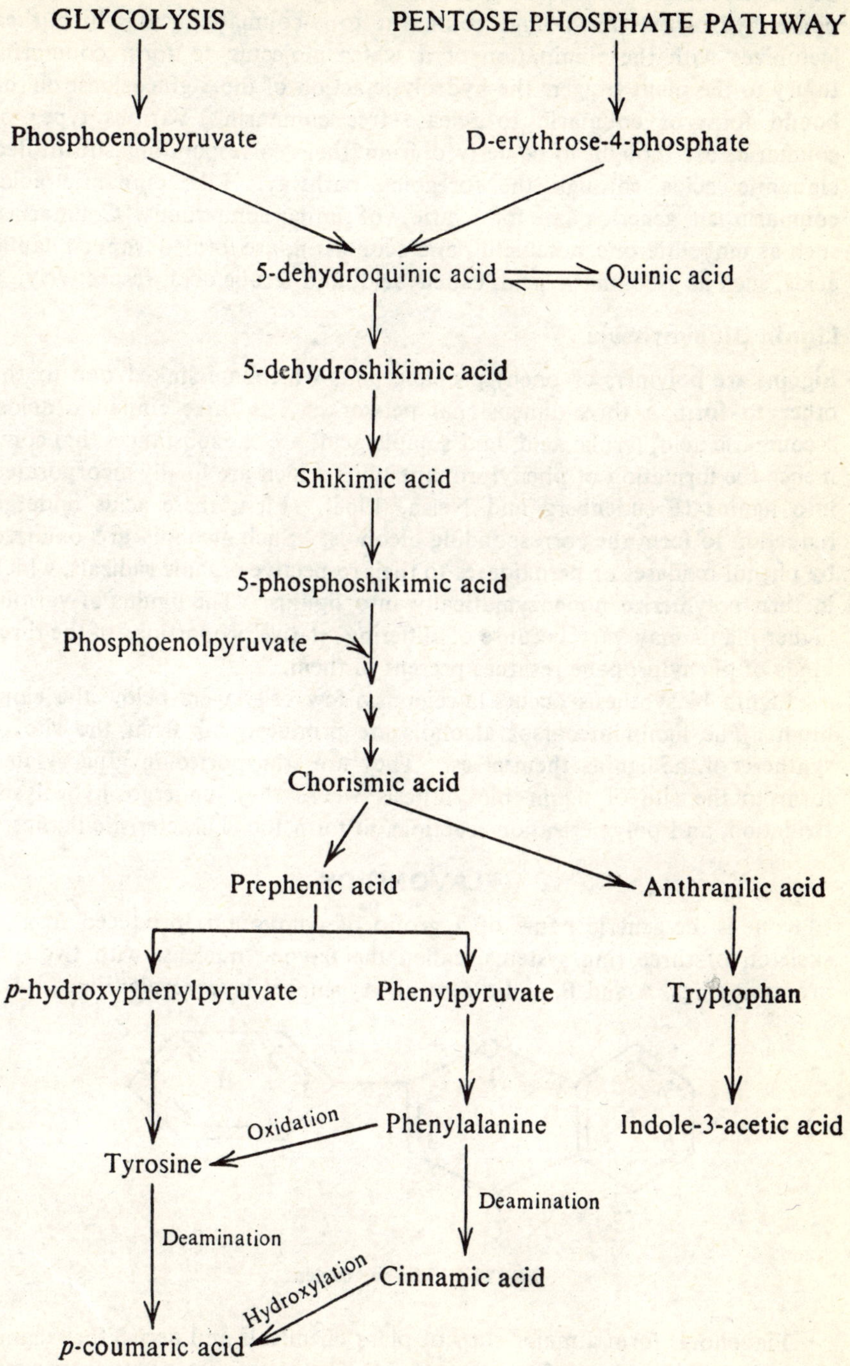

Fig. 14-3 Biosynthetic sequence of formation of cinnamic acid and *p*-coumaric acid in the shikimic acid pathway.

This is the form in which coumarin occurs in plants. The enzyme β-glucosidase hydrolyzes this bound coumarin to *o*-coumaric acid which then lactonizes with the elimination of a water molecule to form coumarin. Injury to the plant triggers the hydrolytic action of the β-glucosidase on the bound form of coumarin to release free coumarin. Various types of coumarins are thought to be derived from their corresponding substituted cinnamic acids through the foregoing pathway. Like cinnamic acid, coumarin is a generic name for a variety of similar compounds. Coumarins, such as umbelliferone, aesculetin, and scopoletin, are formed from cinnamic acids, such as *p*-coumaric acid, caffeic acid, and ferulic acid, respectively.

Lignin Biosynthesis

Lignins are polymers of phenylpropane units which are linked one to the other to form a three-dimensional network. The three cinnamic acids, *p*-coumaric acid, ferulic acid, and sinapic acid, are the substances that commence the formation of phenylpropane units which are finally incorporated into lignins (Freudenberg and Neish, 1968). First, these acids undergo reduction to form the corresponding alcohols. Such alcohols are oxidized by phenol oxidases or peroxidases to their respective organic radicals, which in turn polymerize nonenzymatically into lignins. The lignins of various higher plants may vary because of differing relative proportions of the three kinds of phenylpropane residues present in them.

Lignin biosynthesis occurs in regions a few cell layers below the cambium. The lignin precursor alcohols are produced far from the site of synthesis of the lignins themselves. They are transported in glucosylated form to the site of lignin biosynthesis where they undergo hydrolysis, oxidation, and polymerization reactions to form the characteristic lignins.

FLAVONOIDS

Flavone is the generic name of a group of compounds produced from a skeleton of three ring systems, called the flavone nucleus, with two side aromatic rings, A and B, and a central oxygenated heterocycle (Fig. 14-4).

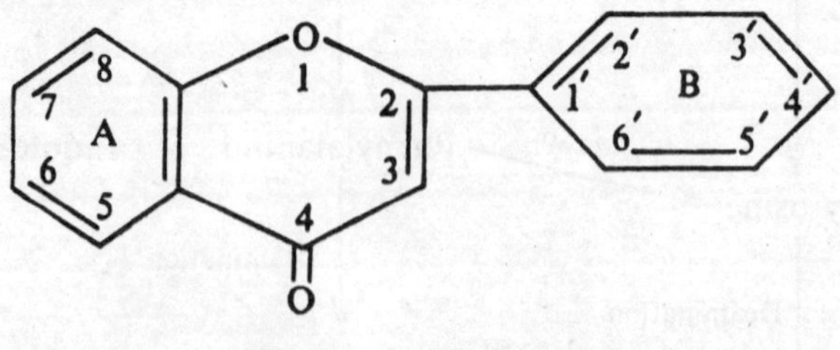

Fig. 14-4 Flavone nucleus.

Flavonoids form a major class of plant chemicals and derive their name from the yellow colour of some of these. They normally occur in various parts of plants, such as flowers, fruits, and leaves, and are commonly found in the form of their glycosides. They are classified on the basis of the oxidation state of the central heterocycle of the flavone nucleus. Each subgroup is further grouped according to the pattern of substitution in the B ring.

Chalcones are not true flavonoids since they lack the characteristic central heterocycle of the flavone nucleus. They too occur as glycosides in plants, especially in floral parts, and are known to get transformed into true flavone derivatives. Butein is a chalcone and its structure is as shown in Fig. 14-5.

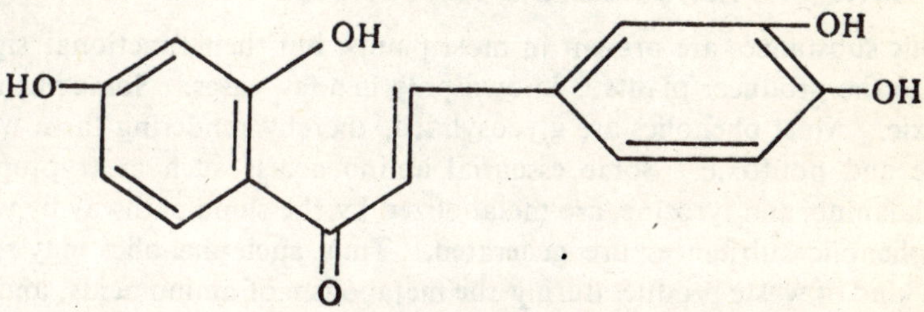

Fig. 14-5 Butein.

Flavones, such as luteolin, have in the central heterocycle two double bonds whereas flavonones have only one double bond. Examples of flavonones are naringenin and eriodictyol which are structurally similar to apigenin and luteolin (both flavones) except for the central heterocycle. Flavonols result from the addition of a hydroxyl group at carbon-3 of the central heterocycle. Kaempferol, quercetin, and myericetin are common examples of flavonols. Some other flavonols, such as flavon-3-ols and flavon-3,4-diols, are very important as they polymerize to produce a large group of tannins, called "condensed tannins". Flavon-3-ol is better known as catechin. Flavonols, like other flavonoids, are also found in plants in their glycosylated form.

ANTHOCYANINS

Anthocyanins are glycosides of anthocyanidins and constitute the red-and-blue colour of many flowers. Sugar-free anthocyanins are anthocyanidins, and differ one from the other according to the pattern of substitution of the flavone skeleton in ring B. Their central heterocycle is different from the central heterocycle of other flavone derivates in that it exhibits an oxonium structure in the acidic medium. The anthocyanidins pelorgonidin, cyanidin, and delphinidin differ only in their substituents at carbon-3, -4, and -5 of ring B.

In addition to anthocyanins, carotenoids and flavone derivatives are constituents of flower colours. Carotenoids mostly impart yellow-red colours whereas flavone derivatives are generally responsible for whitish and yellow colours and, occasionally, blue shades. Anthocyanins commonly (but not always) are red in acid pH and blue in alkaline, but cornflower is blue even though the pH of the cell sap of its floral parts is about 4.8 to 5.0. Anthocyanins are stored in cell saps.

BIOSYNTHESIS OF FLAVONES

Biochemical analysis of the building blocks of anthocyanidin formation has revealed that ring A of the flavone nucleus is built from three acetate units, ring B from phenylalanine, and carbon-2, -3, and -4 of the central heterocycle from cinnamic acid. Methylation of the hydroxyl group in ring B occurs by S-adenosylmethionine. The intermediates involved in biosynthetic reactions have not yet been identified and characterized (Harborne *et al.*, 1975).

FUNCTIONS OF PHENOLICS AND FLAVONOIDS

Phenolic substances are present in most plants, but their functional significance to the producer plants is known only in a few cases. Some phenolics are toxic. Most phenolics are glycosylated, thereby rendering them water-soluble and nontoxic. Some essential amino acids, such as tryptophan, phenylalanine, and tyrosine, are metabolized by the same pathway by which toxic phenolic substances are generated. Thus, such phenolics may represent a kind of waste product during the metabolism of amino acids, and glycosylation might have been a mechanism evolved by plants as a protective device against their toxicity.

Like other secondary metabolites, plants produce a variety of coumarins whose functions are largely unexplored. Among the coumarins whose functions are defined, some are known to inhibit seed germination and cell elongation by controlling the activity of indole acetic acid oxidase enzyme which participates in the regulation of the endogenous level of this phytohormone. Other coumarins function also as antibacterial substances. For instance, dicoumarol is used in the prevention of blood clotting. A few phenolic derivatives seem to act as phytoalexins in defending host plants against fungal or pathogenic attacks (Cruickshank, 1963).

Lignins provide strength to plants and are involved in the cementing, anchoring, and stiffening of cellulose microfibrils in cell walls. The lignification of cellulose microfibrils helps cell walls against any biological, chemical, or physical damage.

Flavonoids seem to be responsible for attracting insects and animals for pollination and feeding. Kaempferol derivatives, being cofactors of the indole acetic acid oxidase enzyme, stimulate the oxidation of indole acetic acid whereas quercetin prevents such oxidation because it is an alternative to and the more preferred substrate of indole acetic acid oxidase enzyme (Galston and Davies, 1969). Quercetin is also supposed to be involved in tendril contraction and coiling in some plants. Other flavonoids, such as pisatin and phaseolin, act as phytoalexins. There are many other secondary plant metabolites whose functions have not so far been determined (Harborne, 1973; Harborne *et al.*, 1975).

ALKALOIDS

The term alkaloid has been derived from 'vegetable alkali' which means bases of botanical origin. It is difficult to define it precisely as it represents a heterogeneous class of natural products. But since most alkaloids exhibit some common properties it should be sufficient for our purpose to define them as basic nitrogen-containing heterocyclic compounds produced by plants as secondary metabolites. Their basic nature results from a ring structure. Some exceptions to this definition are colchicine and ricinine which are not basic, and ephedrine and muscarine which are not nitrogen-containing heterocycles. The inclusion of such exceptional cases among the alkaloids is based mainly on their biosynthetic similarity with the true alkaloids.

Ornithine, lysine, phenylalanine, tyrosine, and tryptophan are the primary metabolites from which the various groups of alkaloids are formed in most cases (Hughes and Genest, 1973). This fact biogenetically separates the parent group of true alkaloids from the terpenoid and steroid alkaloids which are produced from 5-carbon precursors, such as isopentenyl pyrophosphate. Another class of pseudoalkaloids results from the oxidation followed by the alkylation or acylation of certain amino acids. Such alkaloids are called protoalkaloids.

BIOSYNTHESIS

The aliphatic amino acids ornithine and lysine can be transformed into nitrogen-containing five-membered and six-membered ring systems, respectively. The five-membered nitrogen-containing heterocyclic ring is called the pyrrolidine ring system, and the six-membered nitrogen-containing heterocyclic ring is known as the piperidine or pyridine ring system. The two ring systems are the outcome of two alternative pathways. The first pathway produces an amine by decarboxylation reaction, ornithine being decarboxylated to putrescine and lysine to cadaverine; the second pathway involves oxidative deamination of either amino acid. Thereafter, the two pathways merge and the pyrrolidine ring issues from ornithine whereas the piperidine ring likewise arises from lysine. Certain tropane alkaloids are formed from acids such as tropic and benzoic acids which in turn originate from phenylalanine.

Some important members of the pyridine-pyrrolidine and pyridine-piperidine groups of alkaloids are nicotine and nornicotine, found in *Nicotiana tabacum*, and anabasine traced in *N. glauca*. Nicotine and nornicotine are examples of the pyridine-pyrrolidine condensed ring nitrogen. Anabasine is an example of the pyridine-piperidine condensed ring system.

The pyridine ring component of the *N. tabacum* alkaloids comes from nicotinic acid. There are two biosynthetic routes for the formation of nicotinic acid; one route, which functions mostly in animals and fungi, generates it from tryptophan through the intermediate quinolinic acid, and the other route, which operates in bacteria and higher plants, produces it from glycerol and aspartic acid, again involving quinolinic acid as an intermediate. After decarboxylation, quinolinic acid is finally transformed into nicotinic acid mononucleotide which can give nicotinic acid proper directly or indirectly via NAD^+. The pyrrolidine ring is supplied by ornithine, but the detailed biosynthetic process is not yet understood.

The anabasine piperidine component comes from lysine and the pyridine component from nicotinic acid.

The aromatic amino acids phenylalanine and tyrosine are used in the biosynthesis of certain alkaloids in many species belonging to the Amaryllidaceae and Centrospermae and also in the genus *Papaver*. In Amaryllidaceae, tyrosine forms tyramine after decarboxylation, and phenylalanine is transformed into caffeic acid which produces protocatechuic acid after β-oxidation; the next reductive step produces protocatechualdehyde which finally condenses with tyramine to form the parent alkaloid characteristic of the Amaryllidaceae family, viz., norbelladine. Methylation of norbelladine gives rise to the most important alkaloid of the group, called belladine. Further derivatives of norbelladine can result from belladine through other chemical reactions. Colchicine, produced by *Colchicum autumnale*, is also synthesized from the same two aromatic amino acids, but in this case phenylalanine creates sinapic acid instead of caffeic acid; the union of tyramine and sinapic acid generates an intermediate which finally produces colchicine through some unknown steps.

Betacyanins and betaxanthins are the two major groups of alkaloids found in the Centrospermae. Betaxanthins are yellow pigments produced from dihydroxyphenylalanine and proline. The betacyanins are red pigments resulting from the condensation of a pyridine derivative with an indole derivative. In this reaction, both the pyridine and indole nucleus are furnished by dihydroxyphenylalanine. Betacyanins occur abundantly in red beet (*Beta vulgaris*).

Papaverine and morphine are important alkaloids of the genus *Papaver*. An isoquinoline skeleton and a benzyl skeleton which result from dihydroxyphenylalanine are involved in their formation. (In all the cases quoted, the parent source of dihydroxyphenylalanine is the amino acid tyrosine.)

The mode of biosynthesis of the indole group of alkaloids is still uncertain, but it is known that an indole nucleus, often in association with one or two 5-carbon units, is involved in the process. The indole nucleus is derived from tryptophan whereas the 5-carbon units may be furnished by mevalonate or isopentenyl pyrophosphate. Lysergic acid is an example of the indole group of alkaloids. It occurs in some fungi, such as *Claviceps*, and in some higher plants, such as Convolvulaceae. Reserpine and strychnine are other examples of indole alkaloids.

Caffeine, theophylline, and theobromine, which are all methylated purines, are instances of purine alkaloids. They are derived from purine mononucleotide, inosine-5′-phosphate. Both theobromine and theophylline contain two methyl groups (but at different positions) whereas caffeine has three methyl groups. Caffeine is found in coffee, tea, and cocoa plants.

FUNCTIONS

In general, alkaloids have not been observed to perform any vital functions in plants. This contrasts with their significance to animals and man who use them as powerful drugs. In this respect, alkaloids differ sharply from the other two groups of secondary plant metabolites, viz., terpenoids and phenols, some members of which are functionally indispensable to the plants themselves. However, there is reason to surmise that some alkaloids may not be altogether irrelevant in plants since they show rapid metabolic turnover (see Hughes and Genest, 1973). For instance, the *Papaver* plant converts the exogenously supplied morphine endogenously into nonalkaloid components. It has also been suggested that some alkaloids protect plants against phytophagous animals and insects. According to Mothes (1966), the opium alkaloids are essentially excretion products without physiological function. The tropane alkaloids are pharmacologically very important; some of them, e.g., coniine, are highly toxic and were used in ancient Greece to execute criminals. A famous victim of the toxin coniine was the Greek philosopher, Socrates. Coniine is synthesized in *Conium maculatum*, popularly called hemlock. The latex of the opium poppy is the chief source of morphine, codeine, and heroin. The *Cinchona* alkaloid, quinine, is the famous antimalarial drug. *Rauwolfia* plants synthesize a well-characterized alkaloid, called reserpine, which is very effective in the treatment of hypertension. Some alkaloid derivatives of phenylalanine and tyrosine function as hallucinogenic drugs.

Analyses using tissue culture techniques suggest that the formation of root initials parallels the development of the ability to synthesize some of these alkaloids in plants (Street and Cockburn, 1972). When grown in cell suspension cultures, *Datura* does not produce its major alkaloids hyoscyamine and scopolamine. The indole alkaloids, on the other hand, seem to be synthesized in suspension and callus cultures of *Ipomoea violacea*, though the pattern of formation of these alkaloids in such cultures is different from that in the whole plants (see Hughes and Genest, 1973). These instances indicate the possibility of artificially manipulating the factors affecting alkaloid biosynthesis.

NONPROTEIN AMINO ACIDS

Plants produce two major classes of amino acids, viz., the protein-forming amino acids (about 20 different kinds) and the nonprotein amino acids (about 200 different kinds). However, only a few of the nonprotein amino acids are characteristic of any one particular plant species. Nonprotein amino acids such as homocysteine, homoserine, and ornithine are well-established intermediates in the biosynthesis of methionine, threonine, and proline, respectively, but most nonprotein amino acids are metabolically inert and seem to have no apparent use to the producer plants. Some of them are antimetabolites of the protein amino acids, and quite a few seem to function as nitrogenous reserves or in the transport of nitrogen. They are structurally more variable than protein amino acids, but, like the latter, may be acidic, basic, neutral, hydroxyl-containing, sulphur- or selenium-containing, and aromatic.

BIOSYNTHESIS

The formation of most protein amino acids is now well-understood. Carbon skeletons of glutamic acid, alanine, and aspartic acid are derived from alpha-ketoglutaric acid, pyruvic acid, and oxaloacetic acid. These three amino acids are in turn the precursors of three groups of amino acids (see Chapter 10). The carbon skeleton of serine is derived from phosphoglyceric acid, and serine can give rise to glycine and cystine. The carbon skeleton of tryptophan, phenylalanine, and tyrosine comes from chorismate. Glutamic acid is the primary amino acid whose amino group is transferred to a variety of keto acids. This leads to the formation of the respective amino acids by transamination which plays a key role in the biogenesis of nonprotein amino acids, simplifying it to the level of the biogenesis of their corresponding keto acids. However, the mode and the source of origin of carbon skeletons of various nonprotein amino acids are obscure (Fowden, 1973). Alpha-amino adipic acid is implicated as an intermediate in the biosynthesis of pipecolic acid and lysine, and lysine, after a number of steps, gives rise to lathyrine.

We note from Table 14-1 that a plant that produces nonprotein amino acids contains only a few, and often only one, of these amino acids. The limited distribution of nonprotein amino acids is thus a reflection of their biosynthetic diversification and this feature of plants is finding increasing use in the assessment of their taxonomy and phylogeny.

Table 14-1 Distribution of some important nonprotein amino acids in plants

Nature of amino acids	Name of amino acid	Distribution
Dicarboxylic amino acid derivatives	γ-hydroxyglutamic acid, γ-hydroxyglutamine	Liliaceae
	β,γ-dihydroxyglutamic acid	Seeds of lettuce
	cis-α-(2-carboxycyclopropyl) alanine	*Aesculus parviflora*
	3-carboxy-D-phenylglycine, 3-carboxy-L-alanine	*Iris tingitana*
	S-(2-carboxyethyl)-L-cysteine, S-(2-carboxyisopropyl)-L-cysteine	Seeds of *Albizzia* and *Acacia*
	S-methylcysteine	Cruciferae
	S-hydroxymethylhomocysteine	Some algae
	Ethionine	Some bacteria

Table 14-1 Distribution of some important nonprotein amino acids in plants (cont.)

Nature of amino acids	Name of amino acid	Distribution
Imino acid derivatives	L-azetidine-2-carboxylic acid	Liliaceae
	trans-4-hydroxy-L-proline	As a constituent of structural protein in plant cell walls
	3-aminopyrrolidine-3-carboxylic acid (cuçurbitin)	*Cucurbita moschata*
	trans-5-hydroxy-L-pipecolic acid	*Acacia*
Basic amino acid derivatives	α,γ-diaminobutyric acid, β-oxalyldiaminopropionic acid	*Lathyrus*
	Amidine	Leaves of *Indigofera spicata*
	Canavanine	Leguminosae
Aromatic and heterocyclic amino acid derivatives	3,4-dihydroxyphenylalanine	*Vicia faba*
	Mimosine	*Mimosa*
	Lathyrine	*Lathyrus*
Selenium analogues of sulphur amino acids	Selenomethylselenocysteine, selenocystathionine, selenomethylselenomethionine	Selenium accumulator plants of *Astragalus* Selenium nonaccumulator plants of *Astragalus*

FUNCTIONS

Gamma-hydroxyglutamic acid and γ-hydroxyglutamine occur in liliaceous plants but seem to have no obvious functional significance. Gamma-methyleneglutamic acid and the corresponding γ-methyleneglutamine occur in large amounts in the seedlings (but not seeds) of peanuts. Similarly, homoarginine occurs in the seedlings of pea. They have been implicated in the transport of nitrogen in the respective plants (see Fowden, 1973). Canavanine, alpha,gamma-diaminobutyric acid, and homoarginine are stored in large quantities in the seeds of certain leguminous plants and they might represent a kind of readily mobilized pool of nitrogen reserve. Erythro-β-hydroxy-D-aspartic acid is a constituent of a toxic peptide from the poisonous mushroom, *Amanita phalloides*. Mimosine causes hair loss in cattle and inhibits the seedling growth of nonproducer plants. β-(methylenecyclopropyl) alanine, also known as hypoglycin-A, induces hypoglycemia in man. Azetidine-2-carboxylic acid inhibits growth in nonliliaceous plants. These amino acids may help to protect producer plants against phytophagous animals and insects or to compete for the colonization of particular habitats.

Alpha,gamma-diaminobutyric acid and beta-oxalyldiaminopropionic acid are neurotoxins and produce symptoms of lathyrism in animals.

The significance of nonprotein amino acids, such as those of terpenoids and alkaloids, lies in their relation to the chemical systematics of plants (chemotaxonomy). The species *Astragalus* is a good example; a clear-cut biochemical differentiation has segregated them into two groups, viz., the selenium accumulators, which store selenium in methylselenocysteine and selenocystathionine, and selenium nonaccumulators, which incorporate selenium mainly into methylselenomethionine (Shrift, 1969). Genus *Lathyrus* has some species synthesizing both homoarginine and lathyrine and others synthesizing only the former. Homoarginine is an intermediate in the biosynthesis of lathyrine from lysine.

REFERENCES

Cruickshank, I. A. M., Phytoalexins. *Ann. Rev. Phytopathol.*, **1**, 351–74 (1963).

Eisner, I., Catnip: its raison d'etre. *Science*, **146**, 1318 (1964).

Fowden, L., Amino acids. In Miller, L. P. (ed.), Phytochemistry, vol. 2, 1–29, Van Nostrand Reinhold, New York (1973).

Freudenberg, K., Neish, A. C., Constitution and Biosynthesis of Lignin. Springer-Verlag, Berlin (1968).

Galston, A. W., Davies, P. J., Hormonal regulation in higher plants. *Science*, **163**, 1288–97 (1969).

Harborne, J. B., Flavonoids. In Miller, L. P. (ed.), Phytochemistry, vol. 2, 344–80, Van Nostrand Reinhold, New York (1973).

Harborne, J. B., Mabry, T. J., Mabry, H. (eds.), The Flavonoids. Chapman and Hall, London (1975).

Herout, V., Some relations between plants, insects, and their isoprenoids. In Reinhold, L., Liwschitz, Y. (eds.), Progress in Phytochemistry, vol. 2, 143–202, Wiley-Interscience, London (1970).

Hughes, D. W., Genest, K., Alkaloids. In Miller, L. P. (ed.), Phytochemistry, vol. 2, 118–70, Van Nostrand Reinhold, New York (1973).

Johnson, D. F., Bennett, R. D., Heftmann, E., Chloresterol in higher plants. *Science*, **140**, 198–99 (1963).

Mothes, K., Shutte, H. R., Biosynthese der Alkaloide. VEB Deutsher Verlag der Wissenschaften, Berlin (1969).

Muller, C. H., Allelopathy as a factor in ecological process. *Vegetatio*, **18**, 348–57 (1969).

Shrift, A., Aspects of selenium metabolism in higher plants. *Ann. Rev. Plant Physiol.*, **20**, 475–94 (1969).

Silverstein, R. M., Rodin, S. O., Wood, D. L., Sex attractants in grass produced by male *Ips confusus* in ponderosa. *Science*, **154**, 509 (1966).

Street, H. E., Cockburn, W., Plant Metabolism, 2nd edn. Pergamon Press, Oxford (1972).

Walker, J. R. L., The Biology of Plant Phenolics. Edward Arnold, London (1975).

Suggestions for Further Reading

Awapara, J., Introduction to Biological Chemistry. Prentice-Hall, Englewood Cliffs, New Jersey (1968).

Ball, E. G., Energy Metabolism. Addison-Wesley/Benjamin, Reading, Massachusetts (1973).

Bergh, S. G. Van Den, Mitochondria: Biogenesis and Bioenergetics; Biomembranes: Molecular Arrangements and Transport Mechanisms. North-Holland, Amsterdam (1972).

Bidwell, R. G. S., Plant Physiology. Macmillan, Riverside, New Jersey (1974).

Black, C. C., Ecological implications of dividing plants into groups with distinct photosynthetic production capacities. *Adv. Ecol. Res.*, **7**, 87–114 (1971).

Black, C. C., Photosynthetic carbon fixation in relation to net CO_2 uptake. *Ann. Rev. Plant Physiol.*, **24**, 253–86 (1973).

Black, C. C., Campbell, W. H., Chen, T. M., Dittrich, P., The Monocotyledons: Their evolution and comparative biology. III. Pathways of carbon metabolism related to net carbon dioxide assimilation by monocotyledons. *Quart. Rev. Biol.*, **48**, 299–313 (1973).

Boulter, D., Protein synthesis in higher plants. *Ann. Rev. Plant Physiol.*, **21**, 91–114 (1970).

Brown, J. S., Forms of chlorophyll *in vivo*. *Ann. Rev. Plant Physiol.*, **23**, 73–86 (1972).

Calvin, M., Bassham, J. A., The Photosynthesis of Carbon Compounds. Benjamin, New York (1962).

Chichester, C. O. (ed.), Structure and Function of the Chloroplasts. Springer-Verlag, Berlin (1971).

Clayton, R., Molecular Physics in Photosynthesis. Blaisdell, New York (1965).

Clowes, F. A. L., Juniper, B. E., Plant Cells. Blackwell, Oxford (1968).

Cocking, E. C., Plant cell protoplasts—isolation and development. *Ann. Rev. Plant Physiol.*, **23**, 29–50 (1972).

Cohen, G. N., The Regulation of Cell Metabolism. Holt, Rinehart and Winston, New York (1969).

Davies, M., Functions of Biological Membranes. Chapman and Hall, London (1973).

Devlin, R. M., Plant Physiology, 2nd edn. Van Nostrand Reinhold, New York (1969).

Devlin, R. M., Barker, A. V., Photosynthesis. Van Nostrand Reinhold, New York (1971).

Dixon, M., Webb, E. C., Enzymes. Academic Press, New York (1964).

Dowben, R. M., Cell Biology. Harper and Row, New York (1971).

Fogg, G. E., Photosynthesis. The English Universities Press, London (1968).
Fogg, G. E. (ed.), Cell Differentiation. S. E. B. Symposium, No. 17, Cambridge Univ. Press, London (1963).
Fowden, L., The non-protein amino acids. *Endeavour*, **21**, 35–42 (1962).
Garrod, D. R., Cellular Development. Chapman and Hall, London (1973).
Gauch, H. C., Inorganic Plant Nutrition. Dowden, Hutchinson and Ross, Stroudsburg, Pennsylvania (1972).
Gibbs, M. (ed.), Structure and Function of the Chloroplasts. Springer-Verlag, New York (1971).
Giese, A. C., Cell Physiology, 3rd edn. Saunders, Philadelphia (1968).
Goddard, D. R., Bonner, W. D., Cellular respiration. In Steward, F. C. (ed.), Plant Physiology, vol. 1A, 209–312, Academic Press, New York (1960).
Goodwin, T. W. (ed.), Porphyrins and Related Compounds. Academic Press, London (1968).
Goodwin, T. W., Mercer, E. I., Introduction to Plant Biochemistry. Pergamon Press, Oxford (1972).
Greulach, V. A., Plant Function and Structure. Macmillan, Riverside, New Jersey (1973).
Hatch, M. D., Osmond, C. B., Slatyer, R. O. (eds.), Photosynthesis and Photorespiration. Wiley-Interscience, New York (1971).
Heath, O. V. S., The Physiological Aspects of Photosynthesis. Stanford Univ. Press, Stanford (1969).
Hess, D., Jarvis, D., Plant Physiology. Springer-Verlag, Berlin (1974).
Hewitt, E. J., Cutting, C. V. (eds.), Recent Aspects of Nitrogen Metabolism in Plants. Academic Press, London (1968).
Hillman, W. S., Papers in Plant Physiology. Holt, Rinehart and Winston, New York (1970).
Jevons, F. R., The Biochemical Approach to Life, 2nd edn. George Allen and Unwin, London (1968).
Kumar, H. D., Prakash, G., Toxicity of selenium to the blue-green algae *Anacystis nidulans* and *Anabaena variabilis*. *Ann. Bot.*, **35**, 697–705 (1971).
Laetsch, W. M., Relationship between chloroplast structure and photosynthetic carbon fixation. *Sci. Prog.*, Oxford, 57, 323–51 (1969).
Laetsch, W. M., Cleland, R. E., Papers on Plant Growth and Development. Little Brown and Co., Iowa (1967).
Ledbetter, M. C., Porter, K. R., Introduction to the Fine Structure of Plant Cells. Springer-Verlag, New York (1970).
Lehninger, A. L., Biochemistry. Worth Publishers, New York (1970).
Lehninger, A. L., Bioenergetics: The Molecular Basis of Biological Energy Transformations, 2nd edn. Benjamin, New York (1971).
Levitt, J., Introduction to Plant Physiology. C. V. Mosby, St. Louis (1969).
Levitt, J., Responses of Plants to Environmental Stresses. Academic Press, New York (1972).
Lie, T. A., Mulder, E. G. (eds.), Biological Nitrogen Fixation in Natural

and Agricultural Habitats. Martinus Nijhoff, The Hague (1971).

Liu, A. Y., Black, C. C., Glycolate metabolism in mesophyll cells and bundle sheath cells isolated from Crabgrass, *Digitaria sanguinalis* (L.) Scop., leaves. *Arch. Biochem. Biophys.*, **149**, 269–80 (1972).

Meyer, B. S., Anderson, D. B., Bohning, R. H., Fratianne, D. G., Introduction to Plant Physiology, 2nd edn. Van Nostrand, New York (1973).

McElroy, W. D., Cell Physiology and Biochemistry, 3rd edn. Prentice-Hall, Englewood Cliffs, New Jersey (1971).

Nobel, P. S., Plant Cell Physiology: A Physico-chemical Approach. W. H. Freeman, San Francisco (1970).

Nobel, P. S., Introduction to Biophysical Plant Physiology. W. H. Freeman, San Francisco (1974).

Nomura, M., Ribosomes. *Scientific American*, **221**, 28–35 (1969).

Plowman, K. M., Enzyme Kinetics. McGraw-Hill, New York (1972).

Price, C. A., Molecular Approaches to Plant Physiology. McGraw-Hill, New York (1970).

Pridham, J. B. (ed.), Plant Cell Organelles. Academic Press, London (1968).

Rabinowitch, E., Govindjee, Photosynthesis. Wiley, New York (1969).

Ranson, S. L., Thomas, M., Crassulacean acid metabolism. *Ann. Rev. Plant Physiol.*, **11**, 81–110 (1960).

Roller, A., Discovering the Basis of Life: An Introduction to Molecular Biology. McGraw-Hill, New York (1974).

Roman, H. L., Sussman, A. S. (eds.), Topics in the Study of Life: The Bio Source Book. Harper and Row, New York (1971).

Salisbury, F. C., Ross, C., Plant Physiology. Wadsworth, Belmont, California (1969).

Sigel, H., Metal Ions in Biological Systems, 2 vols. Marcel Dekker, New York (1973).

Steward, F. C., Growth and Organization in Plants. Addison-Wesley, Reading, Massachusetts (1968).

Stewart, W. D. P. (ed.), Algal Physiology and Biochemistry. Blackwell, Oxford (1973).

Stiles, W., Cocking, E. C., An Introduction to the Principles of Plant Physiology, 3rd edn. Methuen, London (1969).

Street, H. E., The Repertoire of Plant Cells. Leicester Univ. Press, Leicester (1969).

Street, H. E. (ed.), Plant Tissue and Cell Culture. Blackwell, Oxford (1973).

Street, H. E., Cockburn, W., Plant Metabolism, 2nd edn. Pergamon Press, Oxford (1972).

Street, H. E., Öpik, H., The Physiology of Flowering Plants: Their Growth and Development. Edward Arnold, London (1970).

Tedeschi, H., Cell Physiology: Molecular Dynamics. Academic Press, New York (1974).

Thornton, J. W., Cellular respiration: an example of a module using programmed format. In Creager, J. G., Murray, D. L. (eds.), The Use of Modules in College Biology Teaching, 97–130, Commission on Under-

graduate Education in Biological Sciences, Washington (1971).
Troughton, J., Donaldson, L. A., Probing Plant Structure. McGraw-Hill, New York (1972).
Van Overbeek, J., The Lore of Living Plants. Scholastic Book Services, New York (1964).
Walker, D. A., Three phases of chloroplast research. *Nature*, **226**, 1204–8 (1970).
Watson, J. D., The Molecular Biology of the Gene. Benjamin, New York (1965).
Wilkins, M. B. (ed.), The Physiology of Plant Growth and Development. McGraw-Hill, Maidenhead, England (1969).
Zalik, S., Jones, B. L., Protein biosynthesis. *Ann. Rev. Plant Physiol.*, **24**, 47–68 (1973).
Zelitch, I., Photosynthesis, Photorespiration and Plant Productivity. Academic Press, New York (1971).

Author Index

Subject Index